3rd International Conference on Geoscience and Earth Resources Engineering (3rd ICGERE 2021)

IOP Conference Series: Earth and Environmental Science
Volume 1031

Online
8 – 9 December 2021

ISBN: 978-1-7138-5470-8
ISSN: 1755-1307

Printed from e-media with permission by:

Curran Associates, Inc.
57 Morehouse Lane
Red Hook, NY 12571

Institute of Physics
Dirac House, Temple Back
Bristol BS1 6BE UK

Phone: 44 1 17 929 7481
Fax: 44 1 17 920 0979

techtracking@iop.org

Additional copies of this publication are available from:

Curran Associates, Inc.
57 Morehouse Lane
Red Hook, NY 12571 USA
Phone: 845-758-0400
Fax: 845-758-2633
Email: curran@proceedings.com
Web: www.proceedings.com

TABLE OF CONTENTS

IOP Conf. Series: Earth and Environmental Science 1031 (2022) 011001 doi:10.1088/1755-1315/1031/1/011001

The 3rd International Conference on Geoscience and Earth Resources Engineering: Geosciences and Earth Resources Engineering for Sustainable Future (Preface)

David P. Sahara[1*], Taufiq Hidayat[2], Billal Maydika Aslam[3], Zulfakriza[1], Ridho K. Wattimena[4], Syafrizal[4], Susanti Alawiyah[1]

[1]Department of Geophysical Engineering, Faculty of Mining and Petroleum Engineering, Institut Teknologi Bandung (ITB), Indonesia.
[2]Department of Metalurgical Engineering, Faculty of Mining and Petroleum Engineering, Institut Teknologi Bandung (ITB), Indonesia.
[3]Department of Petroleum Engineering, Faculty of Mining and Petroleum Engineering, Institut Teknologi Bandung (ITB), Indonesia.
[3]Department of Mining Engineering, Faculty of Mining and Petroleum Engineering, Institut Teknologi Bandung (ITB), Indonesia.

*email: david.sahara@gf.itb.ac.id

The issue of IOP Conference Series: Earth and Environmental Science contains the outstanding papers presented at International Conference on Geoscience and Earth Resources Engineering (ICGERE) 2021. Due to Covid-19 Pandemic, the 3rd ICGERE 2021 Conference was held online on 8 – 9 December 2021.

The International Conference on Geoscience and Earth Resources Engineering (ICGERE) is an annual event held by the Faculty of Mining and Petroleum Engineering, Bandung Institute of Technology, Bandung, Indonesia. The 3rd ICGERE 2021, held in collaboration with research centres at ITB.

The conference's theme of ICGERE 2021 was "Geosciences and Earth Resources Engineering for Sustainable Future". Some new insights and techniques in geoscience and earth resources will be shared and discussed during the conference. Energies, environments and infrastructures are the objects that their integrity and sustainability must be contemplated and considered by all members of Faculty of Mining and Petroleum Engineering, in order to ensure the sustainability of earth resources in the future. Furthermore, students will have incredible chances to interact with leading experts from their respective fields.

ICGERE will serve as a meeting point for collaboration, sharing and developing new ideas and technologies. The ICGERE create an opportunity for academics, professionals, and students to promote and discuss the scientific results related to recent development in Geoscience and Earth Resource. Topics to be presented during the conference including earth resources exploration, mining engineering, engineering of drilling, production, and management of oil and gas, reservoir engineering, exploration and engineering seismology, global geophysics, metallurgical engineering, geothermal engineering

There were 6 invited international speakers, 64 contributed speakers with 1 panel discussion and 5 parallel sessions. The interesting topics discussed in the panel discussion with invited speakers from Cambridge University (Prof. Nick Rawlinson), Monash University (Prof. Jian Zhao), Swinburne University (Prof. Akbar Rhamdhani), Seoul National University (Prof. Ki-Bok Min), Kyoto University (Prof. Katsuaki Koike), Sejong University (Prof. W.S Bae).

Documentation:

IOP Conf. Series: Earth and Environmental Science 1031 (2022) 011001 doi:10.1088/1755-1315/1031/1/011001

List of Committees

1. Prof. Ridho K Wattimena (ITB)
2. Dr. Syafrizal (ITB)
3. Dr. Susanti Alawiyah (ITB)
4. Dr.rer.nat. David P. Sahara (ITB)
5. Irwan Iskandar, S.T., M.T, Ph.D. (ITB)
6. Dr.phil.nat. Agus Haris Widayat, S.T., M.T. (ITB)
7. Taufiq Hidayat, S.T., M.Phill., Ph.D. (ITB)
8. Dr. Sendy Dwiki, S.T., M.Eng. (ITB)
9. Dr.rer.nat. Andri Hendriyana, S.T., M.T. (ITB)
10. Dr.Eng. Nurulhuda Halim, S.T., M.T. (ITB)
11. Billal Maydika Aslam, S.T., M.T. (ITB)
12. Muhammad Ichsan, S.T., M.EnergyChange.Adv. (ITB)
13. Betseba br Sibarani, S.T., M.T. (ITB)
14. Angga Bakti Pratama, S.Si., M.T. (ITB)

Editors

1. Dr.rer.nat. David P. Sahara (ITB)
2. Taufiq Hidayat, S.T., M.Phill., Ph.D. (ITB)
3. Billal Maydika Aslam, S.T., M.T. (ITB)
4. Dr. Zulfakriza (ITB)

Reviewers

1. Dr. Fernando Lawrens (ITB)
2. Dr. Andri Hendriyana (ITB)
3. Dr. Eko Januari Wahyudi (ITB)
4. Faridz Nizhar ST.MT. (ITB)
5. Dr. Endra Gunawan (ITB)
6. Dr. Dadi Abdurrahman (ITB)
7. Dona Sita Ambarsari ST.MT. (ITB)
8. Ekkal Dinanto ST.MT. (ITB)
9. Wahyu Hidayat ST.MT. (ITB)
10. Aria W. Baskara ST. MT. (ITB)
11. Dian Kusumawati SSi.MSi. (ITB)
12. Dr.Eng. Widi Astuti (National Research and Innovation Agency of Indonesia)
13. Erik Prasetyo Ph.D. (National Research and Innovation Agency of Indonesia)
14. Silvya Rahmawati Ph.D. (ITB)
15. Ivan Kurnia M.Sc. (ITB)

Peer Review Statement

All papers published in this volume have been reviewed through processes administered by the Editors. Reviews were conducted by expert referees to the professional and scientific standards expected of a proceedings journal published by IOP Publishing.

- **Type of peer review:** Single Anonymous
- **Conference submission management system:** Morressier
- **Number of submissions received:** 59
- **Number of submissions sent for review:** 59
- **Number of submissions accepted:** 33
- **Acceptance Rate (Submissions Accepted / Submissions Received × 100):** 55.9
- **Average number of reviews per paper:** 2
- **Total number of reviewers involved:** 31
- **Contact person for queries:**
  Name: Dr. David P. Sahara
  Email: david.sahara@gf.itb.ac.id
  Affiliation: Institut Teknologi Bandung - Geophysical Engineering

IOP Conf. Series: Earth and Environmental Science 1031 (2022) 012001 doi:10.1088/1755-1315/1031/1/012001

Variants of the flower pollination algorithm for inversion of schlumberger sounding curve

F Raflesia[1] and Widodo[2]

[1] Master Program of Geophysical Engineering, Faculty of Mining and Petroleum Engineering, Institut Teknologi Bandung, Bandung 40132, Indonesia
[2] Applied Geophysics and Exploration Group, Faculty of Mining and Petroleum Engineering, Institut Teknologi Bandung, Bandung 40132, Indonesia

widodo@gf.itb.ac.id

Abstract. Flower pollination algorithm (FPA) is a nature-inspired algorithm that mimics flowering plant pollination behavior. Yang created the FPA in 2012, and it has since proven to be superior to other metaheuristic algorithms. Many FPA variants have recently been developed through modification and hybridization. This paper provides FPA variants consisting of Modified Flower Pollination Algorithm (MFPA), elitism Flower Pollination Algorithm (eFPA), Dimension by Dimension Improvement Flower Pollination Algorithm (DDIFPA) and Flower Pollination Algorithm with Bee Pollinator (BPFPA). Validation of the code that has been created is done through simple model testing. All algorithms provide inversion results that are consistent with true value. The results of two synthetic models indicate that eFPA is the only algorithm that can reach the global optimum. Besides having the best level of accuracy, eFPA also has the best stability.

1. Introduction

To solve geophysical inverse problems, many inversion algorithms have been developed; like the Occam algorithm [1], Genetic Algorithm and Particle Swarm Optimization [2], and Lavenberq-Marquard using SVD [3]. Geophysical inverse issues are usually always ill-posed, meaning that no single solution can be found. This ill-posedness stems from the following typical issues related to geophysical data collecting: intrinsic non-uniqueness (primarily troublesome when employing potential approaches), data collection mistakes, prospective data set inconsistencies, and, most crucially, a lack of data. Even with the most exact data sets (no noise or systematic errors), only limited information about the subsurface can be gleaned, which is usually insufficient for resolving the inverse problem's non-uniqueness.

FPA is one of the newest algorithms. This algorithm is inspired by the process of pollinating flowers. According to [4], FPA is powerful than Genetic Algorithm and Particle Swarm Optimization in simulations. For self-potential data, [5] used FPA in geophysical inversion problems. As a result, FPA is reliable for interpreting SP data and this method does not need an initial model from anomalous sources. [6] proposed a dimension by dimension improvement based on FPA. As a result, this approach commit increase the convergence agility and the aspect of the result effectively. [7] proposed a BPFPA, compared to DE, PSO, FPA and K-Means, BPFPA is not only more accurate but also more stable. [8] popularized an embellished variant of FPA called MFPA. This algorithm is a hybridization of FPA with Clonal Selection Algorithm (CSA). The optimization results show that MFPA is more accurate than SA, GA, FPA, bat algorithm and firefly algorithm. [9] show that adding elitism is better than adding mutation operators and local search components to FPA. eFPA outperforms other strategies in many cases. In the event that the eFPA falls to construct optimal solution, the solution are compose within fair value.

2. Methods

2.1. Variants of the FPA

(a)

1. Establish a switch probability;
2. Begin the randomly generated population;
3. Determine the pollen's objective function;
4. Identify best solution from the first population;
5. **while** *absolute* (min_{RMS}-max_{RMS}) > 10^{-10} && min_{RMS} > 10^{-10};
6. **if** rand <*p;*
7. **for** i=1 : *n* (*n* indicates total of pollen);
8. global pollination;
9. **end for**
10. **else**
11. select *m* best solution from *pop* to make *clonespop*
12. clonning solution at *clonespop*
13. **for** each solution in *clonespop*
14. do local pollination;
15. **end for**
16. **end if**
17. select best population from *pop* and *clonespop* to make *newpop*;
18. Replace *pop* with *newpop*;
19. Find the current best solution *gbest;*
20. if gbest not change for 100 iterations, hold gbest and replace pop with random generation
21. **end while**

(b)

1. Establish a switch probability $p \in [0,1]$;
2. Make *counter vector = zeros(1,npop)*;
3. define *limit*;
4. Begin the randomly generated population;
5. Determine the pollen's objective function;
6. Identify best solution from the firts population;
7. **while** *absolute* (min_{RMS}-max_{RMS}) > 10^{-10} && min_{RMS} > 10^{-10};
8. **for** i=1 : *n* (*n* indicates total of pollen);
9. **if** rand <*p;*
10. global pollination;
11. **if** *counter (1,i) == limit*;
12. make new solution use simplex method;
13. *counter (1,i)=0*;
14. **endif**
15. **else**
16. do local pollination use EBMO;
17. do *crossover* operation;
18. **end if**
19. **if** new solution better;
20. *counter (1,i)=0*;
21. **else** *counter (1,i)=counter(1,i)+1*;
22. **end for**
23. Find the current best solution *gbest;*
24. **end while**

(c)

1. Establish a switch probability;
2. Begin the randomly generated population;
3. Determine the pollen's objective function;
4. Identify the best solution in the first population;
5. **while** *absolute* (min_{RMS}-max_{RMS}) > 10^{-10} && min_{RMS} > 10^{-10};
6. **for** i=1 : *n* (*n* indicates total of pollen);
7. **if** rand <*p;*
8. global pollination;
9. **else**
10. carry out local pollination;
11. **end if**
12. Consider new solutions;
13. Find worst solution and replace;
14. **end for**
15. Examination the most recent best solution*;*
16. **end while**

(d)

1. Establish a switch probability*;*
2. Begin the randomly generated population;
3. Determine the pollen's objective function;
4. Identify the best solution in the first population n;
5. **while** *absolute* (min_{RMS}-max_{RMS}) > 10^{-10} && min_{RMS} > 10^{-10};
6. define *probability*
7. **for** i=1 : *n* (*n* indicates number of pollen);
8. **if** rand <*p;*
9. global pollination;
10. **else**
11. carry out local pollination;
12. **end if**
13. Consider new solutions use DDEIS;
14. **end for**
15. Examination the most recent best solution *gbest;*
16. **end while**

Figure 1. FPA's pseudo-code variants: a) MFPA, b) BPFPA, c) eFPA, d) DDIFPA.

2.2. Forward Modeling Formulation

[10] devised the following formula:

$$\rho_a(s)=s^2\int_0^{\infty} T(\lambda)J_1(\lambda s)\lambda d\lambda \tag{1}$$

In this equation, *s* represents the half-spaced electrode, *J1* represents the first-order Bessel function, and λ represents the integral variable. T(λ), the resistivity transform function, is denoted in such a way:

$$T_i(\lambda) = \frac{T_{i+1}(\lambda)+\rho_i tanh(\lambda h_i)}{[1+T_{i+1}(\lambda)tanh(\lambda h_i)/\rho_i]}, i = n-1, \ldots\ldots,1 \quad (2)$$

n denotes the total of layers, ρ_i and h_i are the actual resistivity and thickness of the ith layer.

2.3. Objective function

In general, appropriateness of model response to observed data instated by an objective function that must be minimized. The search for the smallest objective function is linked to the examination for the best model. The model is tweaked so that the model response matches the data. Throughout operation, it becomes fair that inverse modelling commit only be accomplished if the relationship between the data and the model parameters is accepted. The objective function:

$$E=\sum_{i=1}^{N}\left(rho_i^{cal} - rho_i\right)^2 \quad (3)$$

The objective function commit be defined as the root mean square error (ERMS) with coming equation to have a fairer connotation in equal unit as data (Ωm):

$$E_{RMS}=\sqrt{\frac{1}{N}\sum_{i=1}^{N}\left(rho_i^{cal} - rho_i\right)^2} \quad (4)$$

3. Results

3.1 Synthetic model

Validation of the code that has been created is done through simple model testing. The simple models used are homogeneous models, and two-layer models. All algorithms provide inversion results that match the true value.

Table 1. Results of the validation.

Parameters	True Value	Search Space Min	Search Space Max	MFPA	eFPA	DDIFPA	BPFPA
			Homogeneous Model				
ρ (Ωm)	200	12,61	500	200	200	200	200
			Two-layered Model				
ρ_1 (Ωm)	50	3,153	300	50	50	50	50
ρ_2 (Ωm)	100			100	100	100	100
h_1 (m)	14	0,5	20	14	14	14	14

The appropriateness of present algorithm was tested on synthetic models K-curve (three-layered model with $\rho_1 < \rho_2 > \rho_3$) and HK-type (four-layered model with $\rho_1 > \rho_2 < \rho_3 > \rho_4$) curve. K-type curve uses a maximum iteration of 5000. While HK-type curve uses a maximum iteration of 7000. The results of the FPA variants are contrasted with the standard FPA. We used the same inverted parameters. Table 2 shows the search space as well as the experimental results. The inversion result value which is the same as the true value and the smallest standard deviation shows that eFPA has the best accuracy and stability.

Table 2. Results of the inversions.

Parameter	True Value	Search Space Min	Search Space Max	FPA	MFPA	eFPA	DDIFPA	BPFPA
				K-type curve				
ρ_1 (Ωm)	100			99,99	100	100	99,99	100
ρ_2 (Ωm)	2000	31,53	2804	1763,52	2037,67	2000	807,84	2009,44
ρ_3 (Ωm)	300			299,99	300	300	299,91	300

h_1 (m)	20	0.5	60	19,98	20	20	19,76	20
h_2 (m)	2			2,27	1,96	2	5,36	1,99
Standard deviation RMS of ten runs				0.0025	0.0038	0.0003	0.0117	0.0063
				HK-type curve				
ρ_1 (Ωm)	1000	33,44	3000	1000	1000,01	1000	1000,03	1000.11
ρ_2 (Ωm)	50			126,88	168,48	50	277,66	294,76
ρ_3 (Ωm)	2000			2021,46	2050,12	2000	2455,56	2471,84
ρ_4 (Ωm)	500			500,03	500,1	500	500,84	500,88
h_1 (m)	20	0.5	127	19,80	19,63	20	18,98	18,82
h_2 (m)	2			5,14	6,91	2	12,36	13,22
h_3 (m)	40			39,36	38,59	40	31,08	30,75
Standard deviation RMS of ten runs				0.0231	0.1132	1.8E-6	0.5988	0.0445

Each algorithms run 20 times (10 runs for each model), as illustrated in Figure 2. The eFPA reached the global optimum 11 times out of 20 runs. The FPA, MFPA, DDIFPA, and BPFPA falled in reaching the global optimum in the 20 runs.

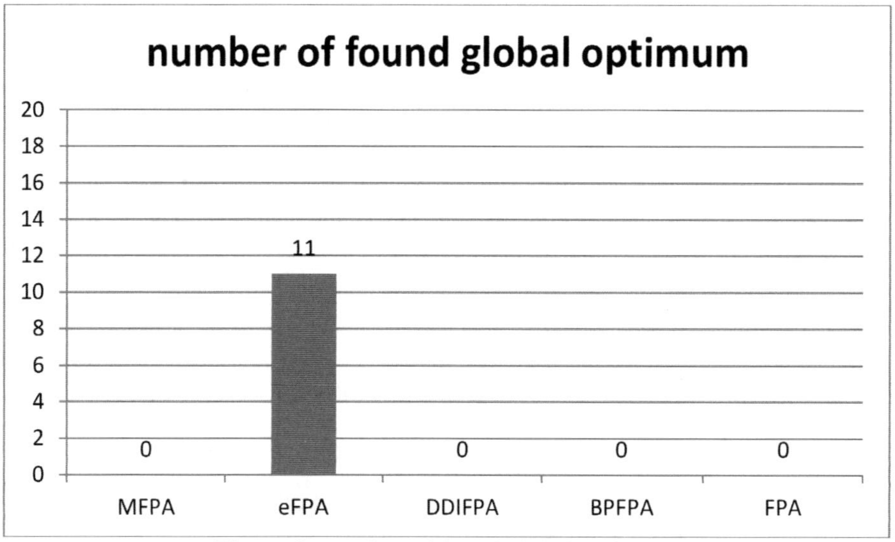

Figure 2. number of found global optimum

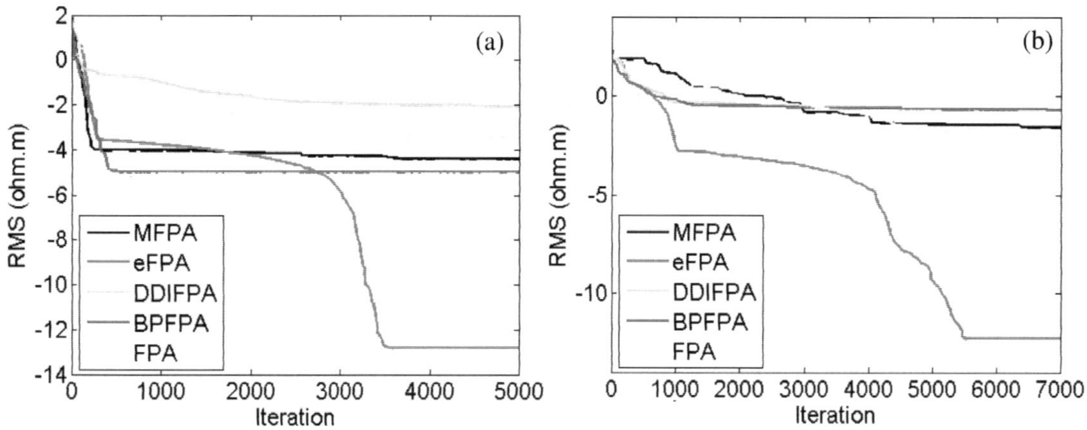

Figure 3. RMS error behavior comparisons (a) RMS error in the iterative operation using the K-curve. (b) RMS error in iterative operation using HK-curve.

Figure 3 reveal that eFPA has better performances than the MFPA, DDIFPA, BPFPA, and FPA in terms of the better RMS error. In Figure 3, it can be seen that eFPA can continue to explore when other algorithms stop at the local optimum (fixed RMS value throughout iterations). In addition to better exploration capabilities, eFPA also has better exploitation capabilities as well. It can be seen from the steep RMS eFPA curve. In contrast to DDIFPA, which has a very gentle RMS curve.

3.2. Field data

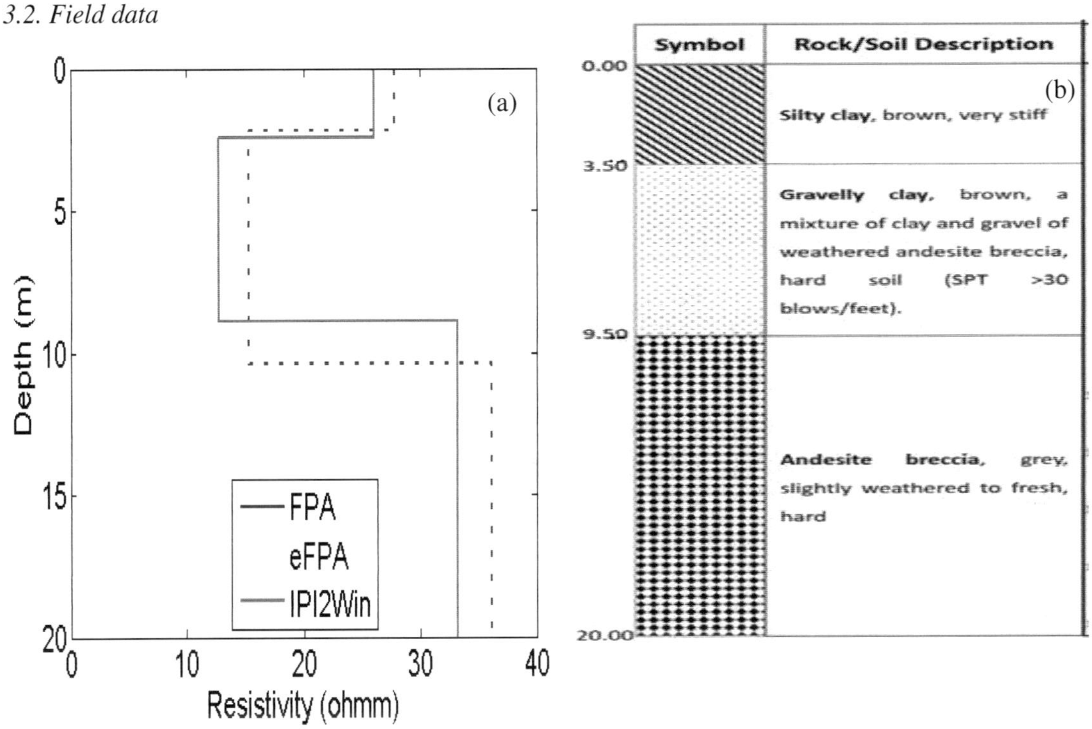

Figure 4. A field data set; (a) inversion results obtained with the FPA and IPI2Win. (b) The borehole data is also shown.

When it comes to field data, the eFPA is pitted against the standard FPA and IPI2Win. Figure 4 depicts the inverse solution of the eFPA for the observed data. The RMS error is 1.9 Ωm (FPA and eFPA) and 2.56 m (FPA and eFPA) (IPI2Win). The inverse result from the FPA, eFPA, and IPI2Win are shown in Figure 4a. The inverse solution is generally consistent with the results collected applying FPA, IPI2Win, and the borehole data. At depths ranging from 0 to 2.18 m, the first layer has a resistivity value 27.81 Ωm. At depths ranging from 2.18 to 10.4 m, the second layer is a low resistivity layer (15.37 Ωm). According to andecite breccia, the third layer has a higher apparent resistivity of 36.21 Ωm. (igneous rock has higher resistivity than clay at first and second layer).

4. Conclusions

The results of two synthetic models indicate that eFPA is the one and only algorithm that can reach the global optimum. Besides having the best level of accuracy, eFPA also has the best stability. Meanwhile, DDIFPA has poor performance, even worse than the FPA standard. In the case of field data, eFPA is better than IPI2Win based on RMS error. However the eFPA has the same results as the FPA standard.

Acknowledgments

This research received financial support by LPDP (Lembaga Pengelola Dana Pendidikan), LPPM ITB, and DIKTI (Directorate General of Higher Education of Indonesia).

References

[1] Constable S C, Parker R L and Constable C G 1987 Occam's inversion: A practical algorithm for generating smooth models from electromagnetic sounding data *Geophysics* **52** 289-300.

[2] Yogi I B S, Widodo 2017 Time domain electromagnetic 1D inversion using genetic algorithm and particle swarm optimization Available from: http://dx.doi.org/10.1063/1.4990901

[3] Widodo W, Saputera D H 2016 Improving Levenberg-Marquardt Algorithm Inversion Result Using Singular Value Decomposition *Earth Science Research* **5(2)** 20

[4] Yang X-S 2012 Flower Pollination Algorithm for Global Optimization. *Lecture Notes in Computer Science* Springer Berlin 240–249

[5] Sungkono 2020 Robust interpretation of single and multiple self-potential anomalies via flower pollination algorithm *Arabian Journal of Geoscience* **13** 100

[6] Wang R and Zhou Y 2014 Flower pollination algorithm with dimension by dimension improvement *Mathematical Problems in Engineering* **2014** 1-9.

[7] Wang R Zhou Y Qiao S and Huang K 2015 Flower pollination algorithm with bee pollinator for cluster analysis *Information Procession Letters* http://dx.doi.org/10.1016/j.ipl.2015.08.007.

[8] Nabil E 2016 A Modified Flower Pollination Algorithm for Global Optimization *Expert Systems with Applications* https://doi.org/10.1016/j.eswa.2016.03.047

[9] Nasser A B Zamli K Z Alsewari A A and Ahmed B S 2018 Hybrid flower pollination algorithm strategies for t-way test suite generation *PLoS One* **13(5)** e0195187.https://doi.org/10.1371/journal.pone.0195187.

[10] Koefoed O 1979 *Geosounding Principle, I: Resistivity Sounding Measurements*. Elseveir Scientific Publishing Co

Manufacture Hollow Glass Microsphere (HGS) in University Laboratory Scale (Phase 1)

B T H Marbun[1*], B Pande[1], S Z Sinaga[1], and B A Purbantanu[1]

[1] Petroleum Engineering Study Program, Bandung Institute of Technology (ITB), Jalan Ganesha 10, Bandung 40132, Indonesia

*Email: bonar.marbun@tm.itb.ac.id

Abstract. Hollow Glass Microsphere (HGS) can be used as an additive of drilling fluid to decrease the density and hydrostatic pressure without changing the chemistry characteristic to make underbalanced drilling condition. HGS is relatively easier to be mixed with the fluid without complicated equipment, can prevent formation damage, and can improve oil and gas production. There are various HGS commercial products that have been developed in the industry for various purposes with a density range 0.125-1.1 gr/cm3, grain size range 2-120 µm, and compressive strength up to 200 MPa. HGS has been successfully used in underbalanced drilling operation in several oil and gas fields in Indonesia. HGS costs less than UBD and the application shows good results with the improvement of oil and gas production. However, the cost of commercial HGS products in the market is very expensive. Petroleum Engineering Study Program Institut Teknologi Bandung (ITB) performed the HGS research in three phases. The first phase reported in this paper was to manufacture HGS in the university laboratory scale. HGS was manufactured based on various chemical compositions and temperature and time variation of the burning process. Scanning Electron Microscope (SEM), Energy Dispersive Spectrometry (EDS), and X-Ray Diffraction (XRD) analysis were performed to identify and characterize the raw chemical material and manufactured HGS. HGS was successfully manufactured with the following characteristics: low density (<0.8 gr/cm3) and low grain size (20-40 µm). These characteristics show that HGS can compete with commercial industry HGS products that have been successfully used in the underbalanced drilling operation. The second phase is to scale-up manufacture HGS from university laboratory scale to industrial scale. The third phase is to test the manufactured HGS with drilling fluid according to the American Petroleum Institute (API) Spec. 13A and API Spec. 13B-1.

1. Introduction

1.1. Background

Hollow Glass Microsphere (HGS) can be used as an additive to drilling fluid. HGS is a spherical-shaped filler material that possesses properties such as low density in the range of 0.29 to 0.63 gr/cm^3, high-temperature resistance, high compressive strength value in the range of 2,000 to 18,000 psi, homogeneous and dimensional-stable [1]. HGS as a drilling fluid additive was initially researched by the U.S. Department of Energy in the late 1990s [2]. HGS can reduce the drilling fluid density down to 5.5 ppg [1] . Application of HGS in drilling fluid can reduce the lost circulation rate from 100 bbl/hr to

6 bbl/hr [3], while the differential sticking [4], formation damage issues, and other possible downhole issues are eliminated [2] [5] [6] [7] [8] [9] [10] [11] [12] [13] [14] [15] [16].

However, the commercial HGS products in the market are very expensive [17]. This study was performed to make a first step in fulfilling the necessity of HGS for drilling new wells in mature fields in Indonesia. Institut Teknologi Bandung (ITB) performed the HGS manufacturing process in the laboratory. In 2012, ITB has successfully performed the HGS manufacturing research [18]. Various chemical compositions and temperature and time variations of the burning process were used and applied in the manufacturing process. The identification and characterization of the chemical material and manufactured HGS were performed using Scanning Electron Microscope (SEM), Energy Dispersive Spectrometry (EDS), and X-Ray Diffraction (XRD). The objective of this research was achieved by manufacturing HGS with characteristics such as low density (<0.8 gr/cm^3) and low grain size (20-40 μm). The material of manufactured HGS costed $136-218.5/sack, which was much cheaper than several commercial HGS products. The test will be performed according to guidelines and procedures in American Petroleum Institute (API) Spec. 10A [19], API Spec. 13A [20], and API Spec. 13B-1 [21].

1.2. Objectives
The study reported in this paper was the first phase with the following objectives:
1. Manufactured HGS in the laboratory.
2. Analyzed the properties of manufactured HGS.
3. Compared the manufactured HGS among several commercial HGS products.

1.3. Study Methodology
Figure 1 shows the methodology of this study.

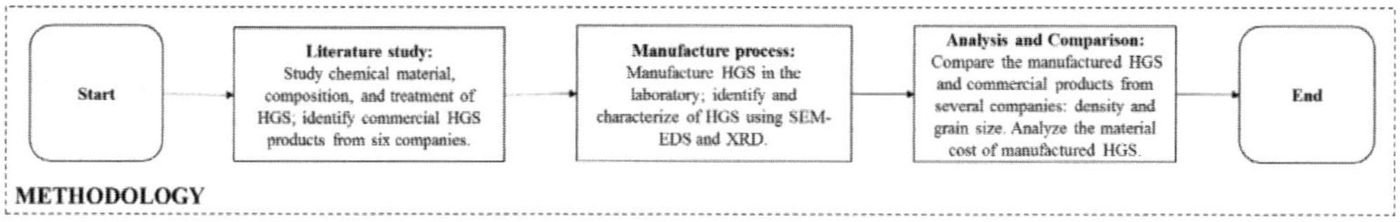

Figure 1. The methodology of the study

Figure 2 shows the manufacturing process of HGS.

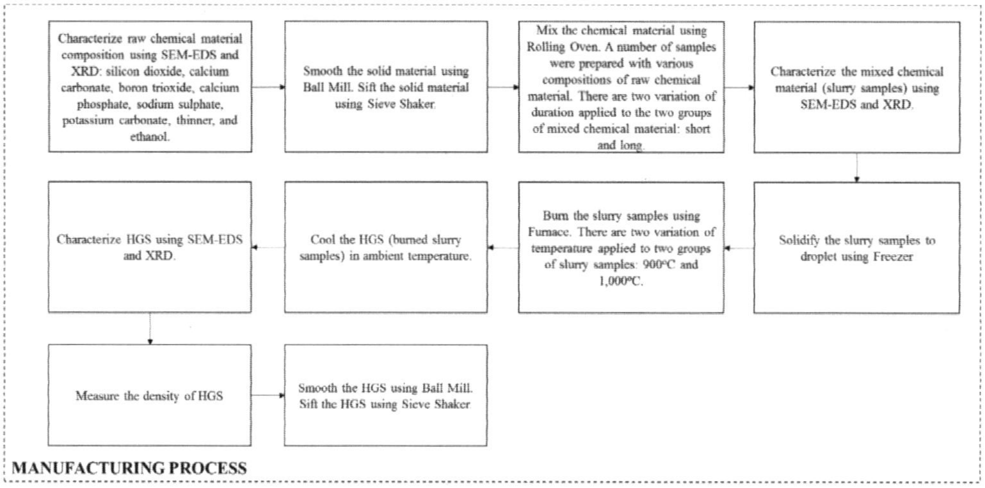

Figure 2. The manufacture process

2. Result and Discussion

2.1. Characteristic of Manufactured HGS

There were 5 samples shown in this study. Figure 3 and Table 1 show the characterization of the manufactured HGS using SEM, EDS, and XRD.

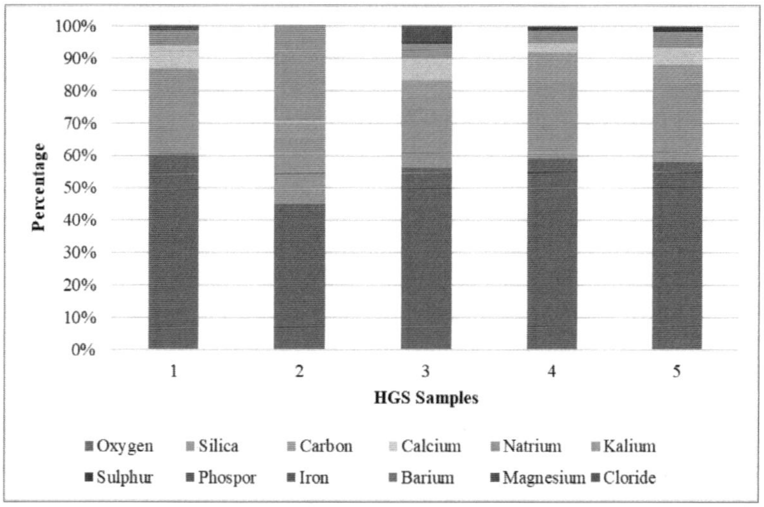

Figure 3. Characterization of the manufactured HGS using SEM and EDS

Table 1. The characterization of the manufactured HGS using XRD

Sample	XRD
1	low Synthetic Tridymite
2	low Synthetic Tridymite and Calcium Carbonate
3	low Tridymite, low Cristobalite, Parawollastonite, and Clinoptilolite
4	low Tridymite, low Cristobalite, and Parawollastonite
5	low Tridymite, low Cristobalite, and Parawollastonite

Table **2** shows the density and grain size measurement of the manufactured HGS.

Table 2. Density and grain size of the manufactured HGS

Sample	HGS Density (gr/cm^3)	Grain Size (μm)
1	0.65	20-40
2	0.6	
3	0.4	
4	0.65	
5	0.55	

2.2. Comparison Analysis of Manufactured HGS with Commercial HGS Products

Table **3** shows the properties comparison of manufactured HGS among commercial HGS products from six companies. The density and grain size of the manufactured HGS can be well compared to the commercial HGS products.

Table 3. Comparison of manufactured and commercial HGS products [22] [23] [24] [25] [26] [27] [28]

Commercial HGS products	MS - VP - A9	Sinosteel	Ceno Series HGT	3M	Hollow Glass Beads Catalogue 19823	Hollow Glass HK20	Manufactured HGS
Manufacturer	NPO Steklopla stik (Russia)	Zhejiang Co, Ltd. (China)	Ceno Technol ogies (USA)	3M Deutschla nd GmBH (Germany)	Polyscience , Inc. (USA)	Zhengzhou Hollowlite Materials Co., Ltd. (China)	ITB
Density (g/cm^3)	0.22 - 0.42	0.25 - 0.6	0.28 - 0.63	0.125 - 0.6	1.1	0.18-0.22	0.4 - 0.65
Average Diameter (μm)	40 - 100	9 - 120	2 - 110	30 - 120	2 - 20	Lower than 120	20 - 40

3. Conclusion

Following are the conclusions of this study:

1. HGS has been successfully manufactured in the laboratory.
2. Based on identification and characterization using SEM, EDS, and XRD, the manufactured HGS has proper properties that can be well compared among the commercial HGS products: low density (<0.8 gr/cm^3) and low grain size (20-40 μm).

4. Further Study

Following are steps for further study:

1. Measure other properties such as compressive and collapse strength, surfacing, pH, and thermal resistance of the manufactured HGS.
2. Scale-up manufacture HGS from university laboratory scale to industrial scale (Phase 2).
3. Test the manufactured HGS with drilling fluid according to the American Petroleum Institute (API) Spec. 13A [20] and API Spec. 13B-1 [21] (Phase 3).

5. Nomenclature and List of Abbreviations

API:	American petroleum institute	$Ca_3(PO_4)_2$:	Calcium Phosphate
EDS:	Energy Dispersive Spectrometry	$CaSiO_3$:	Parawollastonite
HGS:	Hollow Glass Microsphere	$Ca_3(Si_{30}Al_6)O_{72}\cdot 20H_2O$:	Clinoptilolite
HSE:	Health, Safety, and Environment	K_2CO_3:	Potassium Carbonate
SEM:	Scanning Electron Microscope	Na_2SO_4:	Sodium Sulphate
UBD:	Underbalanced Drilling	$Na_2[B_4O_5(OH)_4]\cdot 8H_2O$:	Borax
XRD:	X-Ray Diffraction	SiO_2:	Tridymite
$3H_5OH$:	Ethanol	SiO_2:	Cristobalite
$CaCO_3$:	Calcium Carbonate	SiO_2:	Silicon Dioxide
$CaCO_3$:	Calcite		

Acknowledgements
The authors would like to acknowledge funding from the Institute for Innovation and Entrepreneurship Development (Lembaga Pengembangan Inovasi dan Kewirausahaan) Institut Teknologi Bandung (ITB) through the 2019 Innovation Strengthening Program.

References

[1] Al-Yami AS, Al-Awami M, Wagle V 2015 Investigation of Stability of Hollow Glass Spheres in Fluids and Cement Slurries for Potential Field Applications in Saudi Arabia. In: SPE Kuwait Oil and Gas Show and Conference. Mishref, Kuwait: SPE.
[2] Medley GH, Maurer WC, Garkasi AY 1995 Use of Hollow Glass Spheres for Underbalanced Drilling Fluids. In: SPE Annual Technical Conference and Exhibition. Dallas: SPE
[3] Raju BAT, Pratap KK, Pangtey KS, Trivedi YN, Garg S, Georges GP, et al. 2009 Case Study Using Hollow Glass Microspheres to Reduce the Density of Drilling Fluids in the Mumbai High Basin, India and Subsequent Field Trial at GTI Catoosa Test Facility, Tulsa, OK. In: Middle East Drilling Technology Conference & Exhibition. Manama: SPE
[4] Arco MJ, Blanco JG, Marquez RL, Garavito SM, Tovar JG, Farias AF, et al. 2000 Field Application of Glass Bubbles as a Density-Reducing Agent. In: SPE Annual Technical Conference and Exhibition. Dallas: SPE
[5] Al-Yami AS and Schubert J. 2012 Aerated Underbalanced Drilling System Development. In: SPE Russian Oil and Gas Exploration and Production Technical Conference and Exhibition [Internet]. Moscow: SPE Available from: https://onepetro.org/SPERPTC/proceedings/12ROGC/All-12ROGC/Moscow, Russia/159041
[6] Marbun BTH, Benson P, Putra SK, Zulkhifly S 2011 Evaluasi Penggunaan Glass Bubble Sphere pada Sumur-X *JTM* 1–14.
[7] Marbun BTH, Zulkhifly S, Hariz I, Lumbangaol C 2011 A Methodology of Underbalanced Drilling Design in Depleted Reservoir. In: International Petroleum Technology Conference. Bangkok, Thailand: IPTC
[8] Alajmi SE and Schubert JJ 2003 Optimum Selection of Underbalanced Techniques. In: SPE/IADC Middle East Drilling Technology Conference and Exhibition. SPE
[9] Carver B, Fitryansyah Y, Binyamin AK 2011 Improving Heavy-Oil Well Economics with Hollow Microsphere Cementing Solutions: Case History. In: SPE Western Venezuela Section South American Oil and Gas Congress. Maracaibo, Venezuela: SPE
[10] Chen G, Burnett D 2003 Improving Performance of Low Density Drill in Fluids with Hollow Glass Spheres. In: SPE European Formation Damage Conference. The Hague, Netherlands: SPE
[11] JV Vietsovpetro 2010 Assessment on the Performance of Hollow Glass Microspheres in Low Density Fluids for Workover Programs in Fractured Basement Reservoir at the White Tiger Oil Fields, Cuu Long Basing, Vietnam. Kuala Lumpur: SPE/IADC
[12] Maurer technologi Inc. 2001 Development and Testing of Underbalanced Drilling Products. Houston
[13] McLennan J, Carden RS, Curry D, Stone CS, Wyman RE 1997 Underbalanced Drilling Manual, Gas Research Institute. Chicago, Illinios
[14] Edwin P, Syahriza GR, Yose M, Arga P 2010 Lightweight Water-Based Mud Using Glass Bubbles for Drilling a 6-in. Horizontal Section in a Gunung Kembang Development Well. In: SPE Asia Pacific Oil and Gas Conference and Exhibition. Brisbane, Queensland, Australia: SPE
[15] Talisman Malaysia Ltd. 2013 Prehydrating High-strength Microspheres in Lightweight Cement Slurry Creates Values for Offshore Malaysian Operator
[16] Medley et al. 1997 GH Field Application of Lightweight Hollow-Glass-Sphere Drilling Fluid. San Antonio: SPE

[17] 3M. 3M K37 Glass Bubbles (5 lbs.) [Internet]. Available from: https://www.amazon.com/3M-K37-Glass-Bubbles-lbs/dp/B07PXGS999/ref=sr_1_1?keywords=3M+Glass+Bubbles&qid=1559200138&s=gateway&sr=8-1

[18] Marbun BTH 2012 Studi Hollow Glass Microsphere (HGS) (contract number 001/D00160/2012-S0). Bandung

[19] American Petroleum Institute (API) 2005 API Spec 10A: Specification for cements and materials for well cementing. ANSI/API Recommended Practice 10A. Washington, D.C.: American Petroleum Institute (API)

[20] American Petroleum Institute (API) 2007 API Spec 13A: Specification for Drilling-Fluid Materials. ANSI/API Specification 13A. Washington, D.C.: American Petroleum Institute (API)

[21] American Petroleum Institute (API) 2009 API RP 13B-1: Recommended Practice for Field Testing Water-based Drilling Fluids. ANSI/API Recommended Practice 13B-1. Washington, D.C.: American Petroleum Institute (API)

[22] Simonov-Emel'yanov ID, Trofimov AN, Apeksimov N V., Zubkov SB 2014 The structure formation in polymer composites with hollow glass microspheres *Int Polym Sci Technol.* **41**(1) 21–6

[23] NPO Stekloplastik. Company Information [Internet]. Available from: https://corprussia.com/company/0493385

[24] Zhejiang Co. L. Hollow Glas Microspheres [Internet]. Available from: https://www.glass-bubble.com/products.html?gclid=Cj0KCQjwiILsBRCGARIsAHKQWLMg_z0ldF6r-Yw2bzHOps_FY8yPDQPiLNKGSL1hpd4DoQnuHvRunoIaAkHgEALw_wcB

[25] Ceno Technologies. Glass Microsphere - Ceno Technologies [Internet]. Available from: http://cenotechnologies.com/glass_microspheres.php

[26] 3M. 3M Glass Bubble K1 [Internet]. Available from: https://www.3m.co.id/3M/en_ID/company-id/all-3m-products/~/3M-Glass-Bubbles-K1-15-kg-/?N=5002385+3291919861&preselect=3293786499&rt=rud

[27] Polisciences I. Microspheres & Particles [Internet]. Available from: https://www.polysciences.com/default/catalog-products/microspheres-particles/

[28] Zhengzhou Hollowlite Materials Co. L. Hollow Glass Sphere [Internet]. Available from: https://en.hollowlite.cn/

IOP Conf. Series: Earth and Environmental Science 1031 (2022) 012003 doi:10.1088/1755-1315/1031/1/012003

3D gravity data inversion modeling for subsurface identification of Gondang area, Bojonegoro regency

Mariyanto Mariyanto*, **Reyhan Dzaky Darmawan, Widya Utama**
Department of Geophysical Engineering, Faculty of Civil, Planning and Geo Engineering, Institut Teknologi Sepuluh Nopember, Surabaya 60111, Indonesia.

*Corresponding author: mariyanto@geofisika.its.ac.id

Abstract. Gondang area had experienced an earthquake of a 4.2 magnitude in the mid-2015, showing that tectonics in the area are active. Also, the Gondang area has a mountain categorized as one of the eleven earth thermal prospects in East Java, i.e., Pandan Mountain. It is exciting to study the subsurface structures of the Gondang area. To date, precedent studies only present geological reviews of the surface. This study presents information on subsurface structure conditions based on gravity. The technique used was 3D inversion modeling with Single Value Decomposition (SVD) inversion. The study's primary data were gravity data from 61 points with a study area of 6.9 km x 5.5 km from 3-10 September 2019. Secondary data utilized were Bojonegoro geological map and local geological studies. From 3D modeling results, slicing was conducted for five cross-sections based on density contrast. Then, the subsurface 2D cross-section results were interpreted for each section. Subsurface model results illustrated density distributions in south, northeast, and northwest areas with high densities ranging from 2.8 g/cc to 3.0 g/cc in >500 m depths. Meanwhile, the north, southwest, and southeast areas had low densities ranging from 1.70 g/cc to 2.0 g/cc in >500 m depths. Furthermore, it indicates the continuity of andesite intrusion from the south to the study site in >500 m depths.

1. Introduction

Gondang Area, Bojonegoro Regency, East Java Province is where Pandan Mountain is located with geothermal manifestations of hot springs and mud pools [1]. Heat sources in Pandan Mountain's geothermal system are estimated to correlate with plutonic stone bodies associated with magmatism activities generating the youngest volcanic stones. Based on Bojonegoro Regency's regional geology, Pandan Mountain has intrusions where hot springs originated. One successful method to analyze and interpret subsurface structures is the gravity method with the correlation between Bouguer anomaly and regional geology [2].

The gravity method involves variations of earth's gravity caused by subsurface stone density differences. Subsurface regions have different densities from the surrounding, contributing to gravity value deviations known as gravity anomaly. In geothermal exploration, the gravity method is employed in the preliminary survey stage to understand subsurface geological conditions better.

This study generated subsurface structure delineation in a 3D model surrounding the object interest. The modeling was conducted by 3D inversion against residual anomaly data, combining two inversion methods, i.e., Singular Value Decomposition and Occam inversion. Modeling results were density value distributions to identify the highest anomaly as intrusion stones and the lowest anomaly as the fault zone. Then, the density cross-section model was analyzed based on geological information and previous geophysics study results to generate a more accurate subsurface model interpretation.

IOP Conf. Series: Earth and Environmental Science 1031 (2022) 012003 doi:10.1088/1755-1315/1031/1/012003

2. Method

2.1. Data Acquisition

The study's data were primary data from the gravity method measurement results in the Gondang area, Bojonegoro Regency using Scintrex CG-5. Data collection was carried out in 61 points distributed across the study site with a 6900 x 5500-meter data collection area.

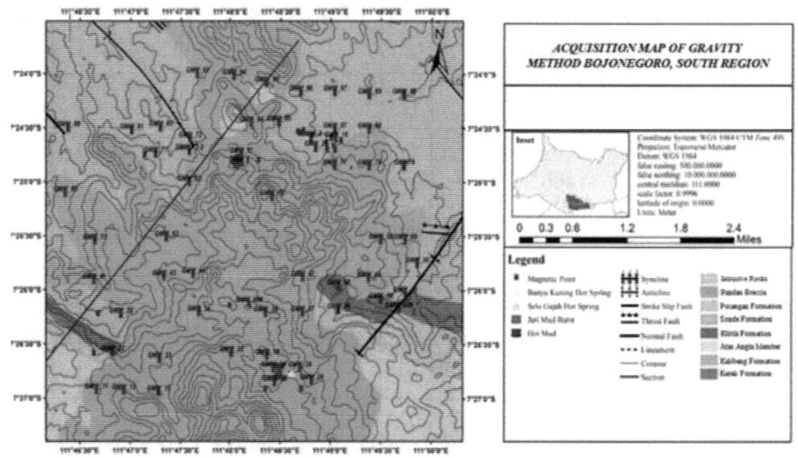

Figure 1. Study Acquisition Map

2.2. Pre-Processing Data

After obtaining field data, the authors applied drift correction, observational gravity value calculation, normal gravity or latitude correction calculation, free air correction calculation, simple Bouguer anomaly calculation, complete Bouguer anomaly calculation, and terrain correction calculation using Mx. Excel. In generating excellent Bouguer anomaly values, one should calculate Bouguer density using the Parasnis method. The Parasnis method was sought by illustrating the graph between free air anomaly (FAA) in the Y-axis and FAC in the X-axis, then obtaining the linear regression value. This regression value is then used as the Parasnis density value. Then, the spectrum analysis was conducted to obtain regional and residual anomalies' depths. This analysis was carried out in Oasis Montaj 8.4 and Ms. Excel on the six measurement paths. Each path produced was spectrum-analyzed in Ms. Excel using the L-curve graph principle, then be plotted. This graph was generated by using complete Bouguer anomaly data or CBA. Then, Fast Fourier Transform (FFT) was applied to CBA data, aided by the Fourier analysis feature in Ms. Excel. After FFT, wave number values (k) and amplitudes (ln A) were calculated.

2.2.1 Filter Moving Average

Filter Moving Average generated the separation of regional and residual anomalies from the CBA map. This separation used window width results from the spectrum analysis.

2.2.2 Three-Dimensional Modeling

In this study, the 3D inversion modeling was carried out using Grablox 2.1 software [3] with inputs of residual anomalies (*.dat) and predetermined initial model (*.inp) to obtain outcomes of two-dimensional (2D) and three-dimensional (3D) density cross-sectional models. The 2D cross-section is a model incision perpendicular to the X- and Y-axis. The 3D model is a model incision perpendicular to the Z-axis (based on the depth) and overall model view in the major block. To display and edit modeling results, Bloxer 1.6 software was employed. The model displayed in Bloxer 1.6 had a format (*.inp), which was the final inversion result in Grablox 2.1 software [3]. Then, both cross-sectional (2D and 3D) were interpreted qualitatively and quantitatively.

IOP Conf. Series: Earth and Environmental Science 1031 (2022) 012003 doi:10.1088/1755-1315/1031/1/012003

2.3. Data Interpretation
The qualitative interpretation was conducted on several 2D cross-sectional models perceived to represent the study site. This interpretation was high, and low-density value distributions associated with the study target. Meanwhile, the quantitative interpretation was conducted on 3D cross-sectional models to obtain density value information in particular depths and locations. Both interpretation results were then analyzed against geological information, both stone types, and topography.

3. Results
The produced map had a value range from 14.8 mGal to 193.3 mGal, as shown in Figure (2), with an area of 38.35 km^2. High anomaly distributions marked red and pink were on the southern study site, while low anomalies marked blue were on the northern region. Cross-section locations in Figure (3) were taken as data samples to calculate the Fourier Transformation. Outputs of this analysis were residual and regional boundary field depth limits.

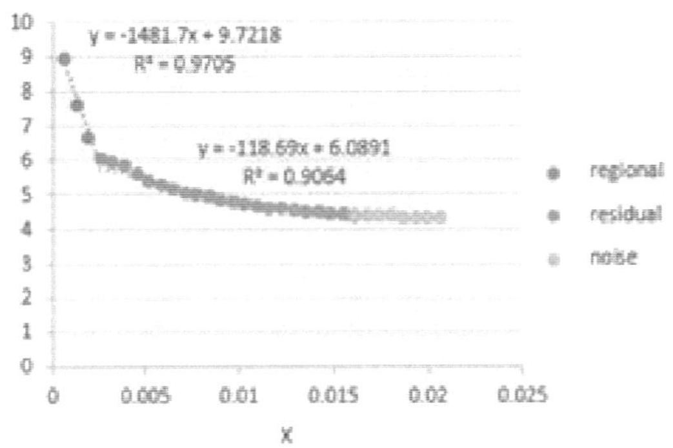

Figure 2. Spectrum analysis results

Table 1. Moving Average depth and window width estimations.

Line	Regional Depth (m)	Residual Depth (m)	kc	N
S1	3360,6	154,5	0,00150	55,656
S2	1481,7	118,69	0,00266	31,433
S3	1611,6	132,77	0,00227	36,957
S4	1043,5	94,643	0,00388	21,586
S5	1272,2	101,76	0,00320	26,145
S6	1560,3	152,91	0,00391	21,405
Average	**1721,65**	**125,88**	**0,0029**	**32,197**

The Red dashed line generated a 1.481 regression value, the regional depth in meter (m) unit, while the grey dashed line generated a 118.69 regression value, the residual depth. Each path from the six paths was calculated for the average regional and residual depths to be used as the 3D modeling initial parameters. Besides generating depth limits, this analysis also estimated the optimum window width to be used during the moving average screening. From Table (1), obtained an overall *k cutoff* (*kc*) value of 0.0029 and *N* value of 32.197.

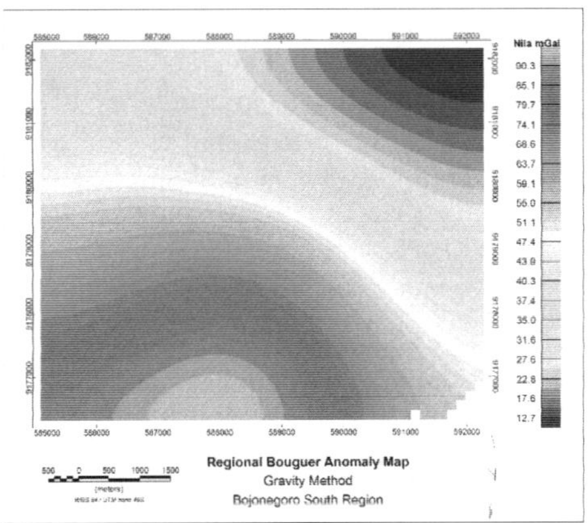

Figure 3. Regional Anomaly Map

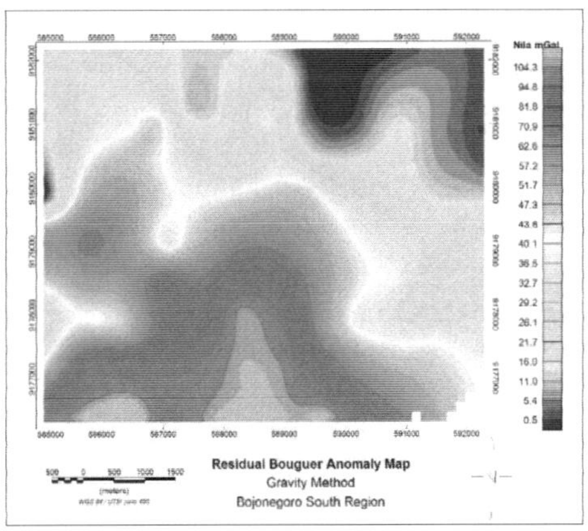

Figure 4. Residual Anomaly Map

IOP Conf. Series: Earth and Environmental Science 1031 (2022) 012003 doi:10.1088/1755-1315/1031/1/012003

From the regional anomaly map, obtained a value range from 12.7 mGal to 90.3 mGal. High anomalies were on the southern study site, while low anomalies were on the northeast. Meanwhile, the residual anomaly map had a value range from 0.5 mGal to 104.3 mGal. High anomaly distributions tended to be on the south region, while low anomalies were on the east region.

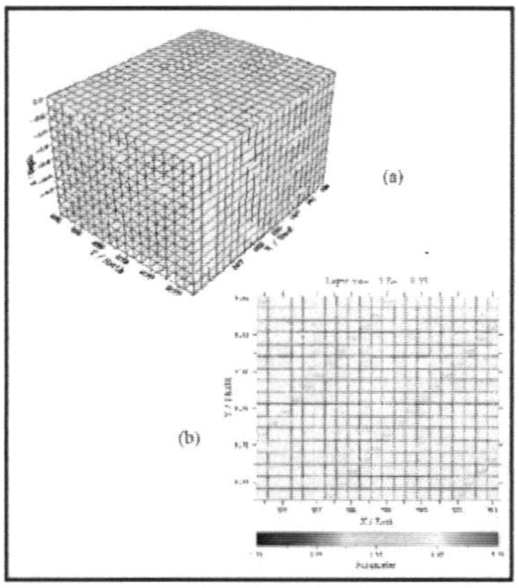

Figure 5. Initial 3D model for initial parameter input in the inversion process. (a) Overall view (b) Top view.

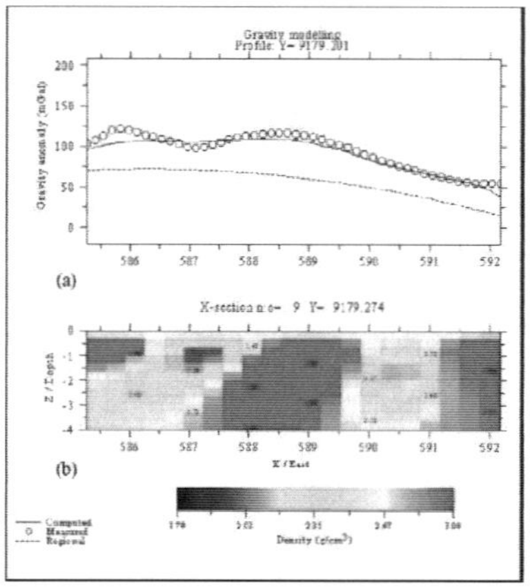

Figure 6. An example of an inversion processing step on a vertical cross-section y=9179. (a) fitting of measurement and calculation data (b) vertical cross-section of density map

The initial model was processed using Single Value Decomposition (SVD) and Occam inversion processes with Grablox software [3]. The input model for data processing uses residual anomaly. From the major block of 21 x 17 x 12, achieved minor blocks for 4284 blocks. Hence, one minor block represents 328.6 x 323.5 x 333.3 (length x width x thickness). Then, the lowest density determination of 1.7 g/cc was caused by the lowest density stone type in the study site of sandstones with densities ranging from 1.61 g/cc – 2.76 g/cc based on Telford et al (1990) [4]. Meanwhile, the highest density stone type in the study site was andesite stones with densities ranging from 2.4 g/cc – 2.8 g/cc. Density parameters were determined using the Parasnis method, obtaining 1.7275 g/cc, in which its value was limited on the geology of the study site rather than using the extremely general earth crust's density of 2.67 g/cc.

4. Discussion

The 2D cross-section interpretation was carried out by selecting formed sections in the initial 3D model production. There were 21 incisions on the X-axis, 17 incisions on the Y-axis, and ten incisions on the Z-axis in this model. Incision orientation directions considered secondary data of geological map having geological structures and intrusion locations. It also considered the study results from Aji et al. (2018) [5] employing modeling surrounding the Pandan Mountain with 2-3 km depth. Then, the model interpretation was strengthened by geological data.

There are two A-A' incisions parallel to the X-axis; one of them was on section 4 with coordinate y=9177.64. The model central region had a density of 3 g/cc, while the western region seemed to clamp central blocks with densities from 1.7 g/cc to 2.0 g/cc. According to Telford et al (1990) [4], andesite stones have a density of approximately 2.85 g/cc, and according to Tan et al. (2003) [6], tuffaceous sandstone density is approximately 1.95 g/cc. Therefore, the central block of this model is interpreted as andesite stones between tuffaceous sandstones. It follows the cross-sectional geology data by

Pringgoprawiro (1992) [7], where pyroxene andesite stones breaking through the Kalibeng formation composed of tuffaceous sandstones and marls.

The B-B' model cross-section is similar to the A-A' model cross-section in the middle of the model with high densities from 2.86 g/cc - 3.0 g/cc. This model is interpreted as a continuation of intrusion from south to north, where this result is consistent with the modeling by Aji et al. (2018) [5] done with different software. Besides, in the geological data in Figure (1), there was also andesite intrusion.

The C-C' path intersected two paths from the south to the north. This cross-section shows a density contrast from a density of 1.76 g/cc to 2.82 g/cc. In the southern part, the cross-section had a lower density than the northern part. It is interpreted as a lithological layer of tuffaceous sandstones to breccias. On the geological map, the formations arranged in the area are the Sonde, Klitik, Kalibeng, and Pandan Breksi formations.

The D-D' path shows the density value from south to center of the study area of 3 g/cc, while the north had a lower density of around 2.68 g/cc. This section is interpreted as a continuation of the andesite intrusion from the south to the study area center. In the northern part of the section with a 1-kilometer depth, there was a lower density of about 1.93 g/cc. This section is interpreted as tuffaceous sandstones. It is consistent with the geological map where the intrusion broke through the southern part of the Kalibeng formation.

The last section, the E-E' path, was located in the eastern part of the study, which cut from south to north. The cross-section shows that the southern part had a low density of around 2.46 g/cc, while the northern part shows a high density of about 2.92 g/cc. The model is interpreted as a continuation of the intrusion zone. Based on the geological map of the low-density areas, the Pucangan formation in the south and Breksi Pandan in the north.

The results of the 3D model inversion show varying density distributions. The rock density values in the model range from 1.7 g/cc to 3g/cc. A hollow block indicates the similarity in value to the surrounding blocks. The high-density blocks of 2.8 g/cc - 3.0 g/cc located in the southern and eastern parts are interpreted as igneous intrusion zone, while densities of 2.35 g/cc - 2.5 g/cc are interpreted as a breached zone. Both of these are following the geological conditions of the study area based on geological maps where the andesite intrusion had densities of 2.4 g/cc - 2.8 g/cc according to Telford et al (1990) [4] breaks through the Kalibeng formation, which is composed of tuffaceous sandstones of 1.95 g/cc according to Tan et al. (2003) [6].

IOP Conf. Series: Earth and Environmental Science 1031 (2022) 012003 doi:10.1088/1755-1315/1031/1/012003

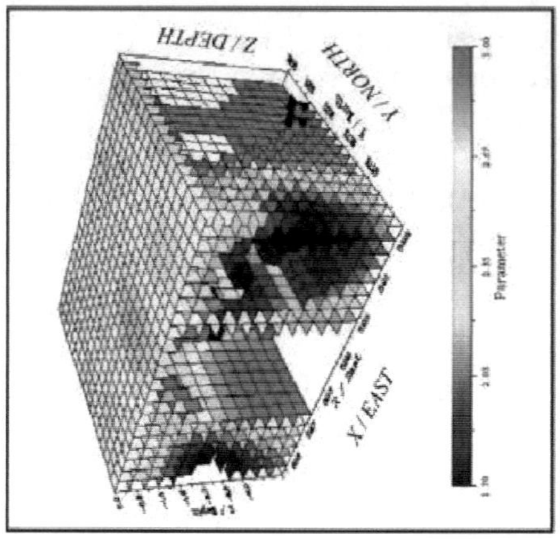

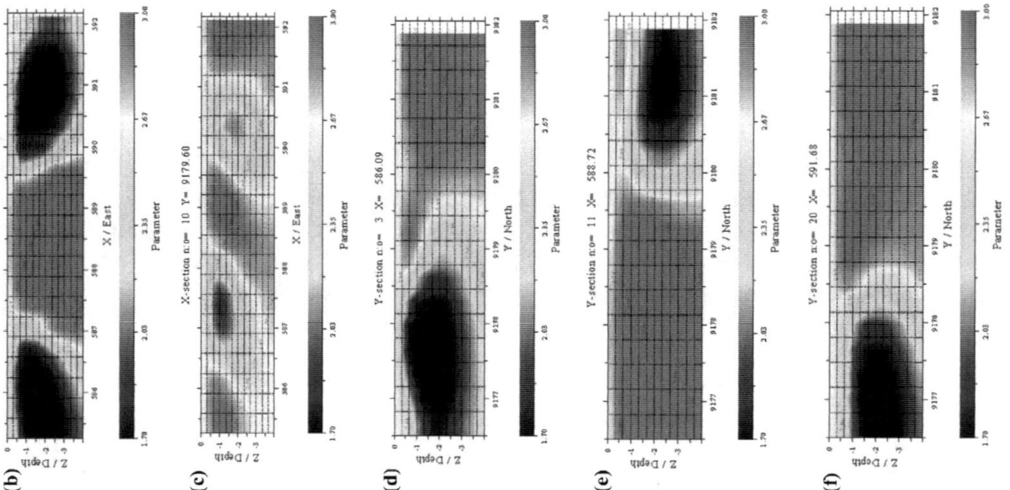

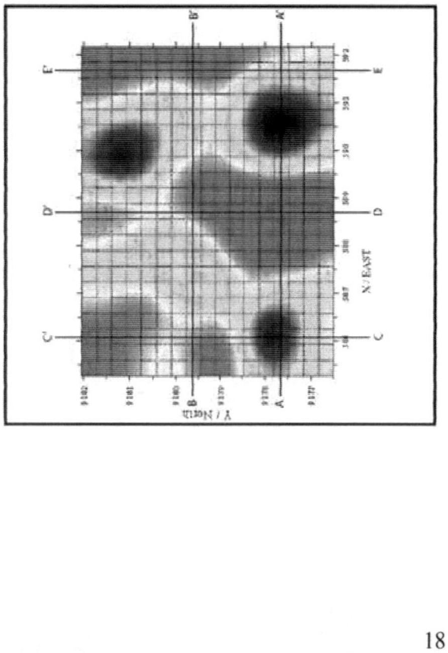

Figure 7. **(a)** 2D cross-section results **(b)** A-A' path of 2D model cross-section **(c)** B-B' path **(d)** C-C' path **(e)** D-D' path **(f)** E-E' path **(g)** 3D inversion model cross-section

5. Conclusion

Based on this study, the distribution of subsurface density in the south, northeast, and west had high densities of about 2.8 g/cc to 3.0 g/cc at >500m depths. Meanwhile, the northern, southwestern, and southeastern parts had low densities of around 1.70 g/cc to 2.0 g/cc at >500m depths. Geologically, it indicates the distribution of andesite rocks in the south, northeast, and west, while in the north, southwest, and southeast, it indicates the distribution of tuffaceous sandstones. Besides, the study area's subsurface geological structure based on 3D inversion modeling obtained continuity of andesite intrusion from the south to the center of the study area with >500m depths.

Acknowledgments
We would like thanks to Institut Teknologi Sepuluh Nopember for funding this research by Hibah Dana Lokal ITS.

References

[1] Pratama O Y, Ramadhani H, WIjayanti H D K and Trisnaning P T 2018 Analisis Tipe Fluida Dan Geotermometer Reservoir Panas Bumi Berdasarkan Data Geokimia Air Daerah Gunung Pandan, Bojonegoro, Jawa Timur. ReTII, 00.
[2] Sihombing J, Lestari W, Mariyanto M and Joni W 2019 Subsurface Analysis Using Gravity Data at Lili Sepporaki Geothermal Area *IOP Conference Series: Materials Science and Engineering* **588** (1) 012009
[3] Parttijavri M 2004 GRABLOX: Gravity Interpretation and Modelling Software Based on 3D Block Model, User's Guide Geological Survey of Finland
[4] Telford W M, Geldart L P and Sheriff R E 1990 *Applied Geophysics 2nd Edition* (Cambridge: Cambridge University)
[5] Aji W, Wahyudi E J, Santoso D, Kadir W G A 2018 Pemodelan Inversi 3-D Data Gaya Berat dengan Perhitungan Iteratif pada Area Gn. Pandan, Jawa Timur, Indonesia, *Jurnal Geofisika* (2018) Vol. **16** No 03 pp 27-33.
[6] Tan T S, Phoon K K, Hight D W and Lerouiel S 2003 *Characterization and Engineering Properties of Natural Soils* (Tokyo: A A Balkema Publishers)
[7] Pringgoprawiro H and Sukido 1992 *Peta Geologi Lembar Bojonegoro* (Bandung: Pusat Penelitian dan Pengembangan Geologi)

IOP Conf. Series: Earth and Environmental Science 1031 (2022) 012004 doi:10.1088/1755-1315/1031/1/012004

Groundwater play in extensional regime: a case study in Cibunar Region, Garut, Indonesia

M Hanif [1,*], W W Parnadi [2]

[1]Research Center for Geotechnology, National Research and Innovation Agency (BRIN), Bandung, Indonesia
[2]Geophysical Engineering Department, Institute Technology of Bandung, Bandung, Indonesia

*corresponding author: hanif.geotek@gmail.com

Abstract. The intermountain Cibunar area of the Garut basin periodically suffered from a drought and thus became a devastating hazard. Its fault structure as an impact of the extensional regime might affect the groundwater play as a transport or barrier system. The tectonic setting and its disruption of volcanic lithologies by faults make the area's hydrogeology very complex. The conducted electrical resistivity tomography coupled with gravity and magneto-telluric studies revealed the aquifer play on the zone is affected by the extensional regime. Tectonically, the drought Cibunar district is close to the top of the horst structure where the uplifted segments might contribute to the limited low resistivity aquifer zone and triggered the flow to the graben zone. The low and high resistivity on the region was related to an unconfined aquifer and unsaturated zone, while the medium zone remains unclear.

1. Introduction

Groundwater in the Cibunar Region, Garut, West Java, Indonesia, as well as the flowing water of Cimanuk stream is suffered from the drought and decreasing water flow in the dry season. This hazard caused a serious problem to the livestock and household activities of the local residents. Moreover, waste due to an increase in population and the growth of industrial activities around the Cimanuk stream could contaminate water reservoirs (called "embung" in local term), which built in 2018. Consequently, these vulnerable circumstances need a more sustainable solution, especially by exploring the unconfined aquifer around the area.

The area is geologically situated on the intermountain Garut basin (Figure 1a), which has an extensional regime [1]. The northern and middle parts of the basin are bordered by Mt. Malabar (2321 m), Mt. Papandayan (2622 m), Mt. Cikuray (2820 m), and Mt. Guntur (1838 m). The lithology of this area (Figure 1b) is dominated by Efflata of Papandayan volcanic (Qhp) which is constituted by volcanic ash, andesite and basalt boulder. Young volcanic deposit from Cikurai Mt. and Guntur Pangkalan Kendang (Qyc and Qypu) consists of eflata and lava flow, mainly andesitic basalt, and undifferentiated volcanic rocks such as volcanic ash, tuff, lapilli, and breccia. The extensional normal faults generally have an important role on the groundwater system. Previous studies were performed on the relation between tectonic structure and groundwater play, which could influence its flow, recharge-discharge relation, and water-table elevations [2–4].

Electrical resistivity tomography (ERT) is a non-invasive geophysical tool and an indirect ground imaging method that can solve some hydrogeological issues. The groundwater feature could be recognized by the resistivity contrasts between saturated/unsaturated formations, such as observed in volcano-sedimentary deposits [5]. Numerous resistivity studies around this basin are mostly focused on the deeper geological scale due to its geothermal potential (e.g. Papandayan and Guntur). However, the near-surface and deep resistivity studies have never been studied to investigate the region's groundwater play and drought factor. This study aims to understand the geologic control on the groundwater, which leads to solving the possible aquifer potential using the resistivity method.

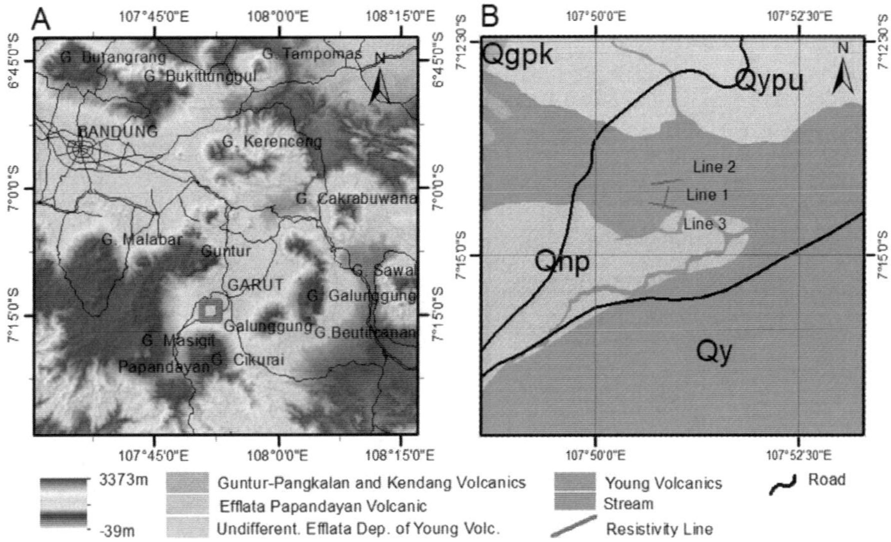

Figure 1 a The Cibunar studied area (red box) overlaid in digital elevation model indicates the intermountain basin which is bordered with Cikurai, Galunggung, Papandayan, Masigit, and Guntur Mountains. **b** Three resistivity lines (red line) are situated around Young Volcanic formation (Qy and Qypu), Efflata Papandayan (Qhp), and Guntur Volcanic Rock (Qgpk). The lithology refers to Garut-Pameungpeuk regional geology map [6].

2. Method

Direct current (DC) resistivity method utilizes a current that is transmitted from a logger through two electrodes. The passage of this current into the ground creates a potential that is measured with two other electrodes [7]. Survey using this method was conducted in Cibunar district, Pakuhaji village by using Lippmann Earth resistivity meter with Wenner electrode configuration at 5 m spacing.

The preferences of this acquisition method and location are based on the geological analysis and drought assessment through the resident dug wells during periods of drought. The resistivity data were collected through 3 lines: 280 m long line 1 in N-S direction, 225 m long line 2 and 200 m long line 3, both in E-W direction as seen in figure 1b. The secondary data such as gravity and magneto-telluric data from published report were utilized to provide a comprehensive and detailed view of geology condition related to the groundwater.

The magneto-telluric (MT) method works based on geomagnetic and geoelectric field variation at earth's surface in the frequency range of 10000-0.00001 Hz. Consequently, it could be used for exploration from near surface to thousands of meter depth. The MT profiles used in this study were obtained from a published paper [1]. The gravity method is also a non-invasive geophysical method that could map the subsurface density distribution, faults, sediment thickness, and groundwater potential. The Bouguer anomaly used in this study was obtained from a published paper [8]. By a deeper penetration and a wider range of MT and gravity method, this region's groundwater play might be clearly understood to find a possible aquifer.

IOP Conf. Series: Earth and Environmental Science 1031 (2022) 012004 doi:10.1088/1755-1315/1031/1/012004

3. Results and Discussions

Two-dimensional resistivity models obtained for 3 lines shown in Figure 2 indicate the subsurface geological structure and might correspond to its groundwater play. On average, the low resistivity value (1-24.5 Ohm.m) is interpreted as the saturated groundwater reservoir, the medium resistivity range (24.5-56.1 Ohm.m) might correlate with the medium saturated zone, while the highest value zone (54.1-90 Ohm.m) is probably related to the unsaturated zone.

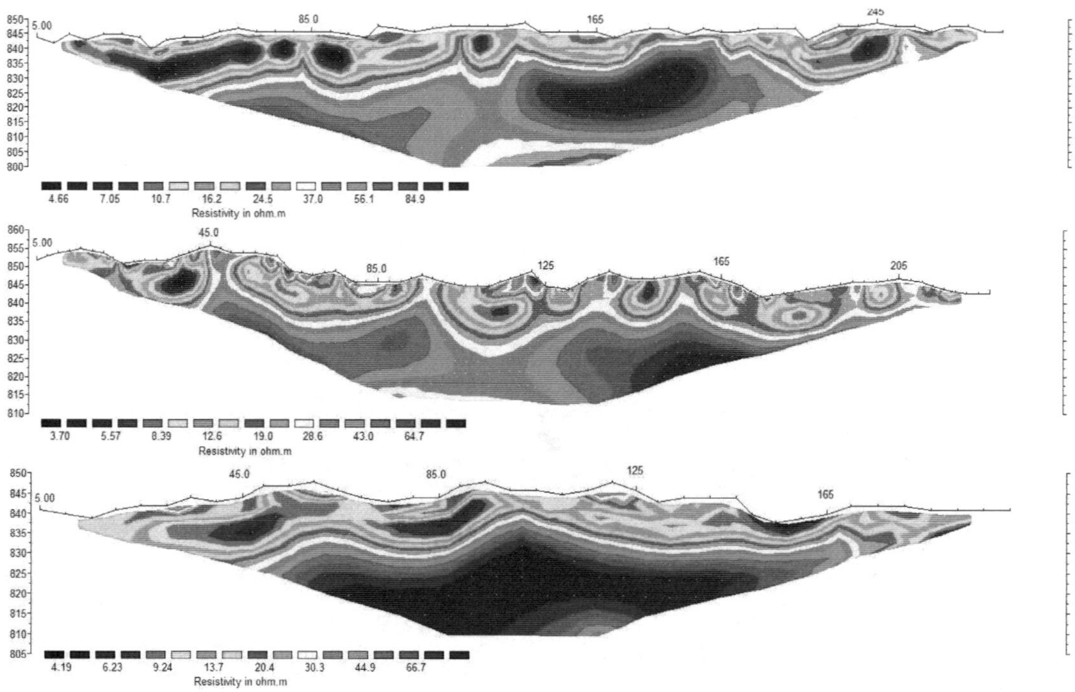

Figure 2 a-c (top to bottom): The resistivity profiles of Line 1, Line 2, and Line 3. The low resistivity (blue to dark green) is interpreted as groundwater reservoir, the medium resistivity range (yellow to orange) is interpreted as medium saturated zone, and the high resistivity zone (red to purple) indicates the unsaturated zone. The map of the line refers to Figure 1.

The low resistivity zone as a saturated water reservoir which has a thickness <20m lies in the depth range of resident dug wells. During drought, it might be expected that this low resistivity zone is an unconfined aquifer that has no impermeable trap to accumulate the groundwater. Besides, the volume is probably vulnerable to the decrease of rainfall rate. The high resistivity zone (red to purple) which seems to be a zone with unsaturated medium, might be interpreted as an impermeable rock dominated on the whole lines (30-40 m). Since a few low resistivity zone remains on the bottom of those profiles, the potential of a deeper confined aquifer might be expected. However, due to the limited penetration (~45 m), a clear confined aquifer could not be delineated.

The magneto-telluric (MT) profiles revealed a broader and deeper resistivity profiles on the region which are concluded as an extensional tectonic regime (Figure 3a) [1]. Both two profiles showed three regions of resistivity which were delineated with horst-graben and normal fault structures. The studied Cibunar area is located around the G03-G05 on the MT line (Figure 3a). Generally, the regional geological background will have no big difference in this distance range (~3 km). The lowest range of resistivity (<100 Ohm.m, red to yellow) with about 250 m thickness might be related to the sediments on the near surface, consistent with the resistivity tomography value range. The thickness indicated a high potential of groundwater on this depth.

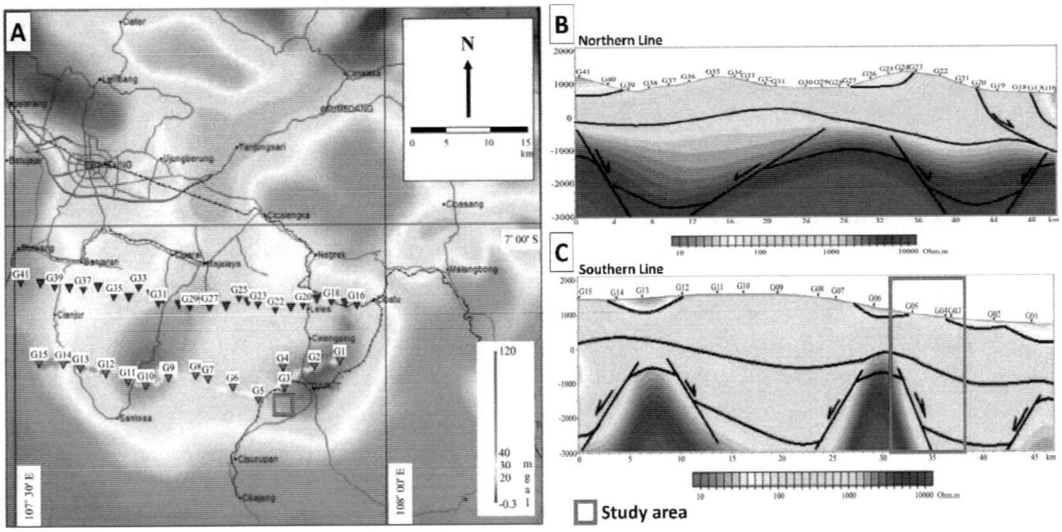

Figure 3 a The Bouguer anomaly map on Garut basin and the magneto-telluric station locations. **b** The northern line resistivity profiles (blue stations) **c** The southern line resistivity profile (red stations). The extensional regime, horst-graben structure, and faults distribution might affect the distribution of groundwater aquifer. These figures were modified from a published paper [1].

The Bouguer gravity anomaly indicated a lowering anomaly to the eastern part of the zone which means a thickening sediment deposition to the eastward and might have more probabilities on the aquifer accumulation. Moreover, the rapid decreasing change on the Bouguer anomaly indicates the fault structures that might exist intensively in the region. This analysis deducts two probabilities on the groundwater play based on the resistivity tomography results (Figure 2). Firstly, the medium zone might be interpreted as a normal fault structure or fracture that plays a conduit to the groundwater. The second alternative, the medium to the high anomaly, is a boulder of impermeable igneous rocks that separates the above groundwater play from the confined one. Consequently, deeper resistivity and geological studies need to be conducted to validate the vicinity of the confined aquifer and the fault role to the groundwater play in this zone.

4. Conclusions

Based on the geological study and electrical resistivity tomography results, the profiles have three distinguished resistivity ranges that illuminate subsurface lithology. The low, medium, and high resistivity values are expectedly correlated with the unconfined groundwater, weathered rock, and igneous rock, respectively. Given the groundwater shortage in the region, the citizens' dug well groundwater source probably has no connected structure to the deeper confined aquifer. The integration with the magneto-telluric and gravity approach revealed the extensional tectonic regime controls the groundwater play in the Cibunar area especially on the horst structure which the prolific aquifer might accumulate in the graben zone. Further research should explore the confined aquifer potential by using resistivity sounding and well bore-based methods.

Acknowledgement
The authors wish to thank Tanoto Foundation and Kemenristek-BRIN (formerly Kemenristekdikti) for the funding under TSRA 2012 and PKM-P grant. Mr. Ilham Togi Sihombing and Social Service Division - Geophysical Engineering Student Association HIMATG TERRA ITB are thanked for the 'Kampung Kita' initiation and field work.

IOP Conf. Series: Earth and Environmental Science 1031 (2022) 012004 doi:10.1088/1755-1315/1031/1/012004

References

[1] Handayani L, Kamtono K and Wardhana D D 2013 Extensional Tectonic Regime of Garut Basin based on Magnetotelluric Analysis *Indones. J. Geosci.* **8** 127–33

[2] Apaydin A 2010 Relation of tectonic structure to groundwater flow in the Beypazari region, NW Anatolia, Turkey *Hydrogeol. J.*

[3] Melchiorre E B, Criss R E and Davisson M L 1999 Relationship between seismicity and subsurface fluids, central Coast Ranges, California *J. Geophys. Res. Solid Earth*

[4] Delinom R M and Suriadarma A 2010 Groundwater Flow System of Bandung Basin Based on Hydraulic Head, Subsurface Temperature, and Stable Isotopes *J. Ris. Geol. dan Pertamb.* **19** 55

[5] Toulier A, Baud B, de Montety V, Lachassagne P, Leonardi V, Pistre S, Dautria J-M, Hendrayana H, Miftakhul Fajar M H, Satrya Muhammad A, Beon O and Jourde H 2019 Multidisciplinary study with quantitative analysis of isotopic data for the assessment of recharge and functioning of volcanic aquifers: Case of Bromo-Tengger volcano, Indonesia *J. Hydrol. Reg. Stud.* **26** 100634

[6] Alzwar M, Akbar N and Bachri S 1992 Regional Geology Map of Garut and Pamengpeuk, Java *Cent. Geol. Res. Dev.*

[7] Knödel K, Lange G, Voigt H-J, Seidel K and Lange G 2007 Direct Current Resistivity Methods *Environmental Geology*

[8] Untung M and Sato Y 1978 Gravity and geological studies in Jawa *Indones. Geol. Surv. Bandung Geol. Surv. Japan, Tokyo*

IOP Conf. Series: Earth and Environmental Science 1031 (2022) 012005 doi:10.1088/1755-1315/1031/1/012005

Subsurface Modeling using Gravity and TDEM in Jiken, Blora Regency

W Aji[1], **E J Wahyudi**[1], **and Warsa**[1]
[1]Institut Teknologi Bandung

Abstract. Geophysics plays an essential role in imaging the subsurface condition of the earth. Gravity and time-domain electromagnetic method is geophysical method that can be used to visualize the subsurface condition. In this research, gravity and TDEM surveys have been carried out in Jiken, Blora Regency. Modeling technique is needed to obtain the subsurface model that bring information related to subsurface condition. Inversion modeling is applied separately to the gravity data and TDEM data, respectively. The density interface inversion modeling is used to gravity data.Subsurface condition is approached by three layers model with different density contrast values in each layer. In these schemes, the modeling process will determine the thickness for every layer. While for TDEM modeling,differential evolution technique is applied to model the TDEM data. The subsurface condition is approached by four layers model. In these schemes, the modeling process will determine the resistivity and thickness for every layer. Finally, the model obtained from gravity inversion modeling is integrated into the model from TDEM inversion data by overlaying the model.

1. Introduction

Earth is composed of various types of rocks that vary both on the surface and subsurface. Rock variations on the surface can be observed directly, while changes that occur below the surface cannot be observed directly. One way to determine subsurface conditions is by conducting geophysical surveys. Geophysics plays an essential role in imaging subsurface conditions to obtain information related to subsurface conditions. Subsurface conditions are characterized by the physical properties of the rocks making up the earth. In this study, two geophysical methods will be discussed, namely the gravity method and time-domain electromagnetic (TDEM). The gravity method can be used to characterize subsurface conditions based on variations in the density of subsurface rocks. The TDEM method can be used to describe subsurface conditions based on differences in the conductivity value of subsurface rocks. To obtain subsurface images from both the gravity method and the TDEM method, a modeling technique is needed. In this study, inversion modeling techniques are applied to gravity data and TDEM data. The density interface inversion technique is used to gravity data, while the differential evolution inversion technique is used to TDEM data. The case study discussed is data from the Jiken area, Blora Regency.

2. Gravity Modeling using Interface Density Inversion

Modeling using gravity data for this work briefly ilustrated as seen in figure 1. For computation of inverse problems we can set-up using simple body as model in the subsurface. Several series of prism (M-series model) are placed below each gravity data (N-series data). For each gravity data in the surface, we have two layers of vertical prism. Two layers in the subsurface will be used to provide sedimentary layer density (ρ_s) and basement density(ρ_b). The densities for sedimentary and basement will be set as homogeneous for each vertical prism.

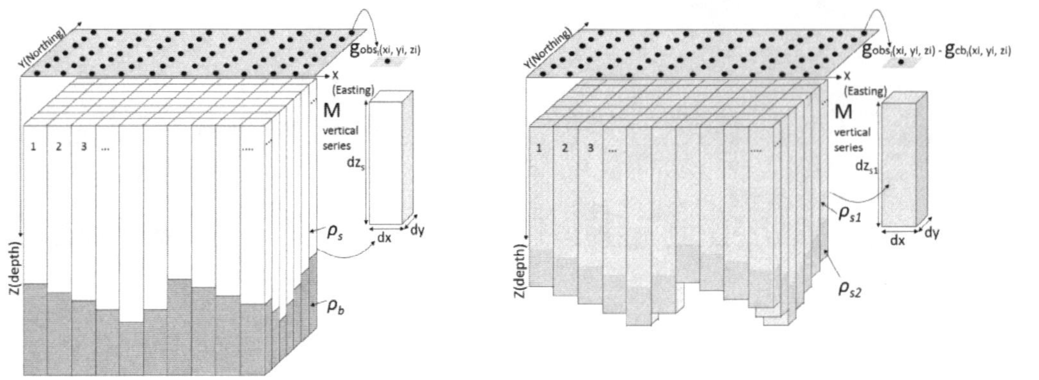

(a) (b)

Figure 1. Illustration of the subsurface model to explain: (a) the interface of homogeneous basement layer (ρ_b) and sediment layer (ρ_s) (flat topography) and (b) the interface between first (ρ_{s1}) & second (ρ_{s2}) layer of sedimentary rock (ρ_s).

Model in the subsurface (as discretization of sedimentary layer and basement) has three dimensional of dx, dy, and dz. Maximum depth will be set-up as as lower bound of basement, while the topography or elevation data will be used as constraint of upper bound of sedimentary layer. The solution of iterative calculation will be targeted for interface layer between the basement and sedimentary layer. The solution will be updated itertively in the scheme of inverse modeling.

Misfit data evaluation will be determined using the differences value between observed (g_{obs}) and calculated (g_{cal}) data. The calculation of calculated data conducted using the equation described by Plouff [1]. The thickness of sedimentary layer ($dz(j)$) will be updated as solution using simple formulation (with modification from previous work [2]), as follow:

$$dz^{(k+1)}(j) = dz^{(k)}(j) + \frac{Rand \times (g_{obs}(i) - g_{cal}(i))}{G\rho_s}, \tag{1}$$

where i, j, and k respectively are mathematical notation for index number of data observation, prism model, and iteration number. Rand is random naumber and G is gravitational constant. In the equation (1) we use the last term of nominator and denominator to provide a rough estimation of updated solution.

We set the maximum iteration number to optimize the misfit value of observed (g_{obs}) and calculated (g_{cal}). After we get the thickness of the sedimentary layer ($dzs(j)$) we continue to divide the sedimentary layer into two layers with different density. The observed data to be fitted in the next iteration (replaced observed gravity) is calculated by substracting our g_{obs} with g_{cb} (gravity calculated of the basement) from the previous iteration. Model designed and formulated using lower bound of depth from the last iteration. The illustration of model design and data to be fitted are shown in figure 1b. From figure 1b, we can see the similarity of the problem, as shown in figure 1a. The interface of the sedimentary layer then calculated with the same scheme as shown in equation (1). Brief procedure of iterative calculation described in the flow-chart in figure 2. Several steps of flow chart will be started with the input and initialization - initial model - iterative process - stopping criteria - output.

IOP Conf. Series: Earth and Environmental Science 1031 (2022) 012005 doi:10.1088/1755-1315/1031/1/012005

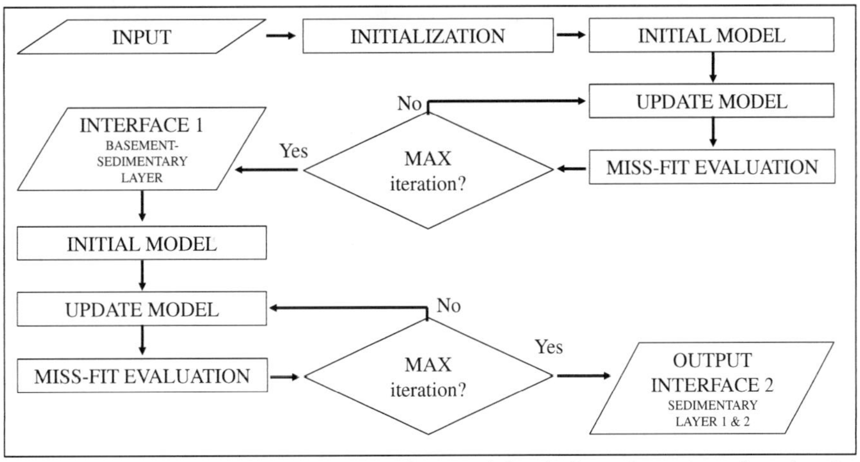

Figure 2. Flow-chart of the program to seek a model solution of two interface density in the subsurface.

Next, we describe the modeling of the study area using iterative calculation, as mentioned previously in the methodology part. Considering preliminary study [2] and well data, the input (density contrast) of the first (upper) layer is -0.10 g/cc and the input (density contrast) of the second (lower) layer is +0.10 g/cc. We calculate with 100 realizations to provide an average solutions from the statistical experiments using a random number. The solution of the interface between the basement-sedimentary layer (first interface) is iteratively computed using 200 times of iteration number for each realization. Vertical prisms dimension (dx and dy) used in the modeling are 250 m. Considering preliminary study [2], residual anomaly and topographic data as input. Residual anomaly variation shows subsurface correlation as well as the seismic data [3]. five kilometers depth is also used as maximum depth of the subsurface model. Total of 441 data observation (residual anomaly) and 441 vertical series (each series consist of two layers) of the subsurface models are calculated. Topographic data with 250 m interval data in the Northing and Easting direction also incorporated in the calculation.

Misfit data changes during 200 iterative calculation of the first interface (basement-sedimentary layer) are shown in figure 3a. Misfit data that are shown in figure 3a is an average value of 100 realizations from the optimization process to find the solution first interface. Until 200 iterative calculation, average misfit data stuck in the value of misfit data around 67 mGal. From these results, we get optimization of the first interface, and the remaining data misfit (with higher frequency or small wavelength) most likely will be fitted with a shallower density interface above the first interface.

Residual anomaly from the first modeling process is used to model two sedimentary layers above the basement. The following input of the upper and bottom layer density will separate the sedimentary layer into two layers, they are -0.15 and +0.05 g/cc (the value is average value from previous iteration for first (upper) layer). For the second optimization process, we also calculate with 100 realizations and iteratively computed using 200 times of iteration number for each realization. Vertical prisms dimension (dx and dy) used in the modeling are 250 m. Varying depth in the grid form of figure 1a (sedimentary layer) will be set-up as constraint second iterative scheme as maximum depth. The calculation of 441 data then replaced with the value from the results of substraction the residual anomaly with basement response (from first scheme iteration).

Misfit data changes during 200 iterative calculation of the second interface (between two sedimentary layers) are shown in figure 3b. After 200 iterative calculation, average misfit data stuck in the value of around 0.5 $mGal^2$. From these results, we have inversion results as shown in the 3D view (figure 4).

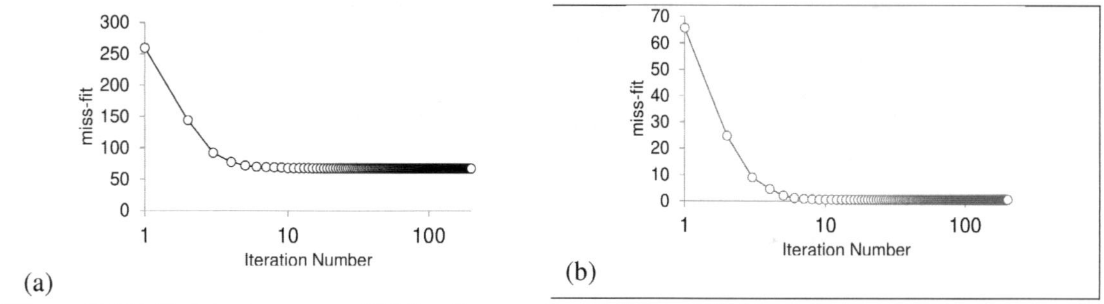

Figure 3. Misfit data changes vs. 200 iteration number (in log scale): (a) misfit data in the average value of 100 realizations in the process of optimization first interface (basement-sedimentary layer) and (b) misfit data in the average value of 100 realizations of the second interface (between two sedimentary layers).

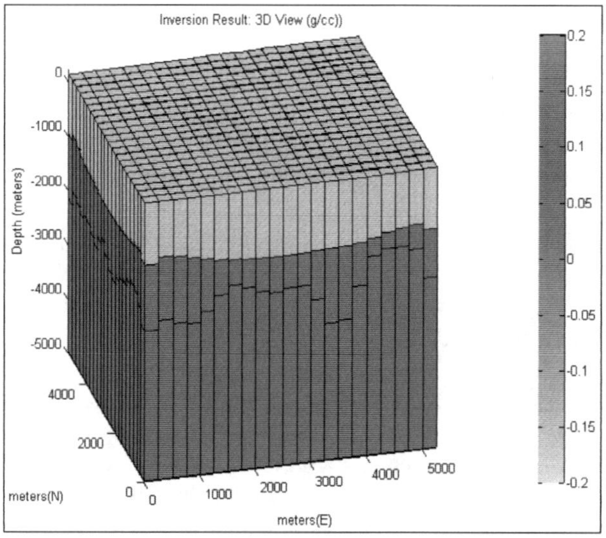

Figure 4. Visualization of the 3D inverse model from 441 vertical series of subsurface model.

3. TDEM Modeling

TDEM method is included in the category of controlled electromagnetic sources [4]. The TDEM method has the principle of injecting current into the ground at certain time intervals and turning off the injection of the current [5]. Model parameters in the TDEM method are the conductivity and thickness of the layer. While the data is the decay of the magnetic field to time. Model parameters and data linked to future modeling. The forward modeling equation for the TDEM method with long-wire configuration [6] is as follows:

$$H_z = \frac{I}{4\pi}\int_{-L}^{L}\frac{y}{R}\int_{0}^{\infty}(1+r_{TE})e^{\mu_0 z}\frac{\lambda^2}{\mu_0}J_1(\lambda r)\,d\lambda\,dx' \qquad (2)$$

Where *Hz* vertical magnetic field (A/m), *I* is electric current (A), *R* is the distance between *transmitter* and *receiver* (m), *x* and *y* are receiver coordinate (m), r_{TE} is reflection coefficient, λ is integration variable of Hankel transform, and *J1* is first-order Bessel function.

In the calculation of calculated data from TEM using the forward modeling equation written in equation (2) will generate data in the frequency domain (Hz). The inverse Fourier transform technique can be applied to transform from calculated data into the time-domain. Mathematically the inverse Fourier transform process can be done by using cosine filters to the vertical magnetic field data (*Hz*) according to the equation formulated in published paper [7].

Next, for the modeling process, we use a differential evolution algorithm (DEA). DEA is population-based global optimization [8]. DEA is consist of two main stages that are initiation and evolution by mutation, crossover, and selection [9]. Brief flowchart of the inverse modeling is shown in figure 5.

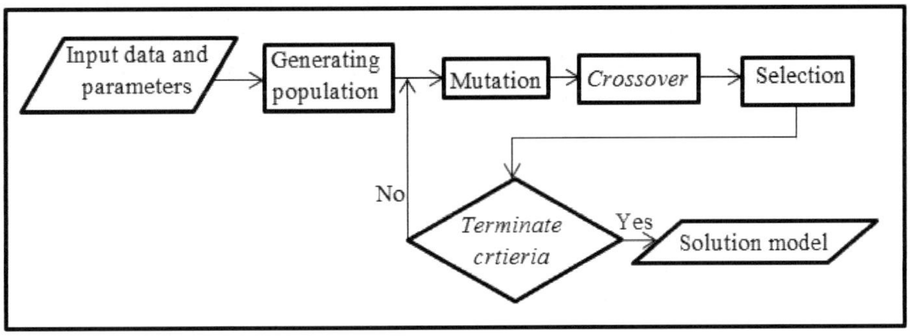

Figure 5. Flow-chart of TDEM inverse modeling.

We modeled 16 TDEM sounding points that form the NW – SE line. The position of the TDEM line can be seen in figure 6. The inversion process is done by 50 iterations. The model approach used is four layers model. Then, the result of TDEM modeling is overlayed by the model from gravity modeling (figure 7).

Based on the model in Figure 7, it appears that the uppermost layer of gravity modeling has varying resistivity values when referring to the results of TDEM modeling. In the middle of the track, the resistivity value is relatively high and is getting lower towards NW and SE. Next, the second layer of gravity modeling has a consistent resistivity value that is relatively low. So it can be said that low resistivity values characterize the second layer. While on the third layer, the resistivity values tend to vary. At some TDEM sounding points, relatively high resistivity values were found, and at other locations, the resistivity values were low.

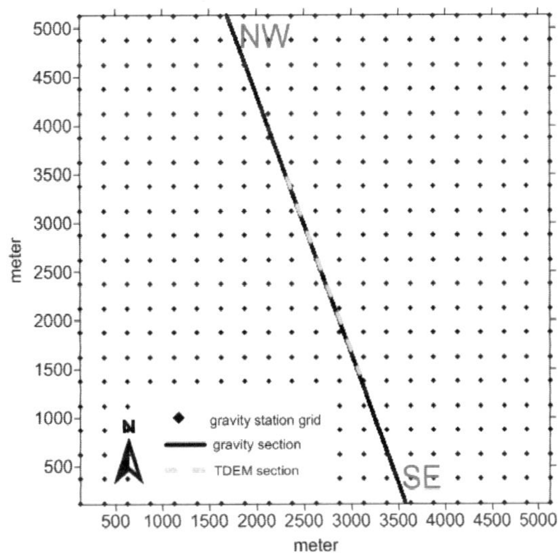

Figure 6. Position of TDEM line and gravity section.

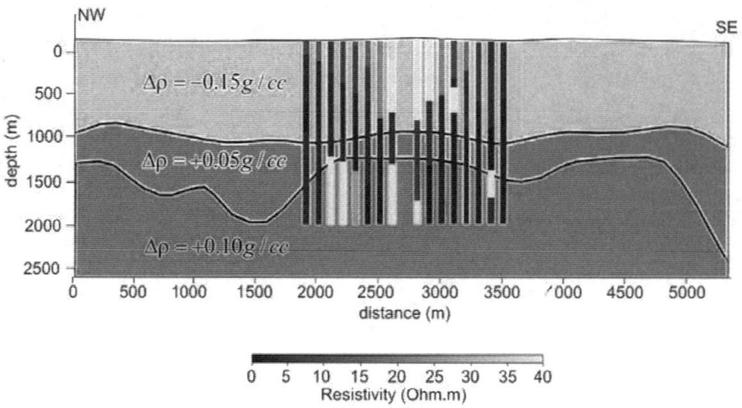

Figure 7. Overlay of the model from TDEM modeling and the model from gravity modeling.

4. Summary

Based on the modeling process that has been done, gravity density interface inversion modeling can produce a model of layer thickness variation based on variations in density contrast. Whereas TDEM inversion modeling can characterize each layer based on resistivity contrast. When modeling is applied to field data in Jiken, Blora Regency, the integration between the results of gravity inversion modeling and TDEM inversion modeling can characterize areas with low resistivity values with contrast density values of +0.05 g / cc. The area is under the layer with a density contrast value of -0.15 g / cc which has varying resistivity values that are higher in the middle of the path, and above the layer with a density contrast value of +0.10 g / cc which also has a resistivity value varies in each TDEM sounding point.

Acknowledgments

The authors acknowledge support from FTTM ITB.

References

[1] Plouff D 1976 Gravity and magnetic fields of polygonal prisms and application to magnetic terrain corrections *Geophysics* **41** pp 727-741

[2] Wahyudi E J and Kadir W G A 2017 Inverse Modeling of Gravity Data with Two Layers Density in Sedimentary Basin Structure *Journal of Physics: Conference Series* **1090** pp 1-11

[3] Tsuji T, Matsuoka T, Kadir W G A, Hato M, Takahasi T, Sule M R, Kitamura K, Yamada Y, Onishi K, Widarto D S, Sebayang R I, Prasetyo A, Priyono A, Ariadji T, Sapiie B, Widianto E, Asikin A R and team G C P 2014 Reservoir characterization for site selection in the Gundih CCS project, Indonesia *Energy Procedia* **63** 6335-6343.

[4] Kanta A, Soupios P, Vallianatos F, Rust D and Barsukov P 2009 The application of time-domain electromagnetic method to characterize the Keritis Basin, in Western Crete, Greece *Proceeding of the 3rd IASME / WSEAS International Conference on Geology and Seismology.*

[5] Srigutomo W, Warsa, Sule R M R, Trimadona, Prasetyo D, Akira S, Nakayama K, Hato M, Prasetyo A and Widarto D S 2015 Time-domain electromagnetic (TDEM) baseline survey for CCS in Gundih area, Central Java, Indonesia *Proceedings of the 12th SEGJ Symposium Tokyo* page 1-5

[6] Ward S H and Hohmann G W 1988 *Electromagnetic theory for geophysical applications in electromagnetic methods in applied geophysics* Vol I 131-311 ed. Nabighian M N SEG Tulsa

[7] Mitsuhata Y, Uchida T, Murakami Y and Amano H 2001 The Fourier transform of controlled-source time-domain electromagnetic data by smooth spectrum inversion *Geophys. J. Int* **144** 123 – 135

[8] Qin A K and Suganthan P N 2005 Self-adaptive differential evolution algorithm for numerical optimization, *IEEE Congress on Evolutionary Computation.*

[9] Balkaya C 2013 An implementation of differential evolution algorithm for inversion of geoelectrical data *Journal of Applied Geophysics* **98** pp 160-175.

Electrical resistivity survey for groundwater investigation in Padalarang, West Java

H Phanjaya, W W Parnadi, dan D Santoso

Geophysical Engineering, Faculty of Mining and Petroleum Engineering, Bandung Institute of Technology, Jln. Ganesha No. 10, Bandung 40132, Indonesia

Abstract. An electrical resistivity survey has been carried out in Padalarang, West Java, to investigate groundwater potential to solve the lack of clean water. The survey consists of three vertical electrical sounding (VES) and four electrical resistivity tomography (ERT) profiles in the study area. The data was processed using inversion method. Wells in the area is available to validate resistivity data. The result of VES shows a low resistivity value (2.82 - 11.72 Ωm) at its fourth layer, which we expected to be a zone with high water content. ERTs result also revealed a low resistivity value (<10.1 Ωm), approximately more than 60 m depth at ERT 3 and 4. We expect this low resistivity zone to have a groundwater potential. A low resistivity zone in the central part of the study area may yield a high groundwater potential. Our result could serve as a guide to engineers for planning on sitting boreholes either for domestic, agricultural, or industrial use.

Keywords: Padalarang, VES, ERT, low resistivity zone, groundwater potential

1. Introduction

Water is essential for humans to live, and one of the safest kinds of water supply is groundwater. Minor water treatment is required for groundwater since it has natural protection against pollution by the covering layers [1]. Lack of access to appropriate water quality for domestic, agricultural, and industrial uses is one of the major limiting factors to socioeconomic development.

West Java has the highest population density in Indonesia, with 18% of the Indonesian population living there. The dry season in West Java always brings drought disasters leading to a lack of clean water for the local community. The main problem of the clean water crisis is the lack of infrastructure to manage or find a new source of clean water. In addition, the amount of population and demand for potable water for domestic, agriculture, and industrial uses in West Java also have increased rapidly.

The resistivity survey is well-known in groundwater studies [2] because of its simplicity, low-cost technique, and efficiency in areas with high contrasting resistivity [3]. We measured ground resistivity by injecting currents between one electrode pair and measured the resulting potential differences at the surface between another electrode pair. Commonly, we apply the one-dimensional vertical electrical sounding (VES) technique to investigate groundwater resources [4]. In some cases, we apply the two dimensional electrical resistivity tomography (ERT) technique where lateral changes or heterogeneity in resistivity exist [5]. Both methods to investigate groundwater resources are recently applied successfully rather than the VES stand-alone technique [6]. The objective of this study is to apply an integrated geoelectrical approach (VES, ERT) to investigate groundwater potential in West Java.

We located the study area in the sub-district of Padalarang, 20 km from the capital of the province of West Java (figure 1). There were four wells available in this area prior to the previous survey. The area was characterized mainly by heavily forested terrain, while the local community used some of the north parts as a farming field.

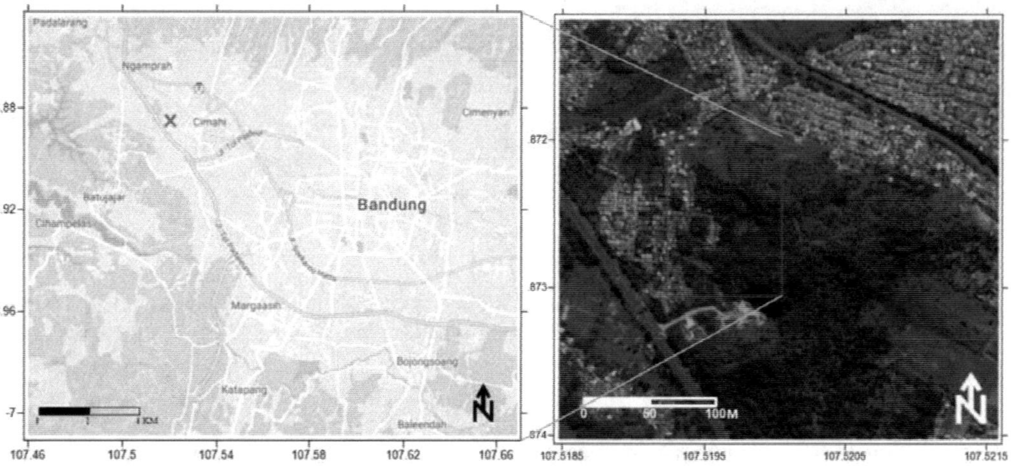

Figure 1. Location of the investigated area.

2. Material and methods

The field resistivity acquisition used in this survey was the ARES II equipment with an array of 48 electrodes. The ARES II system is computer-controlled with direct reading of the apparent resistivity of the grounds in ohm digitally by its LCD monitor. The position of the electrodes was oriented perpendicular to the ground surface to keep signal distribution.

A total of 3 VES by applying Schlumberger array were carried out in the studied area (figure 2,) with the distance between the two current electrodes starting from 2 meters and reaching a maximum of 360 meters. We processed the measured apparent resistivity data by the inversion technique to determine each layer's true resistivity and thickness. In this work, we used MATLAB software to produce depth-sounding curves and inversion of VES data for computer modeling. The algorithm optimizes the fit between the observed and the calculated apparent resistivity data. We initialized the procedure by user-selected initial model parameters and employed the damped least-squares solution with the Singular Value Decomposition (SVD) technique for inversion [7].

ERT data acquisitions also used the Schlumberger array with four lines, which intersect each other (figure 2). ERT 1 and ERT 2 have electrode spacing of 7.5 m and 330 m in length, while ERT 1 and ERT 2 have electrode spacing of 10 m and 430-440 m in length. The acquired resistivity data were then inverted based on the smoothness-constrained least-square method with Res2Dinv software from Geotomo [8]. We added topographic data due to the significant elevation of each profile. We used the root mean square (RMS) value to determine the fitness between observed and calculated data.

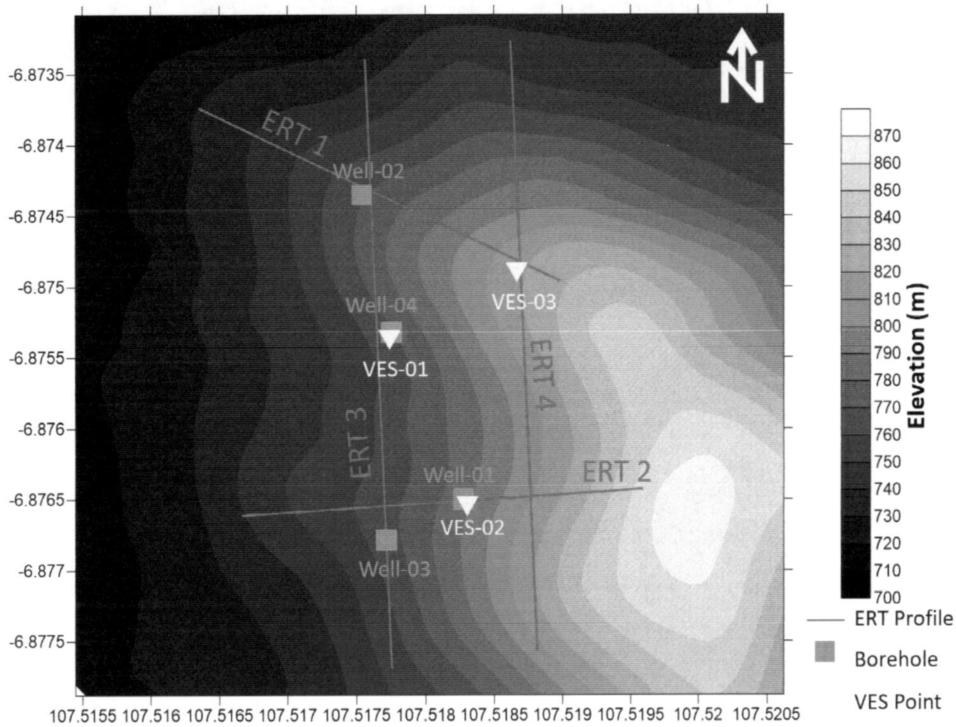

Figure 2. VES station, wells location, and electrical resistivity tomography lines overlaid on digital elevation model derived from the topography map

3. Results

Figure 3 shows the results obtained from VES. Curve type obtained from VES curves revealed a 4-layer with HK type ($\rho 1 > \rho 2 < \rho 3 > \rho 4$). Generally, the RMS value in VES inversion has a relatively low value. Wells in this area were available to correlate the lithological of each layer. Table 1 shows the interpretation of VES combined with wells lithological information.

The result of VES indicate there are low resistivity zone at a depth of around 80 m in VES 3. We also observed the low resistivity value in VES 1 and VES 2 at a depth of around 123 m and 141 m, respectively. Available wells in this area are only limited to 20 m depth, so we could not interpret this low resistivity zone in terms of the lithological unit. We interpret this low resistivity zone (2.8 - 11.7 Ωm) as the zone with high water content.

The ERT 1 (figure 4a), trending NW-SE, has Well 02 for calibration. We interpreted the top layer with moderate resistivity (40 Ωm) as silty clay, while the second layer was relatively low (25.4 Ωm), and we interpreted it as saturated gravelly clay. These two lithological variations fitted well with Well 02 information, where saturated gravelly clay was probably due to irrigation. We interpreted the zone below the Well bottom with high resistivity (64 Ωm) as andesite breccia.

We calibrated the ERT 2 data (figure 4b), trending W-E with Well 1 data. In Well 01, we identified three distinct lithological units (figure 3d). ERT 2 results show clearly that this distinct lithological unit with silty clay has a resistivity value of 25.4 Ωm, gravelly clay has a resistivity value of 32 Ωm, and andesite breccia characterized by a resistivity value of 40.3 Ωm. The higher resistivity means an increase in compactness and fresh andesite breccia. The VES-02 at 160 m shows below this profile around 140 m in depth reveals a low resistivity zone indicating potential groundwater occurrence. Unfortunately, the ERT 2 could only reveal approximately 50 m depth from the surface.

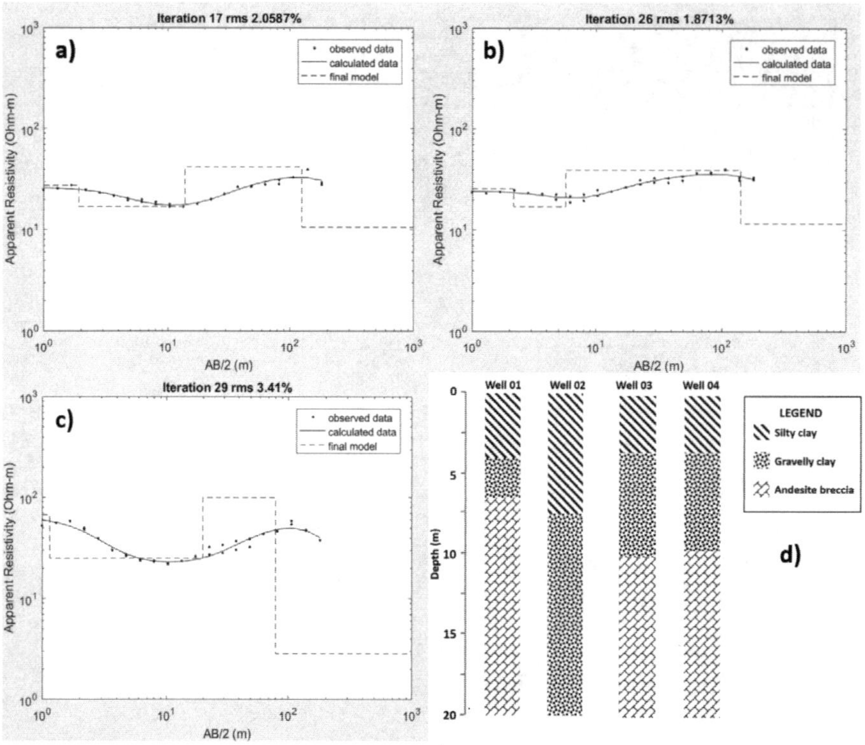

Figure 3. Sounding curves of a) VES-01, b) VES-02, and c) VES-03. d) Wells with lithological information.

Table 1. Interpretation of each layer in VES curves.

VES point	Layer	Resistivity (Ωm)	Thickness (m)	Depth (m)	Lithology	RMS (%)
01	1	27.81	1.91	0 - 1.91	Silty clay	2.06
	2	17.10	11.90	1.91 - 13.81	Gravelly clay	
	3	41.88	109.88	13.81 - 123.69	Andesite breccia	
	4	10.62	∞	123.69 - ∞	High water content zone	
02	1	25.86	2.16	0 - 2.16	Silty clay	1.87
	2	17.14	3.54	2. 16 - 5.7	Gravelly clay	
	3	39.60	135.38	5.7 - 141.08	Andesite breccia	
	4	11.72	∞	141.08 - ∞	High water content zone	
03	1	67.22	1.15	0 - 1.15	Silty clay	3.41
	2	25.18	18.67	1.15 - 19.82	Gravelly clay	
	3	99.66	59.99	19.82 - 79.81	Andesite breccia	
	4	2.82	∞	79.81 - ∞	High water content zone	

Unlike ERT 1 and ERT 2, ERT 3 was trending S-N (figure 4c), which has a better depth of investigation due to greater length. We observed a significant resistivity variation in both near-surface and at greater depth. This variation was probably due to the high location that could lead to noise. Three wells (Well 03, Well 04, and Well 02) are located in this profile to help interpret resistivity results. At a distance of 280 - 320 m, we observed a low resistivity zone (6.35 Ωm) at a depth of 10 - 25 m. We interpreted this zone as a water accumulation. Such zone we also observed at a distance of 240 - 280 m at a depth greater than 60 m. The water accumulation probably flows through the

IOP Conf. Series: Earth and Environmental Science 1031 (2022) 012006 doi:10.1088/1755-1315/1031/1/012006

permeable layer and is stored at a depth greater than 50 m. We observed that the permeable layer was probably gravelly clay at Well 02. VES-01, located at a distance of 210 m, also shows a low resistivity value (10.6 Ωm) at a depth greater than 100 m. We interpreted this low resistivity value to have groundwater potential. We interpreted Andesite breccia as an impermeable layer to hold this accumulated water.

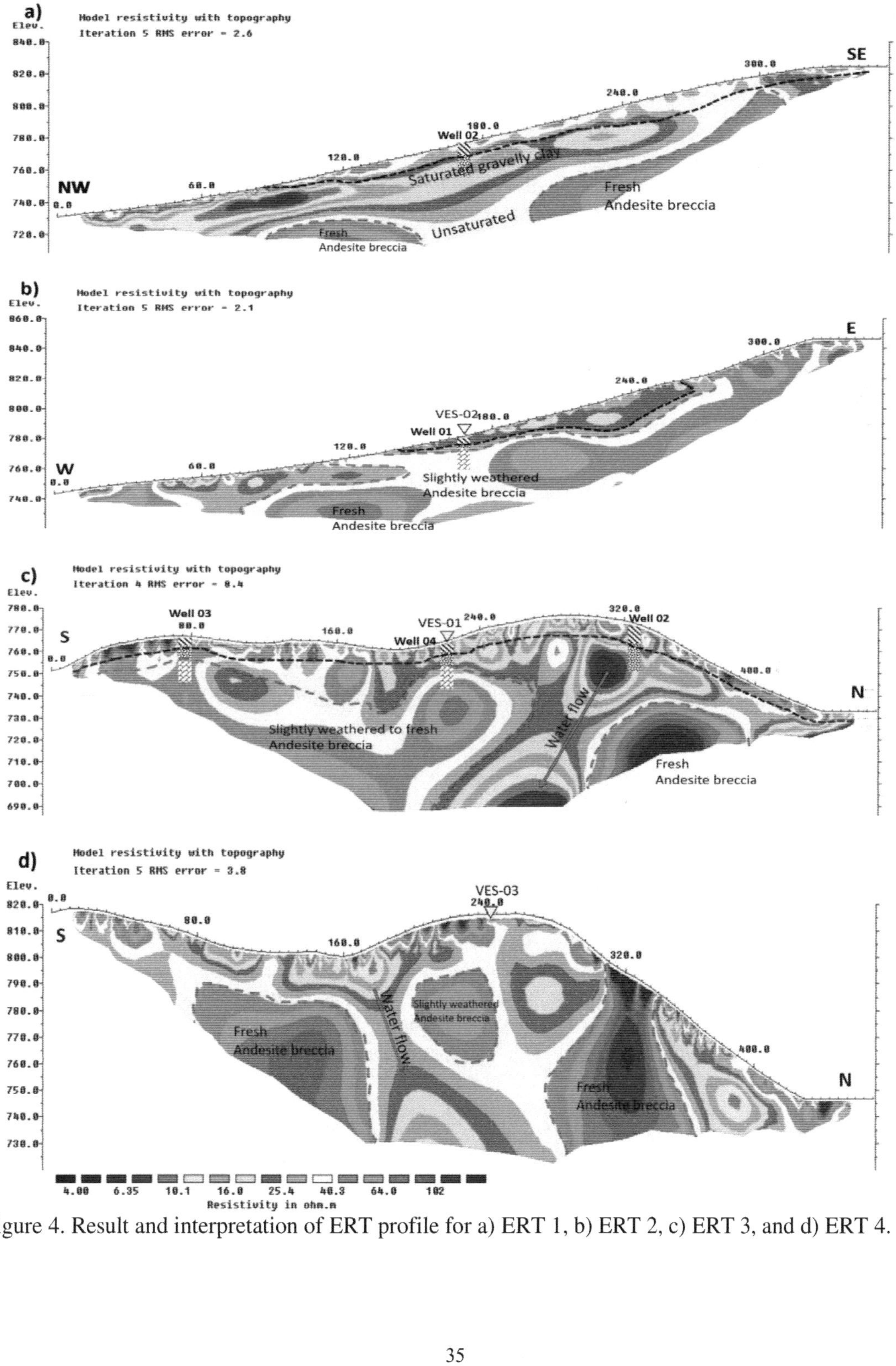

Figure 4. Result and interpretation of ERT profile for a) ERT 1, b) ERT 2, c) ERT 3, and d) ERT 4.

Like ERT 3, ERT 4 also trends S-N and shows a similar pattern (figure 4d). There is no well to calibrate the resistivity result in this profile. Thus, we interpreted this profile using the other profiles. We observed a low resistivity value (10.1 Ωm) at a distance of 140 - 180 m at the top layer until a 15 m depth. The low resistivity value is probably due to high water content. This low resistivity zone was also observed at a distance of 180 - 240 m at a depth of 60 m until a maximum depth of ERT profile, probably due to high water content flowing from the top layer through a permeable layer to a greater depth of the profile. We interpreted the permeable layer as gravelly clay as all wells indicate that gravelly clay is always in the second layer. We found a low resistivity value (2.8 Ωm) at a depth of 80 m at VES-03 located at 240 m. This low resistivity probably accumulated at a depth of greater than 80 m and indicates groundwater potential.

The low resistivity zones observed in ERT 3 and ERT 4 lines indicate that the water flows from higher topography (ERT 4) to lower topography (ERT 3). This indication agrees with VES results where VES-03, located at a higher elevation, encountered low resistivity at 80 m depth, while VES-01 in lower elevation encountered low resistivity at more than 100 depth. We encountered a low-velocity zone in the central part of the study area, which may yield high groundwater potential.

4. Conclusion

This study used geoelectrical methods consisting of VES and ERT to investigate groundwater at Padalarang, West Java. We used available wells to validate in interpreting resistivity data. All VES results revealed four geoelectrical layers and indicated a low resistivity value (2.82 - 11.72 Ωm) at the fourth layer, which we expected to be the zone with high water content. The ERT result provides a lateral and vertical variation of this area also revealed low resistivity values (<10.1 Ωm) at ERT 3 and ERT 4, where the depth was approximately more than 60 m from the surface. We encountered low resistivity values in the central part of the study area, probably yielding a high groundwater potential.

Acknowledgments

Research grants financially supported this research to Widodo Widodo from LPPM (Institute for Research and Community Services), Institute Teknologi Bandung, and RISTEK DIKTI (Ministry of Research, Technology and Higher Education of the Republic of Indonesia).

References

[1] Kirsch R 2009 *Groundwater Geophysics: A Tool for Hydrogeology* (Berlin, Heidelberg: Springer Berlin Heidelberg)

[2] Loke M H, Chambers J E, Rucker DF, Kuras O and Wilkinson P B 2013 Recent developments in the direct-current geoelectrical imaging method *Journal of Applied Geophysics* **95** 135–56

[3] Telford W M, Geldart LP, and Sheriff R E 1990 *Applied Geophysics* (Cambridge: Cambridge University Press)

[4] Adagunodo T A, Akinloye M K, Sunmonu L A, Aizebeokhai A P, Oyeyemi K D and Abodunrin F O 2018 Groundwater Exploration in Aaba Residential Area of Akure, Nigeria *Front. Earth Sci.* **6** 66

[5] Haque S, Kannaujiya S, Taloor A K, Keshri D, Bhunia R K, Champati Ray PK, and Chauhan P 2020 Identification of groundwater resource zone in the active tectonic region of Himalaya through earth observatory techniques *Groundwater for Sustainable Development* **10** 100337

[6] Muchingami I, Hlatywayo D J, Nel J M, and Chuma C 2012 Electrical resistivity survey for groundwater investigations and shallow subsurface evaluation of the basaltic-greenstone formation of the urban Bulawayo aquifer *Physics and Chemistry of the Earth, Parts A/B/C* **50–52** 44–51

[7] Ekinci Y L and Demirci A 2008 A Damped Least-Squares Inversion Program for the Interpretation of Schlumberger Sounding Curves *Journal of Applied Sciences* **8** 4070–8

[8] Loke M H and Barker R D 1996 Rapid least-squares inversion of apparent resistivity pseudosections by a quasi-Newton method1 *Geophys Prospect* **44** 131–52.

Scaling and corrosion formation mechanism in geothermal piping system based on Pourbaix-Lindal diagram-Ryznar index and fluid velocity profile

Sutarno[1,a], E. P. Syuryana[1] and D. B. Saefudin[2]

[1]Metallurgical Engineering Department and [2]Mechanical Engineering Department, Faculty of Manufacturing Technology, University of Jenderal Achmad Yani, Jln, Gatot Subroto, Bandung, West Java, 40285, phone (022) 731274

[a]Corresponding authors : sutarno@lecture.unjani.ac.id

ABSTRACT. Geothermal fluid flow contains scale and corrosion species whose levels differ from each location of geothermal sources. The velocity of scale formation and corrosion mechanism due to specific parameters in case of total key species, flow rate, pressure and temperature of the geothermal fluid based on the Pourbaix-Lindal diagram -Ryznar index. The reduction in pipe diameter due to the scaling formation on the internal surface of the pipe correlates with the flow rate of geothermal fluid into the turbine, which drives the generator to produce electrical energy. This paper describes the mechanism of scale formation and the potential for corrosion on the internal surface of the piping through the momentum transfer approach of geothermal fluids to the turbine. Through this mechanism, curative and preventive measures can be taken to prevent premature degradation of the function and economic life of the geothermal piping system. The mathematical simulation results of fluid displacement momentum, the velocity of geothermal fluid flow, show that the velocity profile at the fluid-internal surface contact of the pipe is zero, which means that the species is at rest on the internal surface of the pipe. Thus, for scale species it will form a scale, while for corrosive species there will be a corrosion process on the internal surface of the pipe.

Keywords: Geothermal fluid, Total key species, Pourbaix diagram, Lindal diagram, Ryznar Index, Fluid velocity profile.

1. Introduction

Geothermal, wind, solar, and biomass are a group of new and renewable energy, environmentally friendly, but their contribution only reaches 12.36% of the total electricity demand [1]. In contrast to wind, solar and biomass energy which requires energy storage in the form of batteries, geothermal energy is abundant and a source of future energy, which can be directly utilized, such as hydropower, steam, coal and petroleum [2].

The geothermal piping system is an integral part of the power generation system which consists of a production well reservoir, a piping system, a pipeline, a turbine and a generator. The function of this piping system is to flow geothermal fluid from the reservoir to the turbine to drive a generator that will produce electrical energy. Geothermal fluid consists of steam containing non-volatile anions such as Cl^-, HCO_3, SO^{4-}, $NH4^-$, F^-, I^-, Br- and cations such as Na^+, K^+, Mg^{2+}, Mn^{2+}, Fe^{2+}, Al^{3+}, and non-condensable gas (NCG), CO_2, H_2S, H_2 and the inert gas He and Ar and brine [3,4,5,6]. Some of these anion and cation species are also influenced by the character of production wells which have a composition of wet vapor, impurity gases, non-condensed gases, and a scale that has the possibility of solid particles.

Scale and corrosion due to contact on the internal surface of the piping system, which is a critical and complex problem in geothermal energy which is well known and disclosed by many researchers and practitioners [7-8]. The main problem that needs to be addressed is why does corrosion and scale occur in flowing geothermal fluids and how is the mechanism of scale formation and corrosion occurring in flowing geothermal fluid. This paper examines the mechanism of scale formation and corrosion based on the pH of geothermal fluids using the Pourbaix-Lindal diagram and Ryznar index approach which is correlated with the velocity profile of geothermal fluid flow. In answering the 2 (two) questions above, there are 2 hypotheses used, namely (1) both the formation of scaling and the corrosion reaction would occur at rest, (2) the Fe on the surface of the internal piping system was oxidized to form Fe^{2+} and Fe^{3+}.

2. Geothermal fluids

2.1. Characteristics parameter

Characteristics of geothermal fluid parameters include key species, pH, TDS, Ω, T, P. During the flow of geothermal fluid in the piping system, there is a decrease in fluid temperature and at the same time there is a total key species interaction with the internal surface of the pipe. The temperature drop occurs due to heat transfer both by convection and conduction through pipe walls and insulators. Lindal diagram maps [9] temperature with fluid phase and its application. For electrical energy, the temperature of the geothermal fluid ≥ 120 °C where the temperature in the range 120-180 °C is known as binary steam, while 180-220 °C is known as superheated steam. A decrease in temperature will result in a decrease in the level of steam dryness, a change in pH value, a decrease in the solubility of key species, and a pressure. The correlation of the temperature decrease with the degree of steam dryness is stated

$$Xw + Xs = 1 \quad (1)$$

Where Xw is the weight fraction of water and Xs is the weight fraction of steam. A decrease in temperature can cause an increase in Xw and a decrease in Xs.

Changes in pH values and decreasing solubility of species as a result of decreasing geothermal fluid temperature have an impact on the tendency of scale formation and corrosion on the internal surface of the piping system. Decomposition of H_2S and formation of metal sulfides (MeS) as scales.

$$H_2S \rightarrow S^{2-} + 2H^+ \quad (2)$$

$$Me^{2+} + S^{2-} \rightarrow MeS \text{ (where Me: Fe, Mn, Cu, Pb, etc)} \quad (3)$$

H_2CO_3 decomposition and $CaCO_3$ precipitation (as scale) take place according to the reaction

$$H_2CO_3 \rightarrow 2H + + CO_3^{2-} \quad (4)$$

$$Ca^{2+} + CO_3^{2-} \rightarrow CaCO_3 \text{ (as scale)} \quad (5)$$

The solubility product, Ksp, is represented by $CaCO_3$ in the form of an equation

$$Ksp = [Ca^{2+}]\,[CO_3^{2-}] \quad (6)$$

Where the value of Ksp <1 means unsaturated; Ksp = 1 means saturated; Ksp> 1 means precipitated. Based on the water hardness classification (as ppm $CaCO_3$), $CaCO_3$ in 15-50 ppm is soft; CaCO3 50-100 ppm medium soft; 100-200 is hard, and $CaCO_3$> 200 ppm is very hard.

The correlation of the descaling and corrosion trends was modeled by Ryznar [10], which was based on the saturation concept of calcium carbonate or calcite. RSI (Reyznar Saturation index) is calculated based on the pH of geothermal fluid and the calculation of pH saturation 2 (pHs) with the equation

$$RSI = 2\,(pHs) - pH \quad (7)$$

Calculation of pHs using equations

$$pHs = (9.3 + A + B) - (C+D) \quad (8)$$

where $A = \frac{(log_{10}[TDS]-1)}{10}$; $B = -13.12\ x\ log_{10}\ (°C + 273.15) + 34.55$;
and $C = \text{Log}_{10}[Ca^{2+} \text{ as } CaCO_3] - 0.4$; $D = \text{Log}_{10}[\text{alkalinity as } CaCO_3]$

2.2. *Pourbaix Diagram*

Pourbaix Diagram is a diagram that correlated pH with chemical potential [11]. As is known, geothermal pipes are generally made of carbon steel and stainless steel with reference to practical recommendations such as API 5L Gr.B, ASTM A53 [12]. In geothermal fluids, corrosion is the reaction between the internal of a pipe surface and a type of corrosive key species in steam. Under acidic, neutral and alkaline conditions the Fe on the internal pipe surface is oxidized or polarized to Fe^{2+} or Fe^{3+} shown in Table 1

Table 1. Corrosion reaction

No.	On the	Acid corrosion	Oxygen corrosion
1	Anodic	$Fe \rightarrow Fe^{2+} + 2e^-$	$Fe \rightarrow Fe^{2+} + 2e^- \rightarrow Fe^{3+} + e-$
2	Cathodic	$2H^+ + 2e^- \rightarrow H_2$	$O_2 + 2H_2O + 4e- \rightarrow 4OH^-$
3	Overall	$Fe + H^+_{(aq)} \rightarrow Fe^{2+}_{(aq)} + H_{2\,(g)}$	$4Fe + 3O_2 + 2H_2O \rightarrow 4FeOOH$

2.3. *Fluid flow velocity profile*

Geothermal fluid flow in the pipe is divided into convective momentum transfer in the direction of the geothermal fluid flow and viscosity momentum transfer in the radial direction shown in Figure 1 [13,14,15]

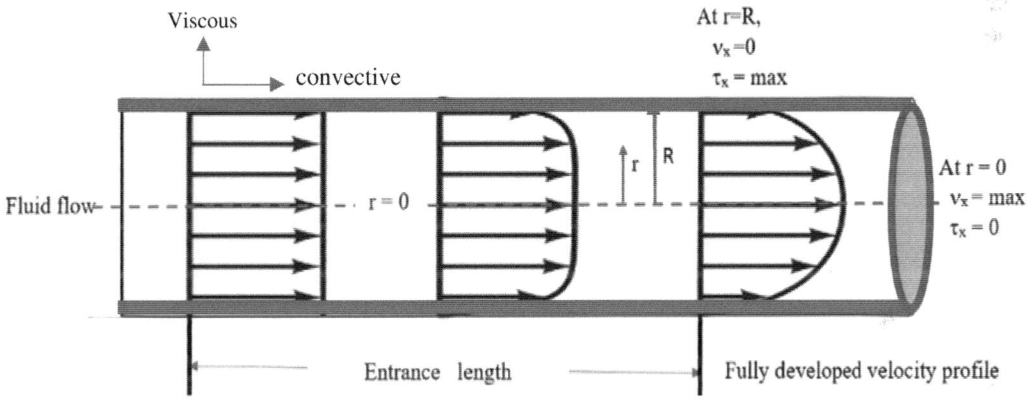

Figure 1. Fluid velocity profile in fully developed conditions

As it is known that the momentum flux is the shear force expressed,τ =-η(dv/dr) so that it is obtained

$$\tau_{rz} = \left(\frac{\Delta P}{L} \mp \rho g\right)\frac{r}{2} \tag{9}$$

In fully developed conditions, the fluid flow velocity profile forms a boundary layer as shown in equation (10) and Figure 1.

$$v_z = \left(\frac{\Delta P}{L} + \rho g\right)\frac{R^2 - r^2}{4\eta} \tag{10}$$

The flow velocity at the part in contact with the pipe surface is 0 (at r = R, v = 0, the fluid is in idle or static condition).

IOP Conf. Series: Earth and Environmental Science 1031 (2022) 012007 doi:10.1088/1755-1315/1031/1/012007

3. Corrosion Mechanism and Scale Formation

Geothermal fluid has characteristic parameters for total key species, pH, TDS, resistivity, and chemical composition. Meanwhile, the geothermal fluid properties include density (ρ), viscosity (η), and flow velocity (v), and pipe diameter (D = 2R). The combination of the interaction of the geothermal-piping system fluid contact interface and the viscous momentum of the radial direction has an important role in the formation of scale and corrosion on the internal surface of the pipe. In simple terms, the approximate parameters and characteristics of geothermal fluid flow are shown in Figure 2.

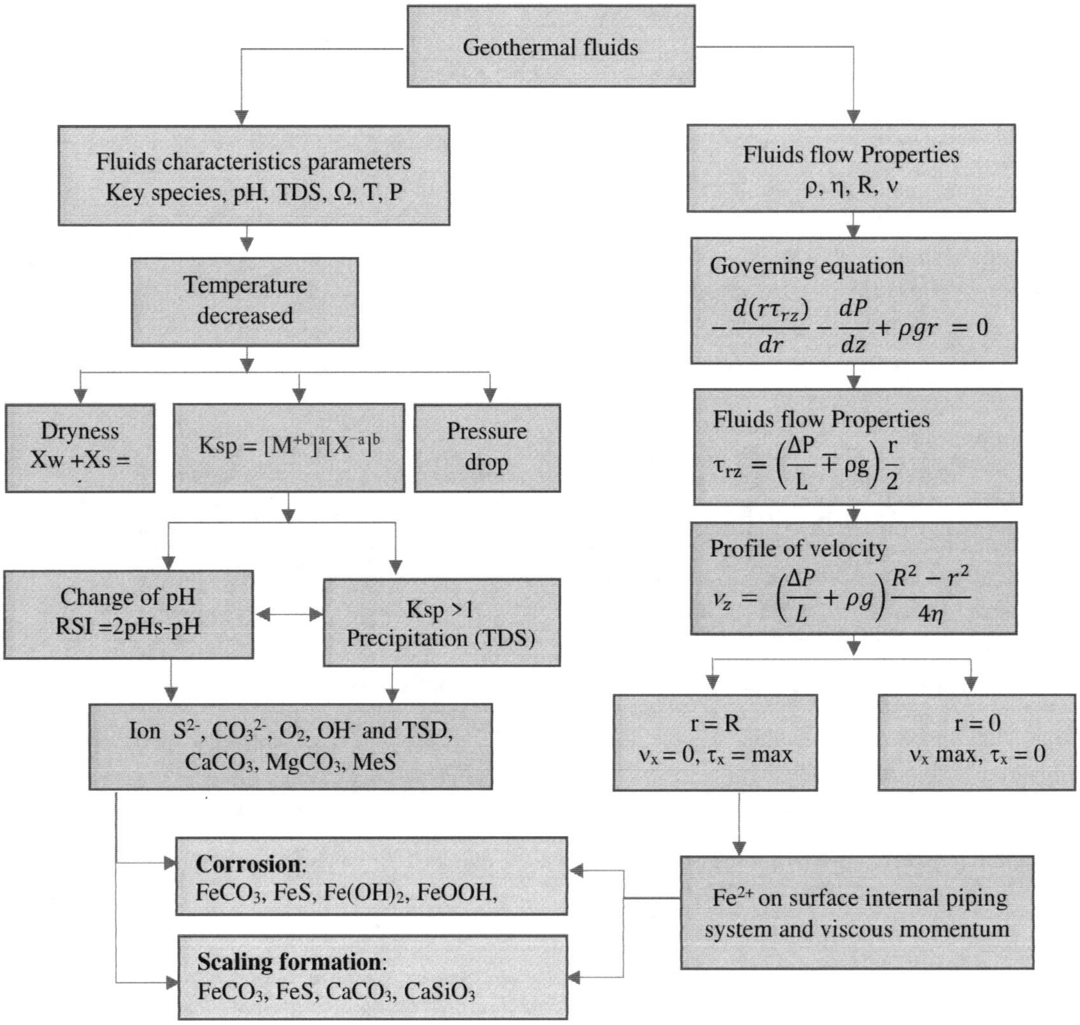

Figure 2. Causal-effect diagram of scale formation and corrosion mechanism

4. Discussion

4.1. Change in pH and RSI value

At RSI ≤ 6 there tends to be scale formation, while RSI ≥ 6 tends to corrode, as shown in Figure 3. The range of RSI and the tendency for scale formation and corrosion is shown in Table 2 [10].

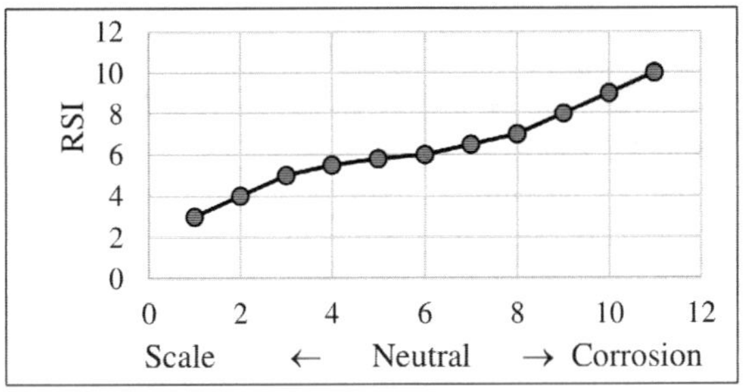

Figure 3. Effect of pH on RSI

Table 2. Scale and corrosion tendency

RSI value range	Interpretation tendency
RSI < 5.5	High Scaling formation
5.5 < RSI < 6.2	Medium Scaling formation
6.2 < RSI < 6.8	No scale and corrosion
6.8 < RSI < 8.5	Medium corrosion
RSI > 8.5	High corrosion

4.2. Corrosion mechanisms and scale formation

During geothermal fluid flows, the internal pipe surface, Fe oxidized to Fe2 +, is on the entire internal surface of the piping system. When it meets CO_3^{-2} it will react to form $FeCO_3$ (siderite), S^{2-} ions to form FeS (pyrite), steam FeOOH, $Fe(OH)_2$, oxygen to form FeO, Fe_2O_3. As for Ca^{2+} with CO_3^{-2}, SiO_2, respectively form $CaCO_3$, $CaOSiO_2$

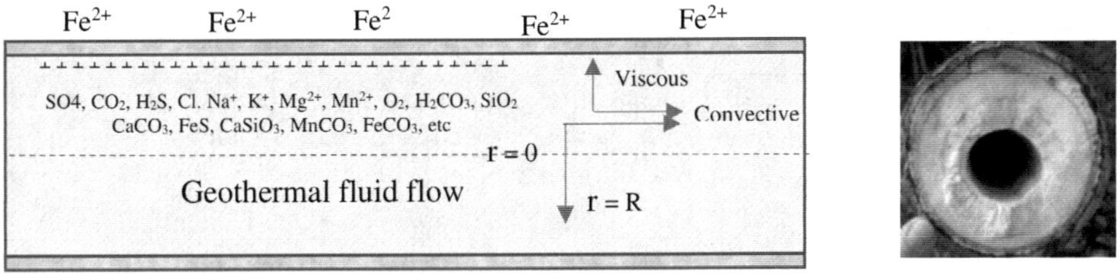

Figure 4. (a). Mechanism for scaling and corrosion in surface internal pipeline; **(b).** Scaling products

The mechanism of corrosion and scale occurs in piping systems in geothermal fluid flow only and only occurs in fully developed flow conditions, at r = R, where flow velocity, v = 0. This has been proven in the research of [13]. using Comsol software, [14] using Ansys software, [16] using the Ansys student version software as shown in Figure. 5. Thus, the conditions for the formation of scale and corrosion or both are due to the oxidation of the steel pipe on the internal surface of the pipe, viscous momentum, Ksp> 1.

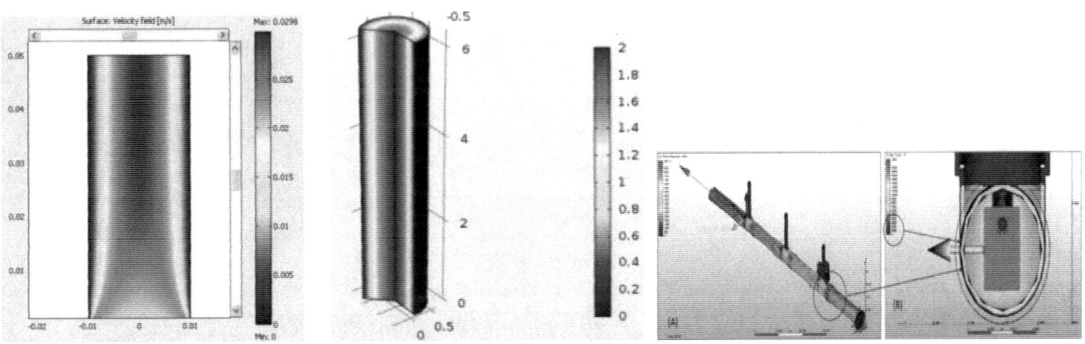

Figure 5. Fluid velocity profile in vertical flow (right) [13, 14], (A) Simulation of flow patterns in pipes; (B) Flow patterns if there is a corrosion coupon or ER / LPR Probe horizontal flow (left) [16]

Corrosion and scale mechanisms occur in piping systems in geothermal fluid flow only and only occur in fully developed flow conditions, at r = R, where flow velocity, v = 0. The Metallurgical Engineering Chemical and Corrosion Laboratory, FTM, UNJANI has developed a closed flow loop simulator (CFLS) to test the corrosivity of fluids in a particular pipe material. The line pipe specifications have been tested, i.e. API 5L X52, API 5L X65, API 5LC (X65 + Ni Base Alloy 825) etc. In the closed-loop simulator design we use computational fluid dynamic (CFD) for velocity profile mapping to describe the visual conditions on the pipe surface and if inside the pipe fitted with corrosion coupons, ER / LPR Probes (Figures A above and below, A and B). [16]

5. Conclusion

Prediction of the tendency of scale formation and corrosion based on pH, temperature, and content of key species using Lindal diagrams, Pourbaix diagrams, and RSI. The mechanism for the formation of corrosion and scale on the inner surface of pipes that flow geothermal fluids tends to occur in fully developed flow conditions where on the internal pipe surface there is oxidation or polarization of Fe to Fe^{2+} and Fe^{3+} and flow velocity, $v = 0$ and this means key species are in contact and at rest on the inner surface of the pipe. The geothermal fluid flow velocity = 0, in the form of a boundary layer at $r = R$, is proven using equation (9) and equation (10) and the simulation shown in Figure 4.

References

[1] Cnbcindonesia.com 2020 Porsi Batu Bara Naik, Target Energi Baru RI Makin Berat Diakses pada 25 Oktober 2020, dari https://www.cnbcindonesia.com/news/20200109164206-4-128910/porsi-batu-bara-naik-target-energi-baru-ri-makin-berat.

[2] Lund and Boyd 2016 Geothermal energy has a large potential for becoming part of the future worldwide energy portfolio, considering its steady growth over the past decades.

[3] Ana Vallejo Vitaller, Ueli M. Angst and Bernhard Elsener 2020 Laboratory tests simulating corrosion in geothermal power plants: infuence of service conditions https://doi.org/10.1186/s40517-020-00163-y.

[4] Niklas Mundhenk 2013 Corrosion and scaling in utilization of geothermal energy in the Upper Rhine graben *Dissertation* Hamburg Karlsruhe German.

[5] Bernhard Köhl, Martin Elsner and Thomas Baumann 2020 Hydrochemical and operational parameters driving carbonate scale kinetics at geothermal facilities in the Bavarian Molasse Basin Köhl et al. *Geotherm Energy* **8** 26 https://doi.org/10.1186/s40517-020-00180-x,

[6] Jamero J and Zarrouk S J 2016 Mineral Scaling In Geothermal Two-Phase Pipelines *Proceedings 38th New Zealand Geothermal Workshop*.

[7] Fridrikssonand T and Thórhallsson S 2015 Geothermal Utilization Geochemical Monitoring of the Reykjanes Geothermal Reservoir 2003 to 2013 *Proceedings World Geothermal Congress 2015* Melbourne Australia 19-25 April 2015.

[8] Stephan S, Andrej L, Paul R and Gerdi B 2016 Operational issues in Geothermal Energy in Europe Status and overview *Geothermal ERA NET Orkustofnun, Grensásvegi* **9** 108 Reykjavík http://www.geothermaleranet.is/ ISBN: 978-9979-68-397-1

[9] Patsa E, Sadiq J, Zarrouk and Zyl D V 2015 The Lindal Diagram for Mining Engineering *GRC Transactions*

[10] Shankar B S 2014 Determination of scaling and corrosion tendencies of water through the use of Langelier and Ryznar Indices *Scholars Journal of Engineering and Technology* (SJET) Vol. **2**(2A) page 123-127.

[11] Pesterfield L L, Maddox J B, Crocker M S and Schweitzer G K 2012 Pourbaix (E–pH-M) Diagrams in Three Dimensions *Journal of Chemical Education* **89**(7) *891–899.*

[12] Kaya T and Hoşhan P 2005 Corrosion And Material Selection For Geothermal Systems, *Proceedings World Geothermal Congress.*

[13] Robert P Hesketh 2007 Flow Between Parallel Plates – Modified from the COMSOL ChE Library module, Chemical Engineering, Rowan University.

[14] Hoai Nguyen, Thang Hoang 2017 Numerical of laminar flow through a pipe using comsol Multiphysics *Intl Journal of Scientific & Eng Research* vol **8** Issue 6 June 2017

IOP Conf. Series: Earth and Environmental Science 1031 (2022) 012007 doi:10.1088/1755-1315/1031/1/012007

[15] Gaskell D R 2013 *An introduction to transport phenomena in materials engineering, 2nd* (New Jersey: Momentum press LLC)
[16] Syuryana E P, Widyanto B and Nasution H 2019 Design and manufacturing closed flow loop corrosion simulator by computational fluids dynamic modeling *Final year project* Metallurgical Eng. Dept. Unjani.

IOP Conf. Series: Earth and Environmental Science 1031 (2022) 012008 doi:10.1088/1755-1315/1031/1/012008

Geochemical Distribution of REE and Grain Size Analysis of Heavy Mineral Associated with Tin Placer-Type Deposit, Bangka

Syafrizal[2], AYA Hakim[2], A Sulastri[1]

[1] Faculty of Mining and Petroleum Engineering, Bandung Institute of Technology, Bandung, Indonesia.
[2] Earth-Resources Exploration Research Group, Bandung Institute of Technology, Bandung, Indonesia.

Email:syafrizal@mining.itb.ac.id, astisulastri@students.itb.ac.id

Abstract. Indonesia is included in the East Malaya Block of the Southeast Asian Tin Belt. Two types of tin deposits in Indonesia are primary and secondary tin deposits (placer). Factors that can cause the formation of placer tin deposits include S-type granite, weathering, transportation, and sedimentation. Factors that control the formation of placer tin deposits include the presence of source rock (granite type S), weathering processes, erosion, transportation and sedimentation, and the presence of basins or valleys where weathering materials are deposited. The transportation process has a major role in the grain shape of a material. This study aims to determine the geochemical distribution and characteristics of heavy mineral grains at the accumulation of placer tin deposits, e.g., colluvial, alluvial, and tailings, in which there are 62 sample points located Bangka Island, Indonesia. The research method used in this study was grain size analysis, radiation level measurement using a scintillometer, and geochemical analysis using ICP-MS.

1. Introduction

The Southeast Asian Tin Belt exists spreadly in a North-South direction along 2,800 km and 400 km starting from Myanmar, Thailand, Malaysia, and the Tin Island in Indonesia. Tectonically, the tin island in Indonesia has included in the East Malaya Block formed during the geological age of Lower Carbon. The characteristic of Indonesian Tin Island is the presence of metasediments with inserts of limestone, shale, schist, chert, and volcanic rock. There are two types of tin deposits in Indonesia, namely primary and secondary (placer) [1].

The genesis of placer tin deposits in Indonesia is mostly the source rock naturally containing primary cassiterite. The liberation of cassiterite in the primary rock does not need excessive comminution. The current mechanical concentration is the breakdown of cassiterite grains where the cassiterite is not consolidated due to mechanical erosion, while other additional factors are control stratigraphy, geological time scale, paleo-environment changes, and climatic conditions [2] (Yim, 2000).

Heavy minerals are generally defined as dense minerals with a specific gravity greater than 2.85. The REE-bearing minerals in placer-type deposits are monazite [(Ce, La, Nd, Th) PO_4] and sometimes xenotime (YPO_4), which are high-density minerals accumulated as heavy minerals. Monazite has been extracted from many heavy mineral placers as a by-product of the recovery processes in the industry, e.g., titanium oxide (ilmenite, rutile), zircon, sillimanite, garnet, staurolite, and others. Xenotime is usually obtained from some alluvial deposits as a by-product of tin (cassiterite) placer mining. Trace elements associated with heavy mineral placers are useful as pathfinder elements, including Ti, Hf, REEs, Th, and U. Reported thickness of economic deposits ranges from 3 to 45 m [3]. Individual ore deposits typically comprise at least 10 million metric tons (Mt) of ore (total size of the individual sand-silt body), with an overall heavy mineral content of 2% to >10% [3].

In Australia, most of the heavy mineral and associated monazite resources are found on the old beach and sand dune deposits, some of which are found on Canning, Perth, Murray, and Eucla basins. In Eneabba (Perth), heavy minerals compose about 6% of the paleoshore sands mined in this district, with monazite composing 0.5 to 7.0% of this heavy mineral. WIM150 mineral sands deposit (Murray Basin) reportedly contains substantial resources of monazite and xenotime, associated with titanium minerals and zircon in the heavy mineral suit, with a deposit thickness of 14 m [3]. The length of that deposit is about 400 km, and the heavy minerals in the sand deposits were formed due to natural weather (storm) and weathering by water (wave).

In Brazil, the monazite placers of the Brazilian coast include elevated paleobeaches, modern beaches, sand dunes, banks, channels, and bars of streams that deposit sediments near the shore. Sources of the heavy mineral are sandstone, strandline deposits, and inland source of detrital monazite, which are amphibolite-granulite rock and sedimentary rock. Some sandstone intervals are rich in monazite, ilmenite, and zircon. These sandstones are eroded and disaggregated by high-tide waves and storm surges. These processes re-deposit sand and heavy minerals into the surf zone, where the heavy minerals are again reworked and sorted by waves, longshore drift, and tides. Thus, the Cretaceous-Tertiary strandline deposits, which occur in slightly higher outcrops near the modern beach, are another source (often containing a very high amount) of monazite. Monazite was recovered as a co-product of the more profitable titanium minerals (ilmenite, rutile) and zircon. In the Guarapari coastline, it is reported that monazite content is 0.83% [4]. Analyses of Brazilian monazites suggest that their average REE oxide content is typically around 57 to 60%, with preferential enrichment in the light REEs.

China has considerable monazite resources within their placer deposits, where monazite is produced as a by-product. Some of China's productive monazite-bearing placer districts are Beihei District, Dianbai District, Haikang District, Nanshanhai District, Sai-Lao, Wuzhang Xinglong, and Xinjiang. The content of heavy minerals in all districts is about 1.5% -6%, which contains minerals including rutile, zircon, monazite, ilmenite, xenotime, magnetite, chromite, anatase, and cassiterite.

Xenotime-bearing alluvial placer deposits in Malaysia were the largest sources of yttrium in the world. It is found as alluvial tin placer deposits containing considerable cassiterite with other minerals, including ilmenite, monazite, and xenotime. Recently, tailings produced from past tin placer mining have been reprocessed to recover monazite and xenotime. Around 350 tons of REOs were produced from Malaysia in 2012 [4].

Modern beach deposits on the northeastern coast of Sri Lanka contain some of the highest concentrations of heavy minerals in the world. These include ilmenite, rutile, and zircon. Sillimanite, monazite, and garnet also exist, with monazite reportedly composing 0.3% of the heavy mineral fraction. In some stretches of beach, the heavy minerals can compose as much as 90% of the sand deposits, while ilmenite, rutile, and zircon formed 65%, 10%, and 10% of the sand deposits, respectively. The Bentota river system carries monazite into seasonal beach sand deposits at Kaikawala and Beruwala, and these have been described as "one of the world's most thorium rich sediments" [4]. The deposits are enriched in LREE to HREE.

Similar to the Malaysian rare-earth resources, the monazite and xenotime deposits in Thailand are recovered as by-products from the upstream process in the production of tin metal in the plant. The host tin placers, enriched with monazite and xenotime, are alluvial deposits usually mined by gravel pumps, and these occur mainly in the southern Thailand area.

IOP Conf. Series: Earth and Environmental Science 1031 (2022) 012008 doi:10.1088/1755-1315/1031/1/012008

The Delta Epsilon SC -133 Portable Scintillation Counter represents a handheld gamma radiation detection device [4]. To determine the mineral's grain size distribution, Udden-Wentworth Grade Scale was used in this study. Krumbein (1934) [5] was introduced as a logarithmic transformation of the scale, which converts the boundaries between grades to whole numbers, and this method was also used later. This scale is known as the Phi Scale: $\phi = -(\log_{10} d)/\log_{10} 2$ [5]. In the present study, cumulative frequency curves (smooth curves) were drawn manually from the cumulative weight of sediment and phi (ϕ) units using graph papers. Phi units, $\phi_5, \phi_{16}, \phi_{25}, \phi_{50}, \phi_{75}, \phi_{84}$, and ϕ_{95}, were determined manually from smooth curves. Then the Median (Md), Mean (M), Standard Deviation (s), Skewness (SK), and Kurtosis (K) of each sediment sample were calculated using graphic method [6], as follows:

$$Median\ (Md) = \phi_{50} \tag{1}$$

$$Mean\ (M) = \frac{\phi_{16} + \phi_{50} + \phi_{84}}{3} \tag{2}$$

$$Standar\ Deviation\ (s) = \frac{\phi_{84} - \phi_{16}}{4} + \frac{\phi_{95} - \phi_5}{6.6} \tag{3}$$

$$Skewness\ (SK) = \frac{\phi_{84} + \phi_{16} - 2\phi_{50}}{2(\phi_{84} - \phi_{16})} + \frac{\phi_{95} + \phi_5 - 2\phi_{50}}{2(\phi_{95} - \phi_5)} \tag{4}$$

$$Kurtosis\ (K) = \frac{\phi_{95} - \phi_5}{2.44(\phi_{75} - \phi_{25})} \tag{5}$$

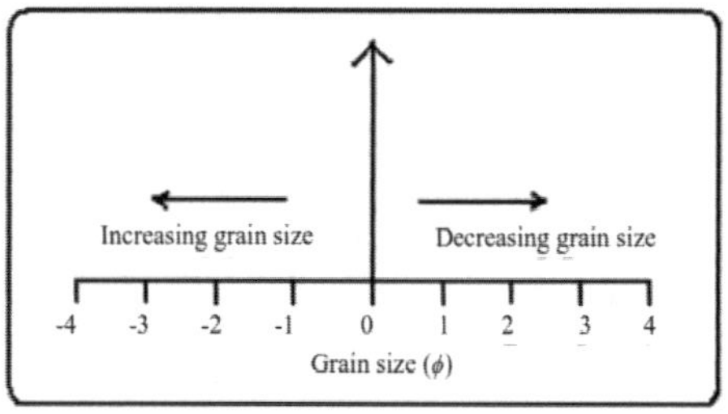

Figure 1. Conventional phi scale showing grain size increasing to the left and decreasing to the right [7].

Particle morphologies are included for at least four concepts: form, sphericity, roundness, and surface feature [8]. Form refers to the gross, overall morphology or configuration of particles. Most measures of form consider the three-dimensional shape of the grains. Roundness measures the sharpness of the corners of grain and is commonly measured in two dimensions only. Surface texture refers to microrelief features, e.g., scratches and pits, that appear on the surfaces of clastic particles, particularly particles that have undergone transport [9].

2. Study Area

2.1. Location

The location of this study is in Bangka Island, Indonesia. The area has a tropical climate with medium rainfall. The area is accessible via asphalt paved road, followed by garden path or small street to reach the location.

2.2. Geomorphological setting

Bangka Island is at an altitude between 0 to > 400 meters above sea level, with slopes ranging from 0 - > 25%. Some parts of the areas with an altitude of > 400 meters above sea level are around the upper reaches of the Mabat River, namely Mount Maras, mostly dominated by flat to hilly areas. Areas with a slope between 0-5% are generally located at the foot of the hill and have a relatively short and small inter-river valley with alluvial plains to the South-East, South, West and North of Bangka Island. [10].

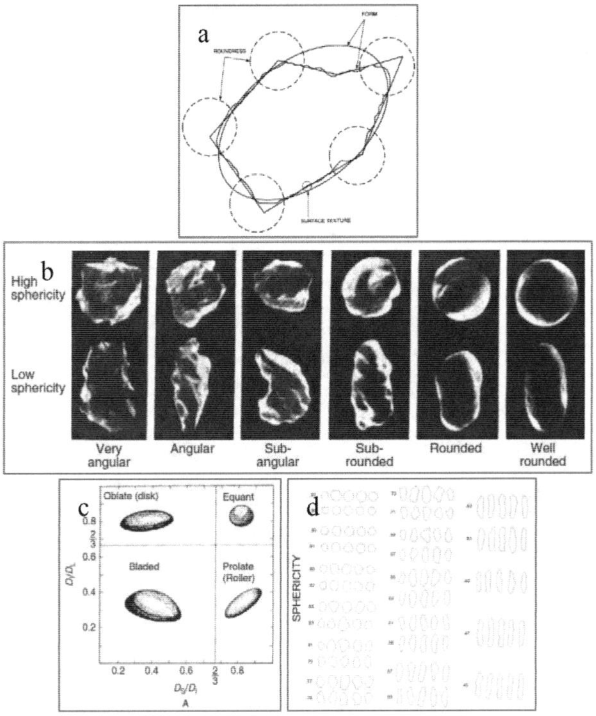

Figure 2. a. The hierarchal relationship of form, roundness, and surface texture [9]. Component aspect to describe particle morphology: b. Roundness [11]; c. Form [9] d. 2D Visually of the sphericity [12].

3. Geological Setting

Samples were collected in the Quaternary-Alluvium (QA), which contained mud, clay, coarse sand, and sediment from swamps, rivers, and beaches; the second was in the Triassic-Sandstone and Claystone of the Tanjung Genting Formation (Trt); the third was the intrusive Late Triassic-Granite Kelabat (TrJkg) which contained biotite-granite, granodiorite, gneissic granite [13] [14]. Regionally, Bangka Island has a strike-slip geological structure, both sinistral and dextral.

IOP Conf. Series: Earth and Environmental Science 1031 (2022) 012008 doi:10.1088/1755-1315/1031/1/012008

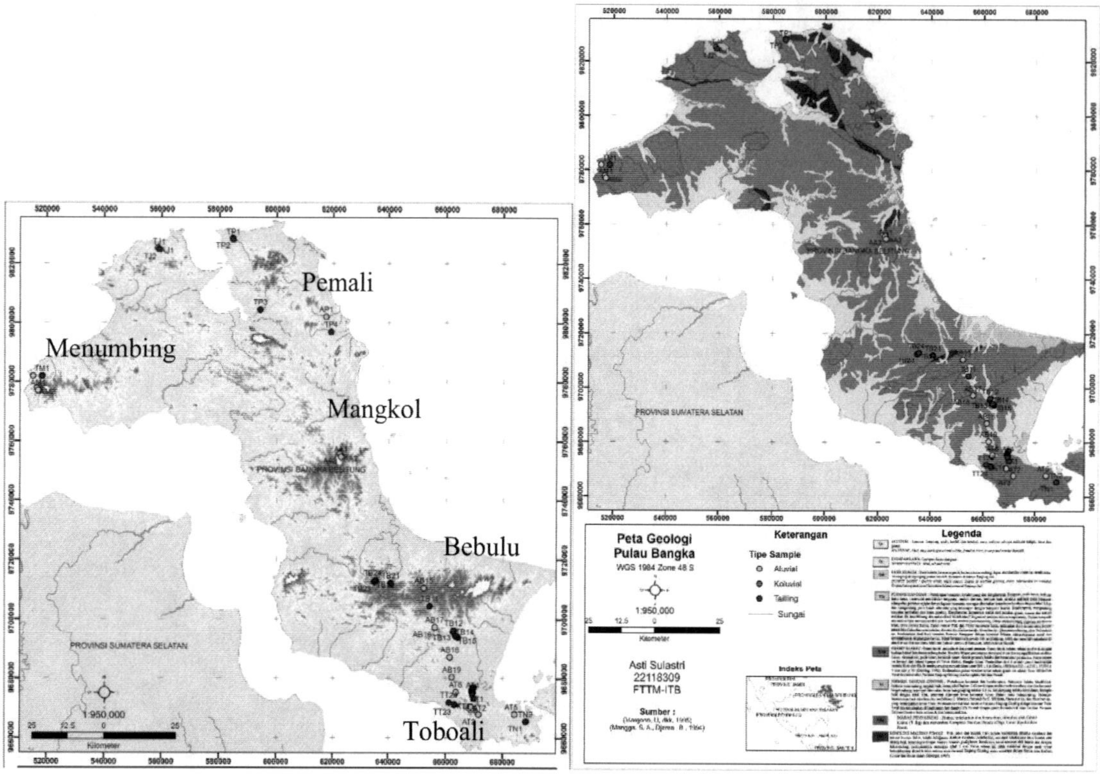

Figure 3. Location and geological map of the alluvial and tailing sample.

4. Material and Method

4.1. Sample and Preparation

In this study, 62 samples were collected, in which 36 were categorized as secondary data while the remaining 26 samples were taken directly from the field. Sample code names were based on sample type, peak location, and serial number. Generally, two types of samples are used: alluvial samples (A) and tailing samples (T). Peak location domain codes were M for Menumbing, P for Pemali, A for Mangkol, B for Bebulu, and T for Toboali.

Alluvial samples were collected from a stream while tailing samples were taken from ex-alluvial-mining locations. Location coordinates were recorded using a handheld GPS device. Samples taken from the location weighed around 2-3 kg were then dried, split using a quartering coning, and sieved into several mesh fractions (#5, #35, #60, #100, and #120) using a standard American mesh sieve. Some ICP-MS samples were crushed using hand mortar then sieved until they passed through the #200 sieve.

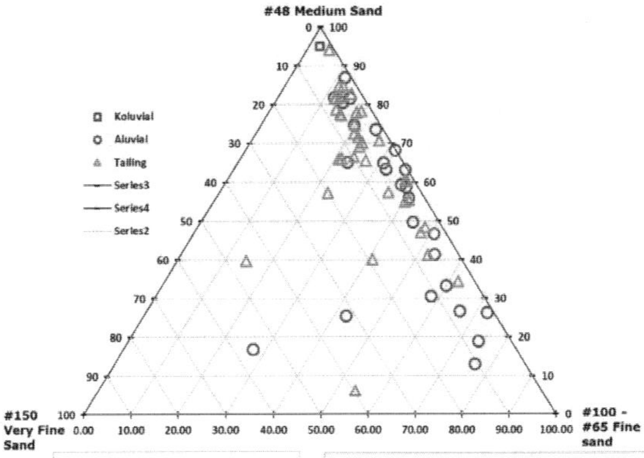

Figure 4. Grain size distribution of the alluvial and tailing sample used in this study. In general, medium sand and fine sand grains dominate the alluvial and tailing samples. However, 2 sand samples have a very fine size, possibly because these samples are the farthest from the peak morphology.

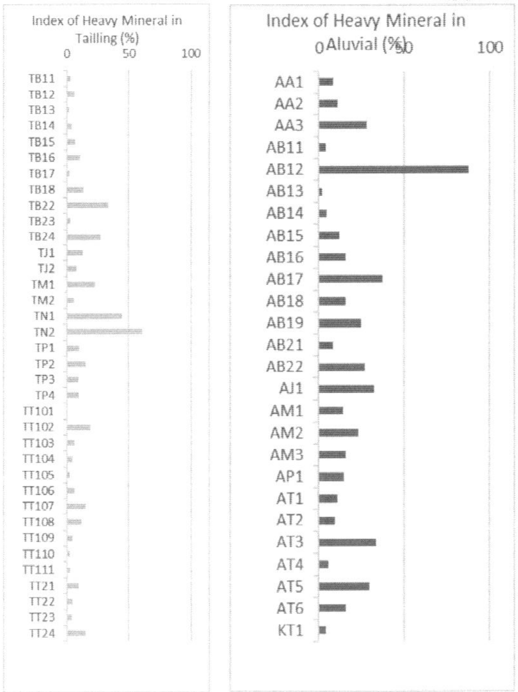

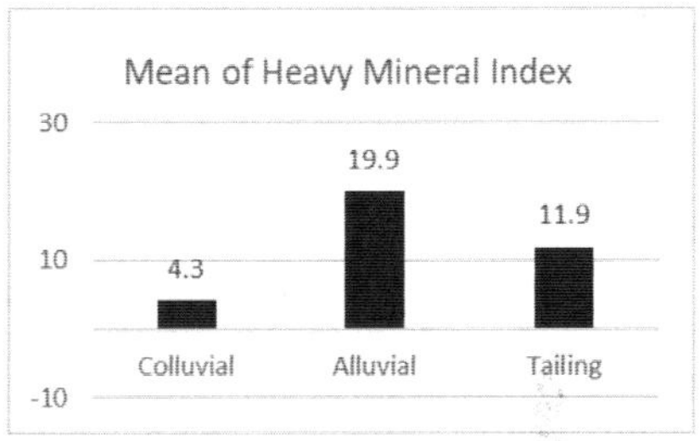

Figure 5. Heavy mineral index in percent. A. Taling, B. Aluvial and Coluvial, C. Mean

4.2. Extraction Procedure and Chemical Analysis

Samples passing through the #5, #35, #60, #100, and #120 sieve sizes were measured for their radiation exposure using an SC-133 portable scintillo counter. The grain minerals and grain morphology of the samples were analyzed using a binocular microscope. Some samples were also subjected to a detailed grain morphological description of a polished incision sample.

The samples were analyzed for their metal concentration (a total of 60 major and trace elements) using ICP-MS. Samples for ICP-MS analysis were taken based on the high radiation values and dominated by the finest fraction (-120 mesh). The ICP-MS analysis was carried out at PT Intertek Laboratory, Indonesia, using 4 solvents, namely HNO_3, $HClO_4$, HF, and HCl.

5. Result

5.1. Mineral Composition of Alluvial and Tailing in Bangka

In general, the light minerals present in the alluvial and tailings samples on Bangka Island are quartz (Qz). Meanwhile, the heavy minerals present include cassiterite (Cass), zircon (Zr), ilmenite (Ilm), rutile (Rtl), xenotime (Xtm), tourmaline (Tour), muscovite (Ms), limonite (Lim), and small amounts of siderite, spinel, magnetite (Mg), and pyrite (Px). Zircon is very abundant in the downstream areas in Bebulu (AB17 and AB19), Jebus (AJ1), and Toboali (AT3 and AT5), presumably since these locations are the farthest locations from each peak of morphology so that heavy minerals are well deposited.

It can be seen in Fig. 6 that the alluvial sample (KT1) tends to have less amount of heavy minerals compared to the alluvial and tailing samples, while the alluvial sample has the largest heavy minerals composition compared to the tailing and colluvial samples (See Fig. 5c). This occurs since the material deposition process is not mature at the alluvial location, and the heavy minerals in the tailings will present at very little concentration. During the early recovery processes, the material is passed into a sluice box to separate the heavy and light minerals, where the light minerals are reported as tailings.

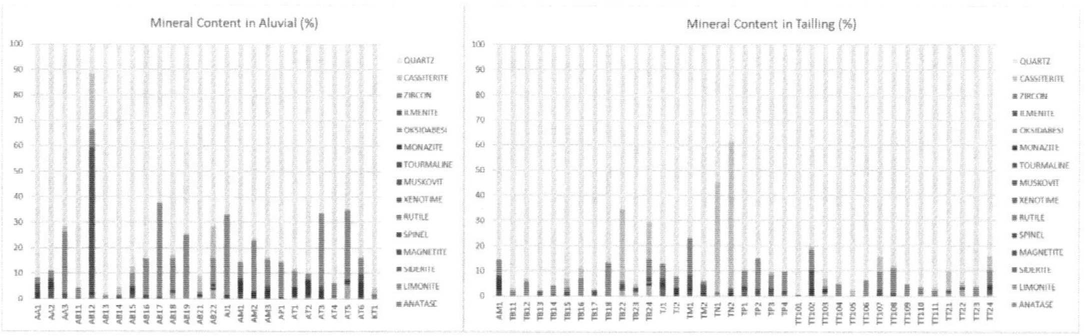

Figure 6. Grain size analysis in alluvial and tailings of Bangka Island sample

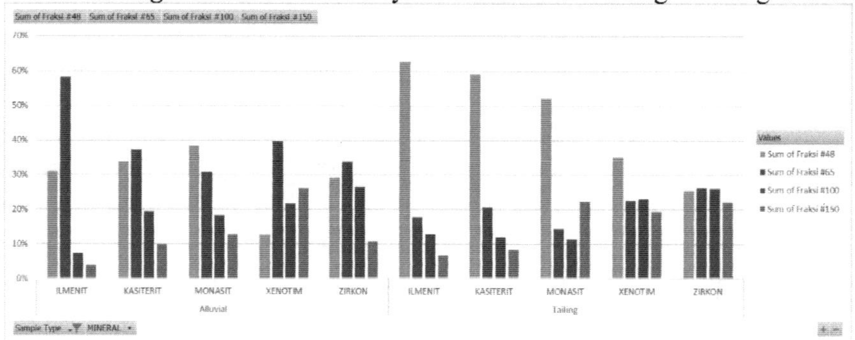

Figure 7. Composition of mineral in each size fraction.

The alluvial samples Ilm, Cass, Xtm, and Zr, were enriched in the fine size fraction (#65), while Mnz was enriched in medium sand (#48) and fine sand (See Fig. 7). Meanwhile, Mnz tailings are rich in medium sand fraction and abundant in very fine sand fraction (#150). Based on Fig. 6 and Fig.7, although the difference in the distribution of Zr in the fraction is not so significant (about 30% in Fig.7), Zr is more enriched in alluvial than tailings samples. This occurs since Zr minerals are recovered to the concentrate stream during the mining process. In contrast Xtm, the highest average Xtm content was in the tailings sample (0.35%), and the tailing Xtm samples were more enriched in medium sand and very fine sand fractions.

5.2. Heavy Mineral Grain Granulometry

5.2.1. Mean of Heavy Minerals
In order to simplify the mesh number, the fraction value in each mesh is converted into phi units [7]. Based on the mathematical calculations [5], the average grain size of heavy minerals in alluvial samples is 1.58 phi, a classification of medium sand grain fraction in the Wentworth grain size classification [9]. The tailing sample has an average grain size of 1.61 phi, similar to alluvial. These results signify that the sample as a whole has the sand grain dominated as medium size (See Fig. 4). Based on similar findings [6], the mean range of 0-4 phi has good sorting characteristics, and the contribution from source rock is quite large.

5.2.2. Sorting of Heavy Minerals

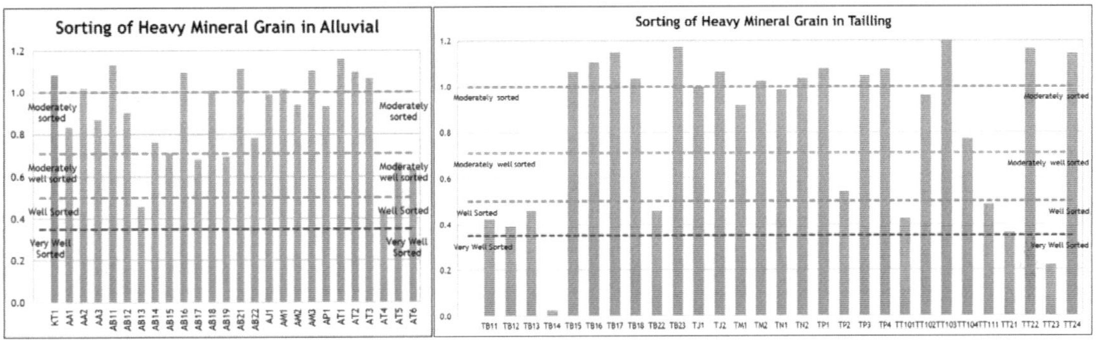

Figure 8. Sorting of heavy mineral grain in aluvial and tailing, Bangka.

In alluvial samples (See Fig. 8), 2 samples are well sorted, 5 samples moderately well sorted, 9 samples moderately sorted, and 10 samples poorly sorted. In alluvial samples as a whole, the sorting value shows a stable trend in the moderately sorted. This can be seen in alluvial samples in the Toboali area (in order AT1, AT2, AT3, AT4, AT5, and AT6) starting from the closest to the furthest from peak morphology, sorting heavy mineral grain sizes from poorly sorted until well sorted. It is evident that the further away from the source, the better the grain size sorting is. This also occurred on the colluvial samples (KT1), which had poorer sorting than the alluvials.

In the tailings sample (See Fig. 8), a similar pattern was found in the samples in Bebulu (TB11-TB23), Jebus (TJ1-TJ2), Menumbing (TM1-TM2), and Namak (TN1-TN2). It is also noticeable that the farther from the source, the better mineral grain size sorting the morphology is.

5.2.3. Kurtosis of Heavy Minerals
Based on the mathematical calculations [15] (See Fig. 9), a good pattern can be found in alluvial samples in the Bebulu (AB11-AB22) and Toboali (AT1-AT6) areas. Meanwhile, the downstream area has a small variation in the size of heavy mineral grains than the upstream area. This indicates that the downstream area is better sorted than the upstream area. On the other hand, heavy mineral tailings predominantly have a very platykurtic value, indicating that the variation in heavy mineral grain size in the tailings sample is quite diverse.

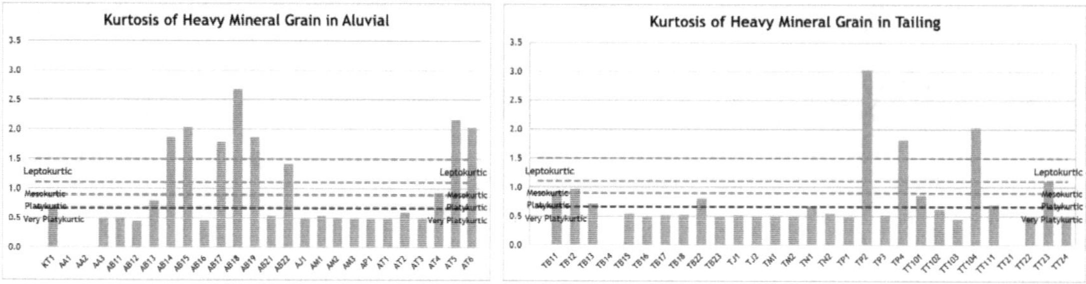

Figure 9. Kurtosis of alluvial and tailing sample in Bangka.

5.2.4. Skewness of Heavy Minerals

Skewness shows the symmetric value of the frequency curve [15]. Fig. 10 shows a good pattern in the samples in Bebulu and Toboali, where the grain size of heavy minerals is becoming coarser in the upstream area. This is quite different from the tailings sample with a random skewness pattern.

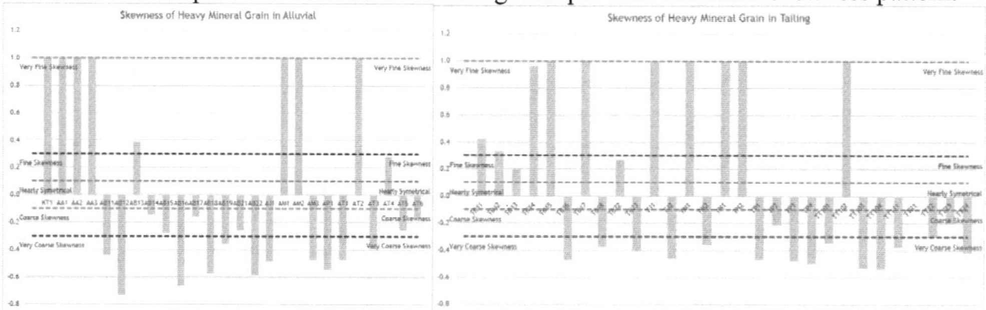

Figure 10. Skewness of heavy mineral grain in alluvial and tailing, Bangka.

5.3. Mineral Grain Morphology

5.3.1. Form of Heavy Mineral Grain

The heavy mineral forms of Cass, Mnz, Ilm, Xtm are predominantly equant, while Zr is predominantly prolate and oblate. The distribution of heavy minerals from upstream to downstream is as follows: Cass oblate to downstream equant, Mnz, Xtm and Ilm in Bebulu from upstream to downstream, changing from equant to oblate, while in the Toboali area, it is consistently oblate. In contrast, it is prolate / oblate to equant from upstream to downstream for Zr.

The predominance of equant and oblate heavy mineral forms is an indication that the transport process tends to occur in a rolling and suspension manner. However, some minerals in prolate and equant forms tend to have bedload transport mechanisms.

5.3.2. Sphericity of Heavy Mineral Grain

The sphericity value of heavy mineral grains was collected in the very equant category. Sphericity is dominated by very elongated to very equant, starting from upstream to downstream. This indicates an erosion factor in mineral grains during the transportation process and indicates the transportation process that occurs, which is the bedload.

5.3.3. Roundness of Heavy Mineral Grain

The value of the roundness of the colluvial sample shows that all minerals are in the sub-angular, angular, and very angular categories. For the alluvial samples, it shows that heavy mineral grains have a roundness ranging from sub-angular to sub-rounded in the upstream area. In contrast, in the downstream area, heavy minerals range from sub-rounded, rounded, to well rounded. This roundness rate is in line with the transport distance of each heavy mineral grain. The more downstream area of the sample, the better roundness is.

5.4. Degree of Radiation Exposure

During the measurement of the radiation exposure, the background value used was 194 CPS. Fig. 11 shows that in alluvial samples, all samples tend to have radiation values above the background. In contrast, in the tailings sample, the radiation exposure values are quite high, including some areas such as Jebus, Menumbing, and Pemali. A significant radiation exposure value is found in the very fine grain fraction sample (-120 mesh). However, the radiation value of the tailings sample tends to be higher than the alluvial sample, namely in the Menumbing area (TM1 at 216.33 CPS), while the highest radiation exposure value in the alluvial sample is in the Toboali area (AT6 at 211 CPS).

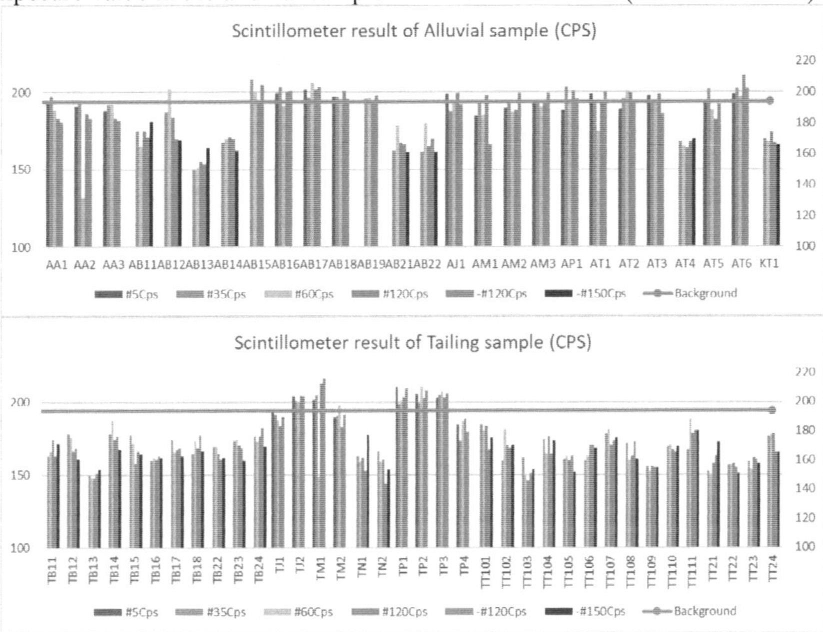

Figure 11. Scintillometer result of aluvial and tailing sample, Bangka.

5.5. Major and Trace Element Content

5.5.1. Relationship into Peak of Morphology

Based on the total REE value, for alluvial, some samples are enriched in REE (REE + Sc + Y> 170ppm) with values ranging from 191.5 ppm - 1860 ppm. These areas are alluvial Mangkol, Bebulu, Menumbing, and Toboali. In the Pemali alluvial, there was no indication of REE enrichment, while in the Jebus area, there was no alluvial location. Meanwhile, there were 2 locations where REE concentrations were high for the tailings sample, namely in Pemali (969.1 ppm) and Menumbing (12,646.5 ppm).

In general, REE is the richest alluvial, especially for the REE. LREE enriches alluvial Bebulu, Menumbing, and Toboali along with Zr, U, Th, Nb, and Sn. The same also applies to Menumbing and Pemali tailings samples. Meanwhile, Mangkol's alluvial is on the LREE.

The high Th content (> 50 ppm) is in Bebulu (AB15 at 60.4 ppm), but it decreases towards the downstream area. The high Th content in the alluvial sample with a value (90-1370 ppm) is the highest in the Toboali alluvial and Menumbing alluvial. At this alluvial location, the value increases downstream.

Regarding the composition of total REE compared to the distance between the alluvial sample and the morphological peak, the pattern cannot be determined with certainty, as not all samples contain REE. However, in general, it can be seen that the greater the distance between the sample and the morphological peak, the total REE concentration increases, as shown in the alluvial of Bebulu, Menumbing, and Toboali.

From the whole sample, the TM2 tailings sample is the sample that has the most significant element composition value, as the content of Total REE, Th, U, Zr, Nb, and Sn is the highest of all samples.

Table 1.Descriptive statictic of ICP-MS and ICP-OES data (Aluvial : 12 sample and tailing: 5 sample)

Element	CODE	DET.LIM	Min Alluvial	Max Alluvial	Mean Alluvial	STD Alluvial	Min Tailing	Max Tailing	Mean Tailing	STD Tailing
Li	PPM	0.1	2.1	69.4	21.6	18.7	8.6	30.9	15.8	8.3
Sc	PPM	1	1.0	12.0	4.6	3.6	1.0	40.0	12.0	14.4
Be	PPM	0.5	0.5	20.0	3.0	5.1	0.5	4.2	2.1	1.4
Na	PPM	20	40.0	1,680.0	618.5	442.2	190.0	1,830.0	680.0	594.6
Mg	PPM	20	20.0	1,090.0	462.3	311.4	70.0	1,180.0	534.0	383.8
Al	PPM	50	480.0	138,000.0	46,160.8	37,397.6	5,540.0	75,600.0	40,662.0	29,835.8
Ca	PPM	50	50.0	1,790.0	504.6	544.6	110.0	250.0	204.0	51.2
S	PPM	50	50.0	350.0	164.6	88.2	80.0	2,900.0	794.0	1,067.3
V	PPM	1	1.0	20.0	11.2	7.4	1.0	28.0	13.2	8.8
Cr	PPM	5	13.0	199.0	94.8	55.7	78.0	301.0	179.0	87.9
Co	PPM	1	1.0	2.0	1.1	0.3	1.0	2.0	1.6	0.5
Ni	PPM	1	1.0	8.0	4.0	2.2	3.0	11.0	6.2	2.9
Mn	PPM	1	18.0	363.0	157.0	104.4	26.0	1,410.0	384.0	517.5
Fe	%	0.01	0.2	2.4	1.1	0.7	0.3	2.4	1.3	0.8
Cu	PPM	1	3.0	26.0	8.5	6.0	3.0	12.0	7.6	3.8
Zn	PPM	1	5.0	79.0	33.8	21.3	6.0	113.0	45.8	36.9
Ga	PPM	0.1	0.2	62.5	21.7	19.5	1.9	218.0	54.3	82.4
As	PPM	1	1.0	73.0	18.2	21.1	1.0	95.0	32.6	34.0
Ge	PPM	0.1	0.4	3.7	1.8	1.1	0.6	24.8	6.2	9.3
Se	PPM	1	1.0	3.0	1.5	0.7	1.0	20.0	5.2	7.4
Ti	PPM	5	449.0	6,970.0	2,998.2	1,935.5	304.0	17,500.0	6,752.8	6,150.5
P	PPM	50	50.0	440.0	151.5	128.8	50.0	2,820.0	652.0	1,086.5
Ag	PPM	0.1	0.1	0.1	0.1	-	0.1	0.4	0.2	0.1
Cd	PPM	0.05	0.1	0.1	0.1	0.0	0.1	0.8	0.2	0.3
In	PPM	0.05	0.1	0.5	0.2	0.1	0.1	1.3	0.4	0.5
Sn	PPM	0.1	1.3	57.3	19.2	17.6	3.0	184.0	51.2	67.1
Sb	PPM	0.1	0.1	3.9	1.4	1.2	0.2	6.6	1.9	2.4
Te	PPM	0.1	0.1	0.3	0.1	0.1	0.1	0.1	0.1	-
Cs	PPM	0.1	0.1	18.5	3.9	5.1	0.8	11.1	3.3	4.0
Rb	PPM	0.1	0.2	372.0	71.4	112.0	7.3	105.0	31.3	37.7
Tl	PPM	0.02	0.0	2.8	0.6	0.8	0.1	1.4	0.5	0.5
Ba	PPM	1	1.0	91.0	27.6	24.1	2.0	96.0	33.6	33.5
W	PPM	0.1	0.3	39.0	10.1	11.4	3.8	71.6	28.4	24.4
Th	PPM	0.05	0.4	262.0	81.6	90.1	2.9	1,370.0	309.9	531.4
U	PPM	0.05	0.2	52.4	14.8	17.0	1.0	307.0	74.4	117.1
K	PPM	20	40.0	24,300.0	5,802.3	8,533.3	580.0	11,000.0	2,928.0	4,045.9
Nb	PPM	0.1	1.1	110.0	30.9	31.3	3.7	214.0	66.3	76.9
Ta	PPM	0.05	0.1	13.1	4.0	4.1	0.2	26.6	8.7	9.4
Pb	PPM	1	1.0	83.0	29.4	27.5	3.0	204.0	53.6	75.9
Mo	PPM	0.1	0.3	3.6	2.0	0.9	1.4	4.4	2.8	1.0
Sr	PPM	0.5	0.7	18.1	8.1	5.2	1.6	12.4	7.0	4.1
Zr	PPM	0.5	4.9	5,300.0	974.1	1,756.8	11.8	31,500.0	9,637.6	12,499.7

Element	CODE	DET.LIM	Min Alluvial	Max Alluvial	Mean Alluvial	STD Alluvial	Min Tailing	Max Tailing	Mean Tailing	STD Tailing
Zr	%	0.01	0.5	0.5	0.5	0.0	1.6	3.2	2.4	0.8
Hf	PPM	0.1	0.2	80.2	20.4	25.9	0.7	340.0	90.0	127.9
Re	PPM	0.05	0.1	0.1	0.1	-	0.1	0.1	0.1	-
Bi	PPM	0.05	0.1	15.3	3.9	4.5	0.4	47.6	11.1	18.3
La	PPM	0.1	0.5	388.0	97.6	127.5	0.8	2,720.0	595.5	1,065.5
Ce	PPM	0.1	1.0	829.0	216.6	268.9	1.9	5,540.0	1,154.8	2,193.6
Pr	PPM	0.05	0.1	88.8	22.8	28.5	0.2	642.0	140.3	251.6
Nd	PPM	0.1	0.4	282.0	74.3	90.8	0.8	1,960.0	432.4	766.8
Sm	PPM	0.1	0.1	55.2	14.4	17.2	0.3	346.0	77.1	135.1
Eu	PPM	0.1	0.1	0.7	0.3	0.2	0.1	1.8	0.8	0.8
Gd	PPM	0.1	0.1	36.5	10.3	11.4	0.4	250.0	57.5	97.1
Tb	PPM	0.05	0.1	4.5	1.3	1.4	0.1	28.3	6.8	10.9
Dy	PPM	0.1	0.2	22.8	6.7	6.9	0.6	138.0	34.5	52.8
Y	PPM	0.1	0.8	100.0	31.3	29.2	2.9	571.0	155.8	216.5
Ho	PPM	0.1	0.1	4.0	1.2	1.2	0.1	24.6	6.3	9.4
Er	PPM	0.1	0.1	14.3	3.9	4.0	0.4	83.0	21.2	31.5
Tm	PPM	0.1	0.1	2.2	0.6	0.6	0.1	12.7	3.3	4.7
Yb	PPM	0.1	0.1	19.0	4.9	5.3	0.5	93.7	24.3	35.4
Lu	PPM	0.05	0.1	2.7	0.7	0.8	0.1	13.5	3.5	5.1
Total of REE			4.8	1,860.4	491.5	594.2	10.3	12,464.5	2,726.0	4,882.1
LREE			2.1	1,643.0	425.8	531.5	4.0	11,208.0	2,400.1	4,410.9
HREE			1.6	206.0	60.8	60.2	5.1	1,214.8	313.1	462.4
LREE/HREE			1.3	9.3	5.2	2.5	0.8	9.2	3.1	3.1
Ce/Ce*			3.8	6.9	4.4	0.8	1.6	4.3	3.5	1.0
Eu/Eu*			0.0	1.0	0.1	0.3	0.0	0.3	0.1	0.1

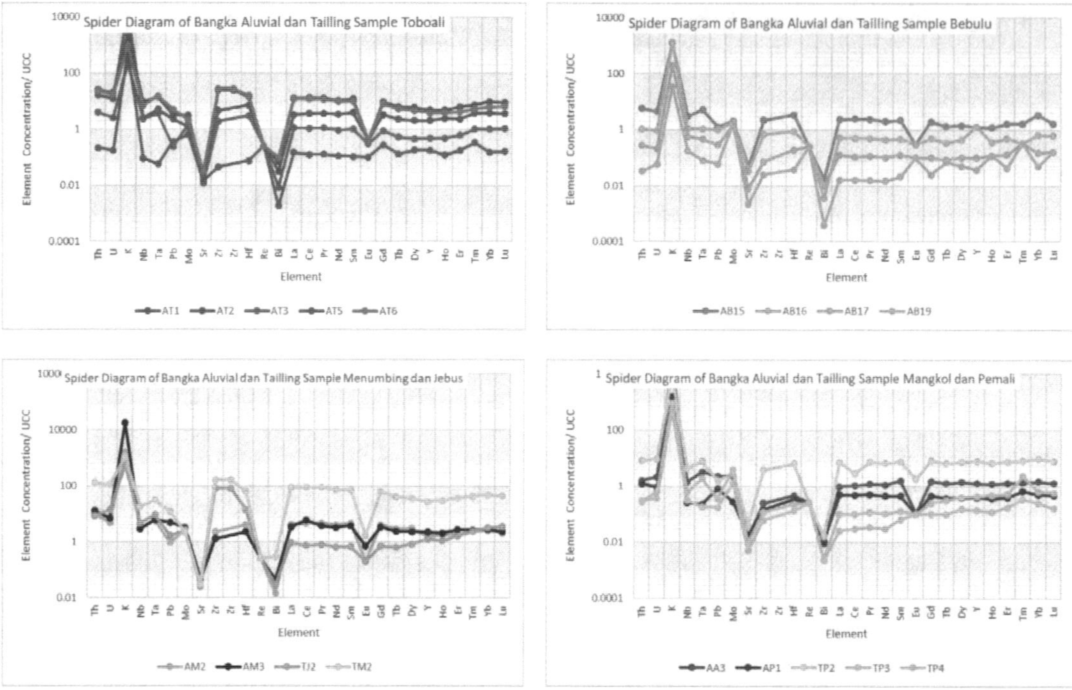

Figure 12. UCC [16-17]-Normalized REE and trace element of alluvial and tailing, Bangka.

The Spider diagram (See Fig. 12) shows that REE in Bangka alluvial is enriched by around 1 - 20 times the Upper Continental Crust level, while REE in Bangka tailing is enriched by about 50 times. There are several anomalies, such as the positive Eu anomaly in the alluvial downstream and Pemali tailings. This occurs since the alluvial Bebulu (AB19) is downstream, and the possibility of sea water contamination affects the high concentration of Eu.

6. Conclusion

1. Based on grain size analysis or granulometry of heavy minerals associated with Bangka Island placer tin deposits, the alluvial sample of the soil is dominated in fine sand grain size, while the tailing sample is mostly in the medium-fine sand grain size.
2. Mathematical analysis of grain morphology of heavy alluvial minerals shows that the grain shape is dominated by equant and oblate (except for Zr prolate, which shows the crystal structure of Zr, namely tetragonal and hardness 7, which is the highest than other heavy minerals). Grain distribution from upstream to downstream is in increasing roundness and stability rate. The transport mechanisms that occur are suspended and bedload.
3. The highest REE enrichment occurred in the Bangka tailing sample, which was 50 times the enrichment of the UCC.

Reference

[1] Schwartz M O, Rajah S S, Askury A K, Putthapiban P and Djaswadi S 1995 *J Earth Science Review* **38** pp 95-293.
[2] Yim W W S 2000 *Handbook Marine of Mineral* Cronan DS (London: CRC Press) p 27.
[3] Sengupta D and Van Gosen BS 2016 *Reviews in Economic Geology* **18** pp 81-100.
[4] Delta Epsilon, SC-133 Portable Scintillation Counters, accessed Dec 2020 (https://deltaepsilon.com/portable-scintillometers/#)
[5] Krumbein W C 1934 *J of Sedimentary Petrology* **4** pp 65-77.
[6] Folk R L and Ward W C 1957 *J of Sedimentary Petrology* **27** pp 3-26.

[7] Hoque M E, Chowdhury S R, Uddin M M, Alam M S and Monwar M M 2013 *Open Journal of Soil Science* **3** pp 71-80.
[8] Folk RL 1980 *Petrology of Sedimentary Rocks* (Texas: Hemphill Publishing Company) p 7.
[9] Bogs S Jr 2009 *Petrology of Sedimentary Rocks* (Cambridge : Cambridge University Press) pp 21-41.
[10] Soegeng W H S, Sucipta 2012 *Proc. Nuclear Geology and Resources* (BATAN).
[11] Power M C 1953 *J of Sedimentary Petrology* **23** pp 117-119.
[12] Rittenhouse G 1943 *Bulletin of Geological Society of America* **54** pp 1725-1780.
[13] Mangga A S and Djamal B 1994 Indonesian Geological Research and Development Center.
[14] Margono U, Supandjono RJB and Partoyo E 1995 Indonesian Geological Research and Development Center.
[15] Dana C D P, Sudirman M R, Noviana A and Hidayat R 2016 *Proc. of the National Seminar on Earth* (Yogyakarta: UGM).
[16] Rudnick R L 2018 *Encyclopedia of Earth Sciences Series.*
[17] Rollinson H R 1993 *Using Geochemical Data* (England : Longman Group UK Limited).

IOP Conf. Series: Earth and Environmental Science 1031 (2022) 012009 doi:10.1088/1755-1315/1031/1/012009

Geological and Isothermal 3-D Model of Blawan-Ijen Geothermal System

Masdukhan Aris, Suryantini, Angga Bakti Pratama

Geothermal Master Program, Institut Teknologi Bandung, Indonesia

masdukhan@gmail.com

Abstract. The Blawan-Ijen is one of the geothermal prospects in the eastern part of Java island, Indonesia. This prospect is situated in the caldera called the Kendeng caldera. The Blawan-Ijen has geothermal manifestations in two different locations, the Ijen crater in the southeast and Blawan in the northern part of the caldera. These manifestations are estimated to have different reservoirs. Different manifestations of characteristic and lithological complexity in the Kendeng caldera make the geothermal system contained in the caldera less known. Therefore, a 3D model was built on understanding the geometries of the Blawan-Ijen geothermal system. In this research, the model is limited to the geological and isothermal model. The data used in building the models are based on the literature and research data that has been done previously. Geological and isothermal models were built using Leapfrog Geothermal software. The 3D conceptual model provides a general overview of the geometries of the Blawan-Ijen geothermal system. The isothermal model shows temperature anomalies and their possible distribution.

1. Introduction

In the initial stages of exploration, the conceptual model of the geothermal system is used to understand the geological conditions and processes that occur in the geothermal system, especially about temperature, pressure, and fluid flow. This conceptual model is generally built based on the integration of geosciences (geology, geochemistry, and geophysics) and well data. This conceptual model will continue to be updated with increasing information or data. Conceptual models are generally limited to 2D models. However, this 2D model has limited information, especially regarding the geometry of the geothermal system. Therefore, the delineation of geothermal systems in 3D models is a way to better understand geothermal systems [1].

The case study in this research was conducted on the Blawan-Ijen geothermal prospect. The Blawan-Ijen is one of the geothermal prospects in the eastern part of Java island, Indonesia. Research in this area has been widely carried out on geology, geochemistry, and geophysics. The Blawan-Ijen prospect is currently in the stage of exploration well drilling. The slim hole was drilled vertically until 2000 m depth and discovered a high temperature of about 283°C [2]. Subsurface 3D models in the Blawan-Ijen area will be built based on data from previous research. This paper only discusses the construction of the geological and isothermal model.

2. Geothermal System of Blawan-Ijen

The Blawan-Ijen geothermal system is associated with active volcanic and a caldera (Kendeng caldera). The Blawan-Ijen geothermal system is controlled by a local (volcanic) structure in the form of the Kendeng Caldera and a regional structure (tectonic) of a north-south fault fracture followed by volcanic

IOP Conf. Series: Earth and Environmental Science 1031 (2022) 012009 doi:10.1088/1755-1315/1031/1/012009

eruption cones directed at the west-east and northwest-southeast faults. The geothermal system in a caldera can be found as long as eruption activity continues. If silicic magma bodies are formed, these resources have a life span of several million years after the last volcanic activity. The amount of a hydrothermal system that can be utilized depends on the level of the permeable zone under the caldera or in the caldera deposit. In most caldera that contain thick and dense tuff deposits, the hydrothermal system is limited to caldera ring fractures and faults that intersect the caldera, which form a permeability zone [3].

The existence of the Blawan-Ijen geothermal system is indicated by the appearance of surface manifestations in the Ijen Craters and Blawan. Manifestations around the Ijen Crater consist of acid crater lakes, acid hot springs, and fumaroles [4]. The acid crater lake has a temperature >30°C and a pH <1 [5]. The acid crater lake has about 1000 x 600 m and 200 m depth [6]. The chemical content in acid crater lakes is controlled by the continuous interaction of meteoric water with crater rocks. The presence of gas vents at the bottom of the lake also controlled the chemical content in the fluid of crater lakes [7]. This crater lake and deeper thermal system are the sources of the manifestation of acid hot springs on the western slope of the Ijen Crater. The hot springs have an acidic pH (pH<1) and a temperature that varies from 20- 30°C [5]. Fluids from both crater lake and acid hot springs have SO_4-Cl type. Fumarole activity can be found in the southeast of Ijen Crater. This fumarole has a high discharge with temperatures varying from 200°C [4,6] at several vents and about 600°C coming out on the silicic dome [8]. The manifestation of Blawan's hot springs is located in Blawan village, in the northern part of Kendeng Caldera. This manifestation has a temperature of around 35-50°C with a neutral pH (pH>~6). This hot spring has the fluids' chemical characteristics of the HCO_3-SO_4 type [9–11].

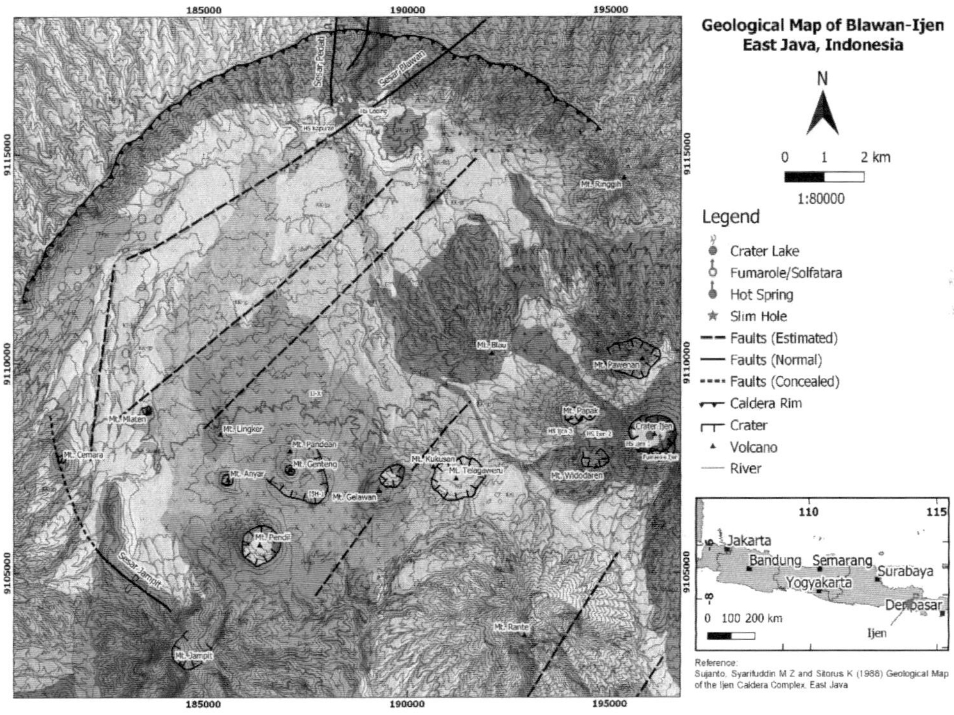

Figure 1. Geological map of Blawan-Ijen modification from Sujanto et al. (1988).

Some wells that have been drilled in the Blawan-Ijen area are slim holes. The ISH-01 slim hole well in the south reach a temperature of 80°C at a 500 m depth. At the same depth, alteration minerals (epidote) are found, estimated to be alteration minerals from extinct systems [9]. The other slim hole (IJ-X) is to the north of the previous slim hole. This slim hole reaches a 2000 m depth and indicates a temperature of 283°C with neutral fluid. High temperatures are also strengthened by the presence of alteration minerals such as epidote found at depths of 1782-2000 m [2]. Figure 1 shows the distribution of the lithological unit and some manifestations in the Blawan-Ijen geothermal area.

3. Data and Method

The data used in this modeling are topography data, geology, geochemistry, and well data. The topographic data used in this modeling is based on DEMNAS with a spatial resolution of about 8 m and bathymetry data from Ijen Crater, as described by Takano et al. [6]. This data is used to build topography in the Leapfrog Geothermal software. A detailed geological survey in the Blawan-Ijen area was conducted by Sitorus (1990) and Sujanto (1988) and produced a surface geological map [12,13]. This surface geology map and cross-section are used as input data to build 3D geological models in Leapfrog Geothermal software. There are about 70 lithology units on this map, but this unit will be simplified by coloring based on rock sources to aid modeling. Subsurface temperature distribution, other than based on the well data, is obtained from the results of temperature estimation given by Saputra (2006) [1]. By this method, the subsurface temperature of the conductive layers can be determined. This method uses surface manifestation temperature data and estimated temperature from the reservoir based on geochemical data. This temperature value will then be interpolated using Leapfrog Geothermal software to provide subsurface temperature distribution in the area. The boundary of the model is limited in the Kendeng caldera area, and the bottom elevation of the model is limited to -1000 masl.

4. Results and Discussion

The results of 3D geological modeling in the Blawan-Ijen area can be seen in Figure 2. The modeling results are arranged based on volcano stratigraphy, as described in the Sitorus (1990). Figure 2 (A) shows a geological model in the early stages of Kendeng caldera formation. Figure 2 (B) is a geological model that shows the Blawan lake sediments and the product of resurgent dome activity (Mt. Blau). Figure 2 (C) is a geological model at the stage of volcano formation along the Kendeng caldera rim. The caldera rim volcano consists of Mt. Jampit, Mt. Ringgih, Mt. Suket, Mt. Rante, Mt. Pawenan and Mt. Merapi (from old to young). Figure 2 (D-F) is a post-caldera stage with several cinder cones formed inside the caldera with an east-west orientation. Figure 2 (G) shows the final 3D geological model of Blawan-Ijen. The 3D geological model shows very complex rock layers near the surface as a result of post-caldera volcanic activity. Complex lithology and lack of structure in the middle part of the caldera may be the cause of the absence of surface manifestations in this area. Based on this model, reservoir rocks are estimated to be the volcanic layer of Old Ijen composite cone with lithology consist of a pyroclastic flow, pyroclastic fall, and lava flow with andesite-basaltic characteristics.

The results of the isothermal model can be shown in Figure 3. The isothermal models are categorized based on their temperature: high temperature (>225°C), moderate temperature (125-225°C), and low temperature (<125°C). Based on the model results, the high-temperature distribution can be seen to tend toward the Ijen Crater. This is possible because Ijen Crater is an active crater found in the Blawan-Ijen area. Besides, the Ijen crater has intense manifestation activity, such as fumaroles. Another possibility is that this high-temperature trend corresponds to the lineament of several cinder cones formed in the caldera. The lineament is likely to have good permeability so that it becomes a magma path to exit and form the cinder cone. This high temperature may be related to the presence of heat sources in the subsurface and was once correlated with cinder cones on the surface. This possibility still needs to be examined in more detail, especially with the lack of data used in this study. Geophysical data and other well data may provide a more detailed overview of the subsurface.

IOP Conf. Series: Earth and Environmental Science 1031 (2022) 012009 doi:10.1088/1755-1315/1031/1/012009

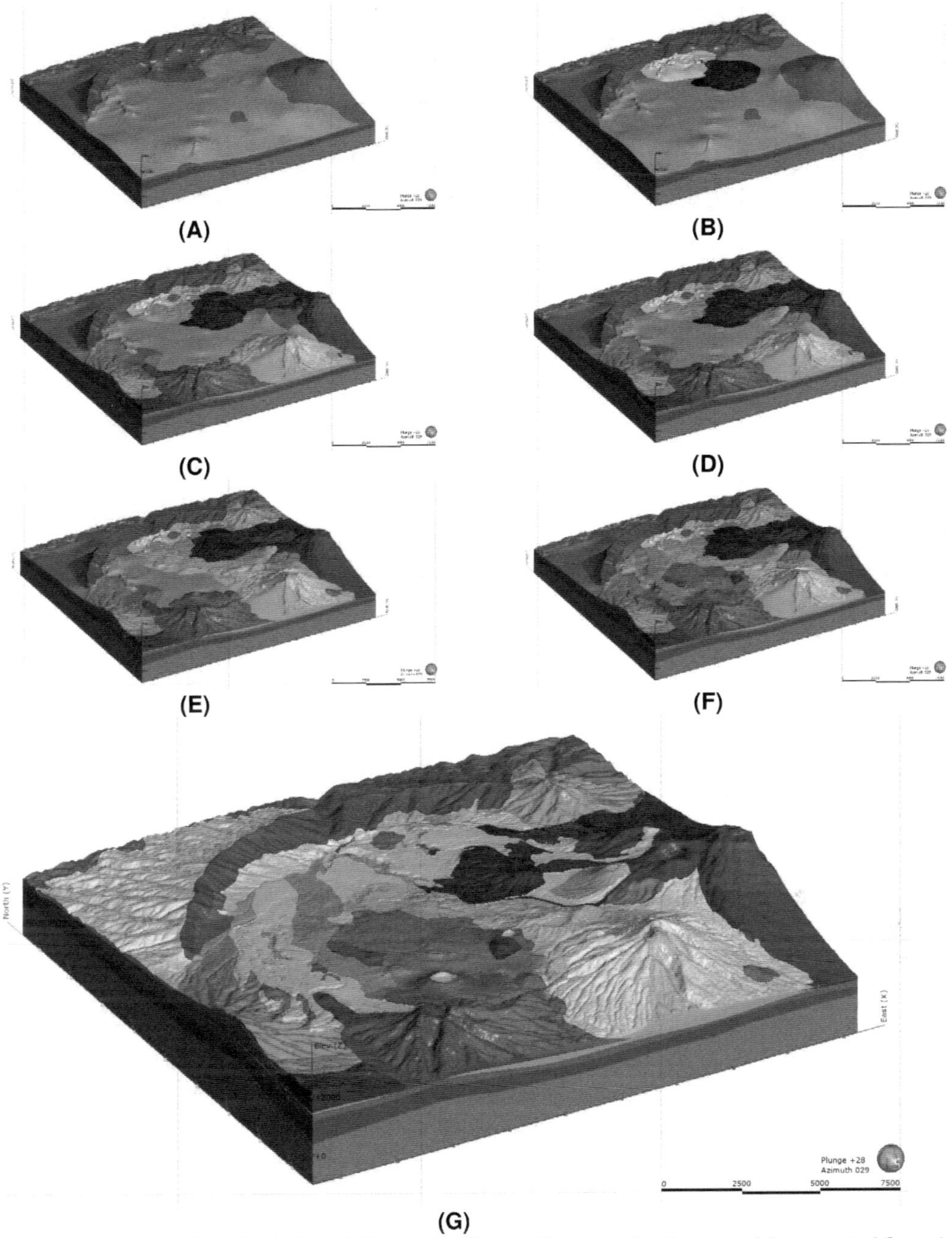

Figure 2. The results of 3D geological modeling in the Blawan-Ijen area, the above models are sorted from A to F based on volcano stratigraphy in Sitorus (1990). A) the initial stage of caldera formation, B) Blawan lake formation and resurgent dome (Mt Blau) activity, C) volcanic formation along the caldera rim, D-F) cinder cone formation in the caldera, and G) final 3D geological model of Blawan-Ijen geothermal area.

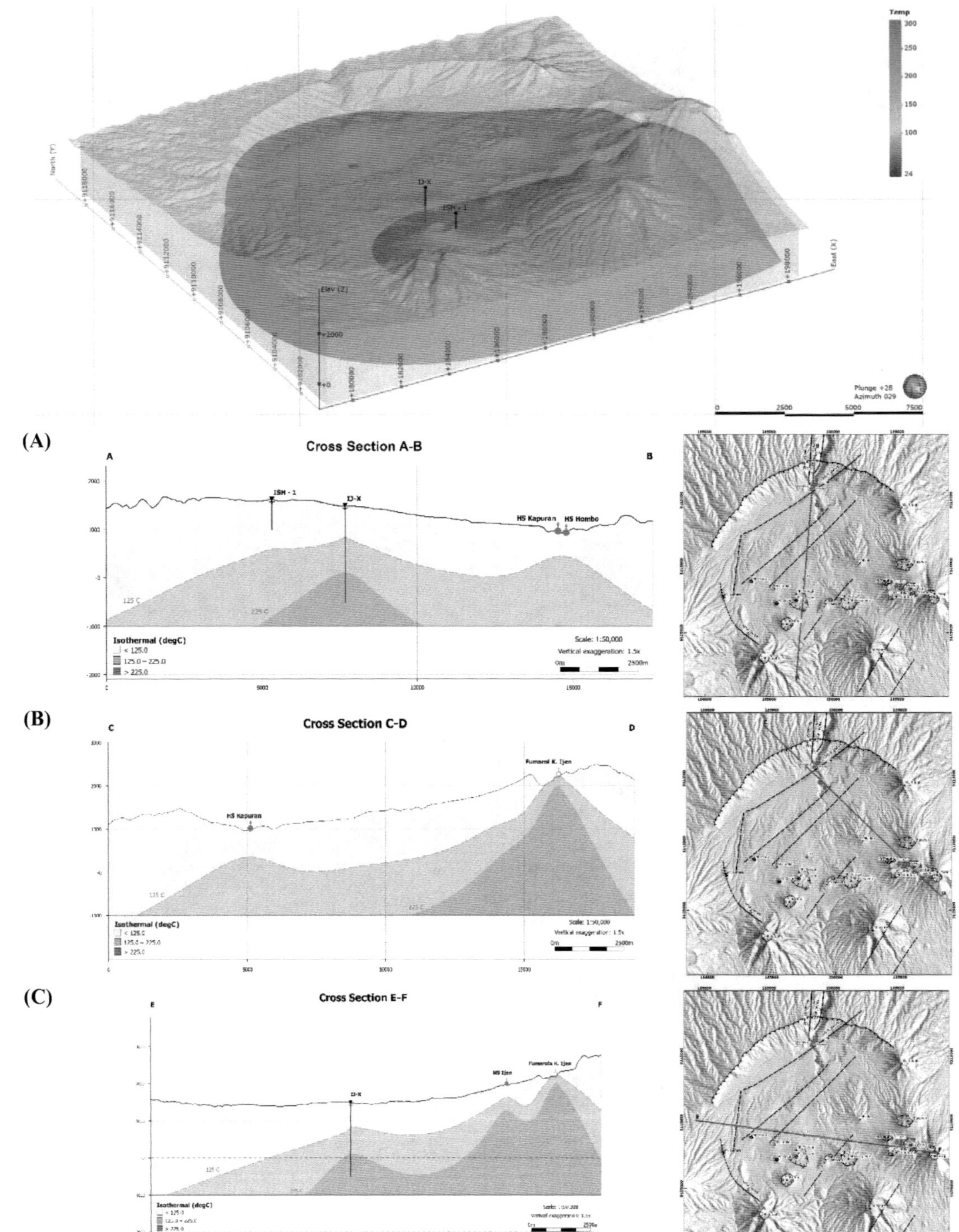

Figure 3. Isothermal model of Blawan-Ijen geothermal area. (A) N-S cross section through Blawan manifestation in the north and Wurung Crater in the south, (B) NW-SE cross section through Ijen crater in the southeast and Blawan manifestation in the northwest, and (C) E-W cross section through Ijen Crater in the west and well IJ-X.

5. Conclusion

The 3D geological and isothermal models show the geometry of the Blawan-Ijen geothermal system and its possible distribution. With limited data used in this 3D model, it is sufficient to provide a subsurface overview of the Blawan-Ijen geothermal system. The heat source of the Blawan-Ijen geothermal system is estimated to be magma associated with volcanic activity on the surface. This interpretation is supported by 3D isothermal model, which shows the correlation of high temperature with volcanic cones on the surface, such as Wurung Crater and Ijen Crater. Based on 3D lithological model, reservoir of this geothermal system is estimated to be old Ijen volcanic products. The absence of manifestations in the center part of the caldera is thought to be due to newer volcanic products with impermeable properties. This 3D model can be updated with increasing data to provide more detailed subsurface information. Further research and the addition of data can provide a more accurate model, so this model can be used as a guide in developing Geothermal fields in the future.

References

[1] Saputra M P, Suryantini, Catigtig D, Regandara R, Asnin S N and Pratama A B 2016 Geological, isothermal, and isobaric 3-D model construction in early stage of geothermal exploration *IOP Conference Series: Earth and Environmental Science* vol **42**

[2] Daud Y, Nuqramadha W A, Fahmi F, Pratama S A, Rahman K R and Subroto W 2017 Discovering "Hidden" Geothermal Reservoir in Blawan-Ijen Geothermal Area (Indonesia) Using 3-D Inversion of MT Data *Proceedings of 1st Geo Electromagnetic Workshop* (Bandung)

[3] Wohletz K and Heiken G 1992 *Volcanology and Geothermal Energy* (Berkeley: University of California Press)

[4] Caudron C, Syahbana D K, Lecocq T, Van Hinsberg V, McCausland W, Triantafyllou A, Camelbeeck T, Bernard A and Surono 2015 Kawah Ijen volcanic activity: a review *Bull. Volcanol.* **77**

[5] Palmer S C J 2009 *Hydrogeochemistry of the upper Banyu Pahit River valley , Kawah Ijen volcano , Indonesia* (McGill University)

[6] Takano B, Suzuki K, Sugimori K, Ohba T, Fazlullin S M, Bernard A, Sumarti S, Sukhyar R and Hirabayashi M 2004 Bathymetric and geochemical investigation of Kawah Ijen Crater Lake, East Java, Indonesia *J. Volcanol. Geotherm. Res.* **135** 299–329

[7] Van Hinsberg V, Vigouroux N, Palmer S, Berlo K, Mauri G, Williams-Jones A, Mckenzie J, Williams-Jones G and Fischer T 2017 Element flux to the environment of the passively degassing crater lake-hosting Kawah Ijen volcano, Indonesia, and implications for estimates of the global volcanic flux *Geol. Soc. Spec. Publ.* **437** 9–34

[8] Mauri G 2009 *Multi-Scale Analysis of Multiparameter Geophysical and Geochemical Data from Active Volcanic Systems* (Simon Fraser University)

[9] Hochstein M P and Sudarman S 2015 Indonesian Volcanic Geothermal Systems *Proceeding World Geotherm. Congr. 2015* 11

[10] Delmelle P and Bernard A 2000 Downstream composition changes of acidic volcanic waters discharged into the Banyupahit stream, Ijen caldera, Indonesia *Journal of Volcanology and Geothermal Research* vol 97 pp 55–75

[11] Palmer S C J, van Hinsberg V J, McKenzie J M and Yee S 2011 Characterization of acid river dilution and associated trace element behavior through hydrogeochemical modeling: A case study of the Banyu Pahit River in East Java, Indonesia *Appl. Geochemistry* **26** 1802–10

[12] Sujanto, Syarifuddin M Z and Sitorus K 1988 Geological Map of the Ijen Caldera Complex, East Java

[13] Sitorus K 1990 *Volcanic stratigraphy and geochemistry of the Idjen Caldera Complex, East Java, Indonesia* (Victoria University of Wellington)

IOP Conf. Series: Earth and Environmental Science 1031 (2022) 012010 doi:10.1088/1755-1315/1031/1/012010

Model Analysis of Phase Transformation on Gadolinium-doped Yttrium Iron Garnet

H Subawi[1], S Soepriyanto[1], A A Korda[1] and B Prijamboedi[2]

[1] Laboratory of Solid Oxide System, Department of Metallurgical Engineering, Institut Teknologi Bandung, Jl Ganesha no 10 Bandung 40132 INDONESIA
[2] Department of Chemistry, Institut Teknologi Bandung, Jl Ganesha no 10 Bandung 40132 INDONESIA

Abstract. The most suitable model to describe the yttrium iron garnet (YIG) sintering reaction has been evaluated in this study. The sintering reaction at the optimum temperature was predicted by the simulation method. When sintering takes place at a temperature of 1200°C, an intermediate compound of YIP is formed, then at a temperature of 1450°C the formation of YIG reaches a steady state. Sintering at 1450°C for 5 hours resulted in a YIG conversion of about 99.7%. The calculation of the activation energy value of the YIG sintering reaction was carried out based on linear regression technique using the Arrhenius plot. SCM modeling has a deviation tolerance of 0.923 and describes a two-step reaction. The SCM modeling was rated the best compared to other models, and was suitable for describing the true Gd-doped YIG sintering reaction. SCM modeling is considered to be able to describe the actual conditions of the YIG and Gd doped YIG sintering reaction.

Keywords: Garnet, Modeling, Phase conversion, Sintering, Srinking core model

1. Introduction

Yttrium iron garnet is needed as a circulator part in radar and telecommunication components. Efforts to improve the performance of YIG have been carried out by adding doping materials, for example, the gadolinium element to form Gd-doped YIG. YIG was formed by the sintering process in solid conditions. So far, research on YIG sintering has been carried out by adjusting the sintering temperature, either conducted at a relatively lower temperature than 1200°C, or above 1500°C. In other case, the raw material composition of Fe_2O_3, Y_2O_3 and Gd_2O_3 was adjusted to obtain a single-phase of YIG [1].

Yttrium iron garnet compounds can be formed through the sintering process using Fe_2O_3 and Y_2O_3 as raw materials with a molar composition of 5: 3. The reactions that occur during sintering can be configured through modeling study. A solid-state reaction mechanism has been described in a model called the shrinking core model (SCM) [2]. This modelling uses the main assumption that there is a process that occurs by diffusion or reaction, at the boundary that separates the two zones in the solid.
The rate of reaction or diffusion rate is defined as a irreversible or reversible reaction which involves the insertion of foreign compounds into the parent structure and results in significant structural modifications

to the parent, for example, the insertion of yttrium in the YIG compound. The shrinking core model has been widely used in process technology to describe gas-solid reactions [3]. The reaction mechanism in YIG sintering at a solid state was also studied using the shrink core model [4].

In this study, several models that can describe the YIG sintering process will be investigated. The indication of intermediate product formation prior to the formation of the final YIG product will also be reviewed in this study. It will also be evaluated when intermediate product formation occurs during the YIG sintering process. The addition of Gadolinium as a doping substance in YIG will also be investigated to see if it shows the same trend.

2. Methodology

The actual condition of the YIG sintering reaction needs to be described by modeling the sintering reaction to confirm the reaction mechanism that occurs. So far, the literature data provides information on the conversion of YIG formation up to a temperature of 1250°C. Calculation and simulation methods were carried out to complement the lack of data as a prediction of the sintering reaction at a practical temperature of 1450°C. Based on the simulation data, the product conversion value, reaction rate, reaction rate constant, deviation tolerance and activation energy value are obtained. The simulation value is then compared with the available data, to determine the suitability of the selected YIG sintering modeling.

3. Result and Discussion

3.1. The governing equation

The sintering reaction of YIG is described simply in terms of the solid-state reaction equation according to:

$$a\,A(s) + b\,B(s) \rightarrow c\,C(s) + d\,D(s) \tag{1}$$

The reaction that occurs between solid oxides is assumed to occur at the particle interface at radius (r), and is described as moving from the outer surface of the particle to its center during the reaction process. A reaction process occurs at the interface of the core and shell, while the ionic diffusion process occurs at the layer between the reaction interface and the outer particle surface. The application of the assumption of shell and core type reactions shows the ease with which the definition of the conversion of the dominant phase product is only related to the position of the shell and core interface. The conversion of the yttrium iron garnet product is given the symbol $X(t)$ formulated in a mathematical equation as follows:

$$X(t) = \frac{n(0) - n(t)}{n(0)} \tag{2}$$

where n is the composition of the reacting solid particles (in mol units).

If the reactants are only present in the nucleus, then the amount of reactant in the particle is proportional to the volume of the nucleus, related to the density of the material. The conversion of the reactant particles can be expressed as a function of the radius, according to the following mathematical equation:

$$X(t) = 1 - \frac{r_i(t)^3}{r_0^{\,3}} \tag{3}$$

where r_o is the particle radius, and $r_i(t)$ is the interface radius between the shell and the core at time t. The definition of an inverse function from equation (3) states that the interface position is a product conversion function.

$$r_i(t) = r_0\left(1 - X(t)\right)^{1/3} \tag{4}$$

Melchiori describes how to calculate the conversion evolution of global particles as a function of a time [4]. The differential equation with respect to time will result in:

$$\frac{\partial X}{\partial t}=\frac{a}{c_A^0}\cdot\frac{3}{r_0^3}\cdot\frac{c_{B,bulk}}{\frac{1}{r_i^2 k'}+\frac{b}{r_i D_{BD}}\left(1-\frac{r_i}{r_0}\right)+\frac{b}{h_B r_0^2}} \quad (5)$$

where c_A^o is the initial molar concentration of A; a, b are the constants of the reaction equation; D_{BD} is the effective diffusion constant B in D; and h_B is the mass transfer coefficient of B.
Finally, by combining equation (4) into equation (5), the variable r_i can be expressed as a conversion function X, so that an equation with only one dependent variable can be obtained as follows:

$$\frac{\partial X}{\partial t}=\frac{3a}{c_A^0}\cdot\frac{c_{B,bulk}}{\frac{r_0}{k'}(1-X(t))^{-2/3}+\frac{b.r_0^2}{D_{BD}}\left((1-X(t))^{-1/3}-1\right)+\frac{b.r_0}{h_B}} \quad (6)$$

Equation (6) above in view of the shrinking core model is an equation which states that the particles are assumed to be spherical. The ion concentration in equilibrium between particles can be expressed in terms of a shrinking nucleus model, which is described by the parameter $c_{B,bulk}$ as a function of time. If the parameter $c_{B,bulk}$ is constant, that is, a condition in which a high flow rate of ions is fed to the surrounding particles at a constant composition, then the mathematical equation has the following analytical solution:

$$t=\tau_K.\left[1-(1-X)^{1/3}\right]+\tau_D.3\left[1-(1-X)^{2/3}-\frac{2}{3}X\right]+\tau_M.X \quad (7)$$

where the parameters τ_K, τ_D, τ_M are the characteristic times associated with kinetic resistance (τ_K), diffusion (τ_D) and external mass transfer (τ_M), respectively.

The value of the characteristic time is related to the importance of each resistance in the reaction process. For example, this means that a high diffusion value τ_D cannot be ignored, it can even be associated with a low diffusion coefficient. It is useful to understand which type of reaction controller is decisive in determining each characteristic time. If the concentration of reactant B on the pellet surface is constant in time, as well as the temperature of the system, then the parameter τ_M in equation (7) can be neglected. In this mathematical equation, τ_K and τ_D correlate with the importance of kinetic and diffusion resistance. The results of the rearrangement into the form of a non-dimensional mathematical equation, the above equation can be expressed as:

$$\theta(X)=\frac{\tau_K}{\tau_K+\tau_D}.\left[1-(1-X)^{1/3}\right]+\frac{\tau_D}{\tau_K+\tau_D}.3\left[1-(1-X)^{2/3}-\frac{2}{3}X\right] \quad (8)$$

Next, define another term, namely Thiele's modulus, $Ø^2$, which is an equation that expresses a direct relationship between two characteristic times τ_K and τ_D.

$$\phi^2=\frac{18\,\tau_D}{b\tau_K} \quad (9)$$

The term of Thiele's modulus describes the correlation between catalytic activity and particle size [5]. When the particles have a large size and the reaction rate is very fast, then the diffusion force is only able to carry the product away from the particle surface. Therefore, only the promoter reactant surface will undergo reaction. Based on the parameters of Thiele's modulus, three criteria can be distinguished for a catalytic activity: diffusive, kinetic, or a transition between the two. In the diffusion regime, the reaction rate is considered faster than the diffusion rate in solids [6]. If the value of Thiele's modulus is sufficently large, it can be noted that the solution of the shrinking core model follows equation (8) with $\tau_K = 0$. In this case, the solution for the srinking core model model can be expressed in non-dimensional numbers.

$$\theta(X) = 3\left[1-(1-X)^{2/3}-\frac{2}{3}X\right] \quad (10)$$

It should be noted that the diffusion regime ($\O^2 >> 1$) means that the value of the kinetic constant as well as the expression of the reaction rate is not important. In the other hand, in the kinetic regime ($\O^2 << 1$), the rate of diffusion in solids is faster than the rate of reaction. The kinetic regime is achieved when the diffusion coefficient of D_{BD} is sufficiently large, while compared to the kinetic constant. In the kinetic regime where $\tau_D = 0$, the shrinking core model solution can be expressed in terms of a non-dimensional number.

$$\theta(X) = \left[1-(1-X)^{1/3}\right] \quad (11)$$

3.2. The shrinking core model

When the sintering reaction is not at optimum conditions, the formation of the YIG product is accompanied by the appearance of a secondary phase, namely yttrium iron perovskite (YIP). The YIP phase can decrease or disappear, when the sintering conditions are carried out at a higher temperature of 1450°C or the duration is extended to 5 hours, but does not exceed that time interval. This shows that the YIG sintering reaction mechanism is through the YIP intermediate product formation process. The YIG formation reaction takes place in two stages, so that modeling is chosen according to the actual conditions. The evaluation of the formation of the YIP intermediate product as an intermediate product of the YIG sintering reaction was reported by Buscaglia. YIP and YIG phases will be formed during the sintering process when Y_2O_3 and Fe_2O_3 are pressed together at high temperatures [7]. The cross-sectional evaluation results of the sintering product between Y_2O_3 and Fe_2O_3 indicate that the YIG phase is formed on the Fe_2O_3 side and YIP is formed on the Y_2O_3 side.

The Fe_2O_3 component reacts and adheres firmly to the Y_2O_3 substrate. Two reaction products detected were the $YFeO_3$ layer and the $Y_3Fe_5O_{12}$ layer. The $YFeO_3$ coating adheres well to both end components. The growth of the $YFeO_3$ secondary phase layer is not in contact with the Y_2O_3 phase. In other words, the compound Y_2O_3 does not react directly with $YFeO_3$. So that there are two faces of the $YFeO_3$ layer, one grows on the Y_2O_3 substrate and the other grows on the $Y_3Fe_5O_{12}$ substrate. The reaction product is solid, and no impurities have been detected. Observation of differences in Gd-doped YIG and YIG samples showed a finer and denser crystal structure compared to the pure YIG image [8].

In general, sintering up to 1100°C is dominated by the YIP intermediate product formation reaction. Up to 1100°C has not shown significant YIG formation [9]. Evaluation of the Ginstling-Brounshtein (GB) model shows that the *kt* value tends to be relatively low compared to the Jander model (JDR) and the shrinking core model (SCM). Based on the five selected models, SCM modeling shows the best suitability to describe the yttrium iron garnet sintering reaction compared to other evaluated models. The literature data data plot at 1450°C is shown as 'figure 1(a)'.

'Figure 1(b)' illustrates the shrinking core model to plot data of YIG reaction rates with and without gadolinium dopant. The YIG phase transformation rate was chosen to follow the diffusive regime. In the actual conditions before the reaction reaches steady state, an intermediate product of YIP is still formed in the first hour of the sintering reaction, so it takes time for the complete conversion to occur. In the curve above, it can be seen that the actual YIG conversion shows a lower value than the simulation. It is possible that in the first 60 minutes, the reaction mechanism was still dominated by the YIP intermediate-product formation reaction. Then in the next hour, YIP will be converted into YIG products. This is shown at the fifth hour where the YIG conversion approached the simulation result value. The use of the SCM model shows a curve with a steady position at around 0.94. This value looks the best compared to other models with a much lower value.

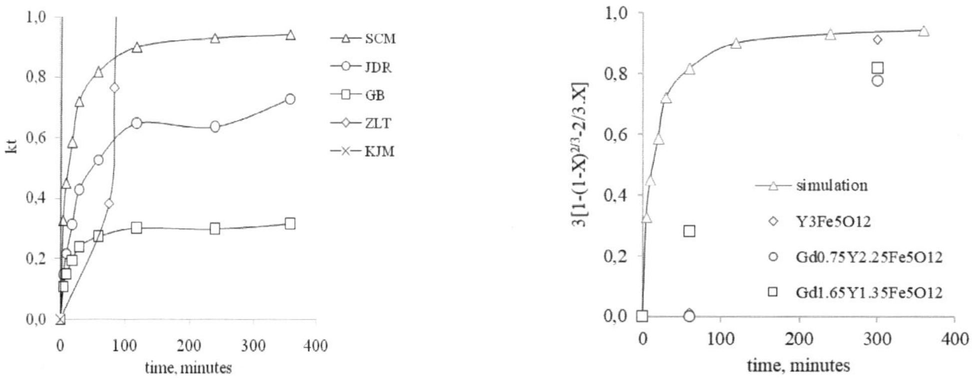

(a). Measured reaction kinetics (b). SCM based reaction kinetics

Figure 1. The measured and model based reaction kinetics of garnet at 1450°C

Evaluation of the five models shows that two out of five models show a single reaction, meaning that they cannot represent the real YIG sintering reaction. The other three show a two-stage reaction. Based on the Arrhenius linear regression plot, modeling the two-stage reaction mechanism in the shrinking core model, shows the highest deviation tolerance (R^2 = 0.923). Therefore, the shrinking core model is considered capable of describing the phase transformation conditions in the Gd-doped YIG sintering reaction.

The SCM model assumes that the reactants diffuse through the particle interface layer. The reversible reaction occurs at the shell and core interface and depends on the total composition of the reactants and products, as illustrated in 'figure 2'. The interface layer can shift towards the center of the particle, if the global reaction rate is positive. The application of SCM modeling is in accordance with the assumption that changes in the concentration profile of the reactants take place very quickly when compared to changes in the position of the reaction interface.

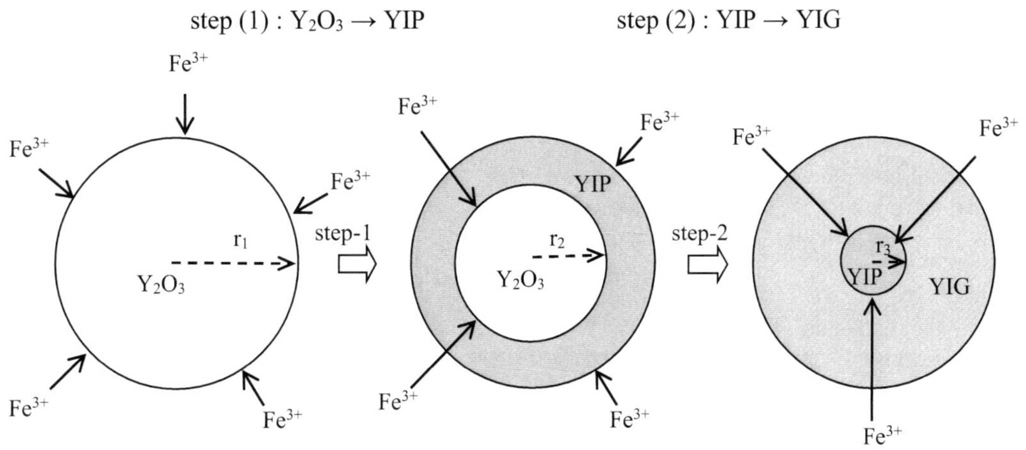

Figure 2. Scheme of shrinking core model of YIG

The study of the sintering reaction rate was carried out by observing the sintering process data from 1000°C to 1250°C [10]. Furthermore, the simulation is carried out at a temperature of 1450°C to predict the sintering conditions at that temperature. The application of KJM modeling is corrected based on the plot of kt against t to ln (1/(1-α)) against t (in minutes). Thus, the total value of k will be greater than zero (positive). The shrinking core model (SCM) shows a good correlation factor (R^2 = 0.923).

By using the reaction rate curve (kt) of each model against t, the values of k_1, k_2 and k_{total} are obtained. The value of k_1 is obtained from the regression of the points at the beginning of the reaction, while k_2 is obtained from the regression of points that are close to steady state conditions. Furthermore, an Arrhenius plot between ln k and 1/T is made to get the value of activation energy (Ea). Table 1 shows the comparison of the activation energy and the deviation tolerance of YIG sintering models.

Table 1. Activation energy at temperature of 1450°C

Model	Ea_1 (kJ/mol)	Ea_2 (kJ/mol)	Ea_tot (kJ/mol)	R²_tot
SCM	378	380	380	0.924
JDR	573	113	594	0.859
GB	146	44	226	0.763
ZLT	386	0	386	0.963
KJM	399	0	399	0.817

Although the Zhuralev-Lesotkin-Tempelmen (ZLT) model showed a better R^2 value (0.963), ZLT modeling could not be used because it did not configure the YIG formation reaction that took place in two stages. Likewise, the Kolmogorov-Johnson-Mehl (KJM) model obtained a value of R^2 = 0.816, which is considered unable to explain the YIG sintering reaction mechanism which takes place in two stages. The ZLT and KJM models both do not show a second reaction mechanism. It does not represent the actual reaction stage and cannot be used to describe the YIG sintering reaction mechanism.

4. Conclusion

The YIG sintering reaction takes place in two stages, through the formation of the YIP intermediate product. The formation of the yttrium iron perovskite phase reaches its maximum value at a sintering temperature at 1200°C, while the formation of yttrium iron garnet reaches its maximum value at a sintering temperature of 1450°C. Sintering carried out at 1450°C for 5 hours resulted in a very significant YIG conversion value of up to 99.7%. The YIG phase transformation rate is predominantly determined by the diffusive regime rather than the kinetic regime. The application of the SCM model is considered to be the best for representing the actual conditions of the yttrium iron garnet, and the gadolinium doped yttrium iron garnet sintering reaction.

References

[1] Wan Ali F F, Othman M A, Ain M F, Abdullah N S, and Ahmad Z A 2013 Studies on the formation of yttrium iron garnet (YIG) through stoichiometry modification prepared by conventional solid-state method *J. Eur. Ceram. Soc.* **33** pp 1317–24

[2] Yagi S and Kunii D 1955 Studies on combustion of carbon particles in flames and fluidized beds *The 5th Symposium on Combustion* New York USA pp 231-6

[3] Gbor P K and Jia C Q 2004 Critical evaluation of coupling particle size distribution with the shrinking core model *Chem. Eng. Sci.* **59** pp 1979-87

[4] Melchiori T 2014 Study of non-catalytic gas-solid reactions- development of single particle model, Disertation Universita Degli Studi di Padova

[5] Thiele E W 1939 Relation between catalytic activity and size of particle *Industrial and Engineering Chemistry* **31** pp 916–20

[6] Wittrup K D 2007 *Chemical and biological reaction engineering* Course materials for 10.37 MIT Open Course Ware

[7] Buscaglia V, Caracciolo F, Bottino C, Leoni M and Nanni P 1997 Reaction diffusion in the Y_2O_3–Fe_2O_3 system *Acta Metall.* **45** pp 1213–24

[8] Akhtar M N, Yousaf M, Khan S N, Nazir M S, Ahmad M and Khan M A 2017 Structural and electromagnetic evaluations of YIG rare earth doped (Gd, Pr, Ho, Yb) nanoferrites for high frequency applications *Ceramics International* **43** pp 17032 - 40

[9] Zhao C, Li Y, Shen X, Cao Z, Cao Z, Wen Z, Yuan X, Ma C and Cao Y 2020 *Studies on highly dense pure YIG polycrystalline ceramics fabricated by tape-casting method* Renmin University of China Beijing 100872

[10] Wan Ali W F F, Othman M A, Ain M F, Abdullah N S and Ahmad Z A 2015 The investigation of the phenomenological YIG phase formation within 1000°C to 1250°C, a kinetic approach *J. Am. Ceram. Soc.* **99** pp 1-9

Study of Water Balance for Tropical Peatland Restoration Case Study: Sinarwajo Village, Jambi Province, Indonesia

S.N. Fauziah[1], L.E. Widodo[2*], H. Ramdan[3]

[1)] Groundwater Engineering, Bandung Institute of Technology, Indonesia
[2)] Research Group on Earth Resources Exploration, Bandung Institute of Technology, Indonesia
[3)] Research Group on Bioresources Management, Bandung Institute of Technology, Indonesia

[*)] Corresponding author: lew@mining.itb.ac.id;

Abstract. Efforts to restore degraded peatlands that have declining groundwater levels are carried out to prevent disasters such as forest fires and carbon emissions into the atmosphere. At present, peatland restoration is an essential topic for researchers, especially regarding hydrological restoration (the re-wetting). Water balance calculation is needed in providing hydrological information and supporting re-wetting of degraded peatlands. In this paper, a study of water balance was conducted for peatland re-wetting in Sinarwajo Village, the Province of Jambi. The result showed that the evapotranspiration budget is 56% from the rainfall. Besides, run-off and groundwater storage is about 32% and 12% respectively. This information can be a primary goal of re-wetting, so the efforts of re-wetting should increase more groundwater storage and limit the water discharge from peatland.

Keywords: water balance, hydrology, tropical peatland, restoration, re-wetting.

1. Introduction

Peatland restoration efforts to prevent fires and reduce carbon emissions have become an essential agenda item over the five years period from 2016 to 2020 in approximately two million hectares of degraded and burned peatlands in Indonesia [1]. These areas are spread across seven provinces: Riau, Jambi, South Sumatra, West Kalimantan, Central Kalimantan, South Kalimantan, and Papua [2]. Hydrological restoration or re-wetting through groundwater level managing has succeeded in reducing about 1.5 Mt of CO_2 per year in an area of 590 km^2 [3]. Another study states that a 30-cm increment of groundwater level (GWL) from the initial condition of 60 cm below ground level can reduce carbon emissions by around 49% [4].

Re-wetting is an effort to increase GWLs on degraded peatlands. By increasing GWLs, land fires are prevented, because the combustible peat material is kept wet [5]. One of these efforts is canal blocking [6]–[8]. By limiting the rate of water coming out of drainage canals, the stream will be blocked and thus able to re-wet peat soil near canals [9], [10].

Research in the field of hydrology supporting re-wetting is growing rapidly. Water balance studies have been conducted in both boreal peatlands (e.g., Thompson et al. [11]) and Borneo's tropical peatlands (e.g., Kumagai et al. [12]; Ritzema and Jansen [13]). Tropical peatlands in Indonesia differ from temperate or boreal peatlands in both the condition of water sources [14] and in physical properties [15]. Also there are more specific studies on the water balance components, such as

IOP Conf. Series: Earth and Environmental Science 1031 (2022) 012011 doi:10.1088/1755-1315/1031/1/012011

precipitation [16], evapotranspiration [17], and run-off [18]. There is a re-wetting plan in Sinarwajo Peatland, Jambi Province, Indonesia. Water balance calculation is needed in providing hydrological information and supporting the re-wetting of degraded peatlands there. This paper shares the results of the study of hydrological conditions based on water balance calculation.

2. Data and Methods

2.1. Data Collection
Data in this study can be divided into two parts: hydrological data and geological data. The former consists of (1) monthly average weather data from the Sultan Thaha Jambi Meteorological Station (WMO ID: 96195) and in situ measurements of rainfall and temperature in Sinarwajo Village, (2) Sinarwajo Canal and Mendahara River data obtained from field surveys, and (3) GWL data measured from 21 observation points. The geological data consists of (1) land cover images from Landsat satellite imagery, (2) data on the hydraulic properties of soil from literature.

2.2. Study Site and Field Survey
We conducted the following activities during the field survey (on January 2018): observed and measured some parameters consist of: GWL measurements at 21 observation points (figure 1); geometry, discharge water, and fluctuations of water level in canals and rivers; ground-check topographic elevation; land use observation; vegetation observation; and soil sampling.

2.3. Water Balance and Water Budget
Water balance is the balance of the input and output of water mass in a system in the form of components of the water cycle. In this study, water balance is calculated using the Mock method [19]. In general, the balance of water balance in one year can be written as in equation (1):

$$P = ET + I + Ro \pm \Delta S, \tag{1}$$

Where P is precipitation (mm), ET is evapotranspiration (mm), Ro is run-off (mm), I is infiltration (mm), and ΔS is storage (mm). Evapotranspiration in the Mock method (1973) was calculated by Penman's (1984) formulation, which combines energy balance and aerodynamic effects that are influenced by the removal of water vapor by wind movement. This water balance calculation was needed to analyze the water budget.

3. Case Study Results and Discussion

3.1. Results of Area Study and Field Survey
The peatland studied is located in Sinarwajo Village, in the western region of Tanjung Jabung Timur District, Jambi Province, Indonesia. This peatland is a part of the Mendahara River–Batanghari River Peat Hydrological Unit (KHG) and was burned in September 2017. From the geomorphological perspective, this peatland has tidal swamp landscapes with a sloping topography of 0–25 masl, and the slope range is between 0–3%. Two rivers flank this peat area: the Mendahara, with a length of 49.5 km, and the Lagan in the east, with a length of 40.9 km [20]. Both rivers flow into the Berhala Strait. There is a transverse canal with a length of 8 km and a depth of 1.8 m (figure 1).

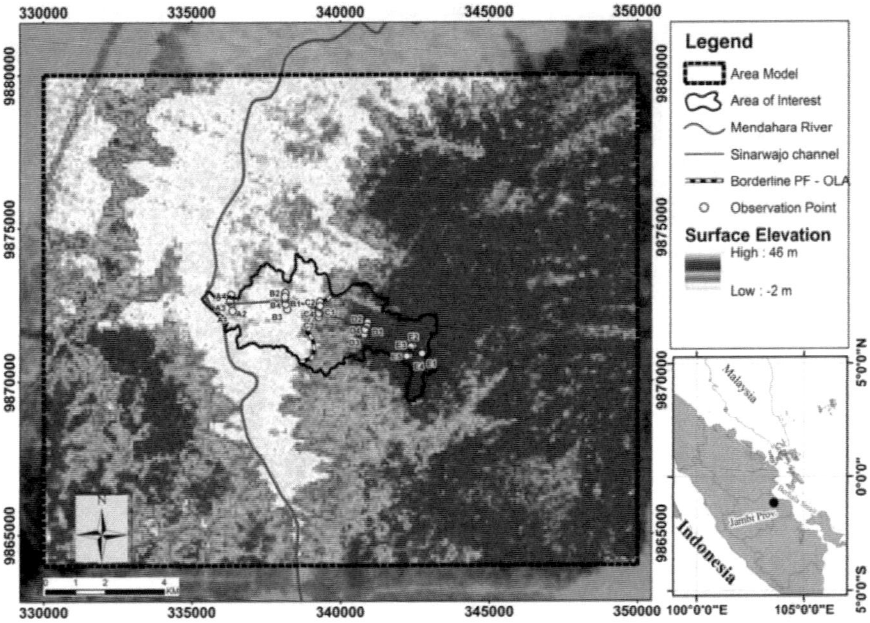

Figure 1. Map of Sinarwajo Village peatlands, with the Canal crossing from the east (upstream) to the west (downstream). The yellow circles are the GWL measurement points.

Based on Jambi climatology data, the study area has a tropical climate, with a rainy season from October to May and a dry season from June to September. Rainfall in this area has a bimodal equatorial pattern, with two rain peaks in different months. Air temperature ranges from 26.2°C to 32°C. Based on temperature measurements obtained in the field, the temperature in degraded peatlands was about 1 to 2 °C higher than at the Sultan Thaha Meteorological Station. Based on the Jambi Geological Sheet Map, the area is composed of three formations: (1) alluvium (Qa), consisting of greed, gravel, sand, silt, and clay; (2) swamp deposits (Qsw), consisting of sand, silt, clay, mud, and peat, and the (3) Kasai Formation (QTk), consisting of tuffaceous sandstone and tuffaceous claystone.

In the Sinarwajo peatlands, land use is divided into two functions. In the eastern area, there is a protected forest (PF), while the western part features other land use (OLU). The PF area is dominated by shrubs, while the vegetation in the OLU area consists of oil palm trees, coconut trees, and areca trees and shrubs. Peat thickness distribution shows that the eastern part of the study area can reach more than 4 m in thickness, which decreases to the west. Overlaying the land use information with GWL measurements reveals that the GWLs in the PL area are quite shallow with a value higher than -40 cm. Conversely, the GWL in the oil palm plantation is more in-depth, less than -40 cm; the GWL can even reach -125 cm (Figure). The reason supposed because the different types of vegetation affect the evapotranspiration rate and water requirement for each plant. Hence, oil palm plantation seemingly needs more water in the evapotranspiration process, thus causing deeper GWL in oil palm areas.

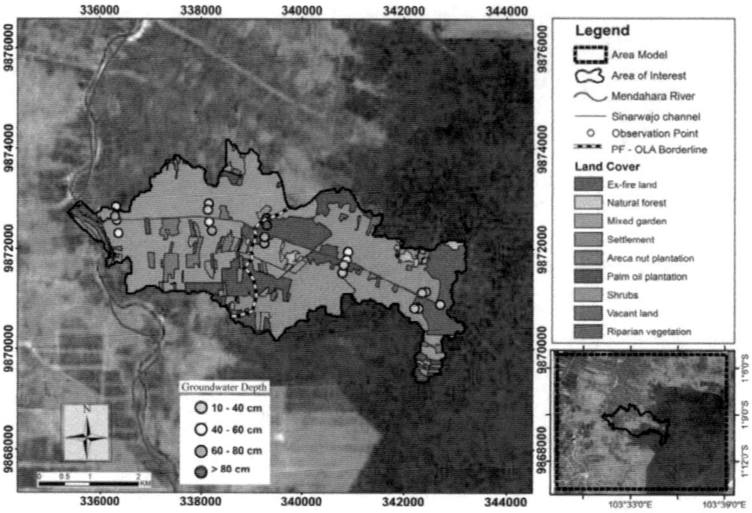

Figure 2. Classification of land use and land cover vegetation and groundwater level from survey measurement.

3.2. Water Balance and Water Budget

The water balance reported in Table 1 shows that the peak of the dry season occurred in August, with rainfall beginning to decline in May, and that the rainy season peak occurred in November. The annual rainfall of 1711 mm is lower than the annual rainfall in other parts of Indonesia, where average annual rainfall reaches more than 2500 mm [21]. The lower annual rainfall may be a reason for the area's drought and land fire vulnerabilities. Rainfall in Jambi is influenced by the Indian Ocean Dipole (IOD) phenomenon, so there should be further research on the possible effects of the IOD on peatlands in the study area.

Monthly average evapotranspiration in the study area reached 80 mm; annual evapotranspiration was 961 mm (56% of rainfall). This evapotranspiration value is in the middle of oil palm plantations (851 mm or 43% of rainfall) and swamp forests (1093 mm or 69% of rainfall) in the Sarawak region of Malaysia [17].

Table 1 shows that run-off pattern is more fluctuate than rainfall and groundwater storage patterns. The percentage of rainfall that became run-off was 32%, but only 12% becoming groundwater storage. It is expected that the percentage of groundwater storage could be increased through hydrological restoration efforts. As a comparison, the run-off budget in the peatlands of Johor, Malaysia, can reach 62% of rainfall [18], while in the Pennine blanket peatlands in the UK, the run-off was up to 40% [22].

In short, either by evapotranspiration or run-off exceed, the decrease of GWLs is inevitable. Therefore, if the degradation of peatlands is not immediately reversed, then the area will be more vulnerable to drought during the dry season and prone to flooding in the rainy season, due to the loss of peat's hydrological function that is supported by Miettinen's study [23] that urgent to take efforts to mitigate peat drainage, while reducing carbon emission or fire.

Table 1. Water balance in the study area.

	Jan	Feb	Mar	Apr	May	Jun	Jul	Aug	Sep	Oct	Nov	Dec	Sum	Percen-tage
Rainfall	145	162	140	186	120	97	99	86	103	132	223	217	1711	100%
ET	82	77	95	77	90	73	73	70	83	82	86	73	961	56%
Run-off	46	61	33	79	22	17	19	11	14	36	98	104	540	32%
Groundwater storage	52	15	16	17	18	15	13	11	9	9	14	21	210	12%

4. Conclusion

Peatland restoration efforts become an essential agenda in Indonesia. One kind of restoration is re-wetting, which focuses on restoring the hydrological condition. Water balance calculation is needed in providing hydrological information and supporting the re-wetting of degraded peatlands. The result showed that the evapotranspiration budget is 56% from the rainfall. Besides, run-off and groundwater storage is about 32% and 12% respectively. This information can be a primary goal of re-wetting, so the efforts of re-wetting should increase more groundwater storage and limit the water discharge from Peatland.

Acknowledgment

This study was funded by the Indonesian Endowment Fund for Education (LPDP). Some data was obtained from the ITB research team in collaboration with the Peat Restoration Agency (BRG). We thank these parties for their support of this study. The authors would also thank for the Research Institution and Community Services of Bandung Institute of Technology (LPPM-ITB) for P3MI 2018 and 2019 research support.

5. References

[1] Safitri D M A 2016 *Peatland Restoration in Indonesia* presented at the Japan Pavilion of COP 22 Marrakech Morocco 2016
[2] Wardhana B 2016 *BRG's Roadmap for Peatland Restoration* Bangkok 2016.
[3] Jaenicke J, Wösten H, Budiman A and Siegert F 2010 *Planning hydrological restoration of peatlands in Indonesia to mitigate carbon dioxide emissions* Mitigation and Adaptation Strategies for Global Change **15** no 3 pp 223–239 doi: 10.1007/s11027-010-9214-5
[4] Astiani D, Burhanuddin, Gusmayanti E, Widiastuti T and Taherdjadeh M 2018 Enhancing water levels of degraded, bare, tropical peatland in West Kalimantan, Indonesia: Impacts on CO2 emission from soil respiration *Biodiversitas* **19** doi: 10.13057/biodiv/d19022
[5] Wösten J H M, Clymans E, Page S E, Rieley J O and Limin S H 2008 Peat–water interrelationships in a tropical peatland ecosystem in Southeast Asia *CATENA* **73** no 2 pp 212–224 doi: 10.1016/j.catena.2007.07.010.
[6] Dohong A 2017 *Bolstering Peatlands Restoration in Indonesia through 3Rs Approach* presented at the Workshop: Developing International Collaborations to Address Fire and Other Conservation Issues in Central Kalimantan Indonesia
[7] Dohong A 2016 *An Assessment of The Restoration Efforts of Degraded Peatland in Central Kalimantan Indonesia* The University of Queensland
[8] Ritzema H and Jaya A 2006 *Water management for sustainable wise use of tropical peatlands* p. 15
[9] Fitriati U 2017 *Canal Blocking to Maintain Groundwater Level at Peatland Central Kalimantan* p 7
[10] Suryadiputra N *et al.* 2005 *A Guide to the Blocking of Canals and Ditches in Conjunction with the Community* (Bogor: Wetlands International - Indonesia Programme and Wildlife Habitat Canada)
[11] Thompson D K, Benscoter B W and Waddington J M 2014 Water balance of a burned and unburned forested boreal peatland: Water balance of boreal peatlands subject to wildfire *Hydrological Processes* **28** no 24 pp 5954–5964 doi: 10.1002/hyp.10074
[12] Kumagai T *et al.* 2005 Annual water balance and seasonality of evapotranspiration in a Bornean tropical rainforest *Agricultural and Forest Meteorology* **128** no 1–2 pp 81–92 doi: 10.1016/j.agrformet.2004.08.006
[13] Ritzema H and Jansen H 2015 *Assessing the water balance of tropical peatlands by using the inverse groundwater modelling approach* p 5

[14] Sumawijaya N 2006 Hydrology and peat-land development in Indonesia *Tropics* **15** no 3 pp 279–284
[15] Andriesse J P 1988 The Main Characteristics of Tropical Peats *Nature and Management of Tropical Peat Soils.* **59** Rome Italy
[16] Li W, Dickinson R E, Fu R, Niu G-Y, Yang Z-L and Canadell J G 2007 Future precipitation changes and their implications for tropical peatlands *Geophysical Research Letters* **34** no 1 Jan doi: 10.1029/2006GL028364
[17] Musin 2017 Evapotranspiration from Tropical Peatland of Sarawak, Malaysia
[18] Katimon A, Shahid S, Khairi Abd Wahab A and Ali Md H 2013 Hydrological behaviour of a drained agricultural peat catchment in the tropics. 1: Rainfall, runoff and water table relationships *Hydrological Sciences Journal* **58** no 6 pp 1297–1309 doi: 10.1080/02626667.2013.815759
[19] Mock F J 1973 *Land Capability Appraisal Indonesia: Water Availability Appraisal : Basic Study* Food and agriculture organization of the United Nations
[20] Jambi Provincial Government 2014 *Buku Data Status Lingkungan Hidup Daerah Provinsi Jambi Tahun 2014* (Jambi: Jambi Provincial Government)
[21] Lee H 2015 General Rainfall Patterns in Indonesia and the Potential Impacts of Local Season Rainfall Intensity *Water* **7** no 12 pp 1751–1768 doi: 10.3390/w7041751
[22] Evans M G, Burt T P, Holden J and Adamson J K 1999 Runoff generation and water table fluctuations in blanket peat: evidence from UK data spanning the dry summer of 1995 *Journal of Hydrology* **211** no 3–4 pp 41–161
[23] Miettinen J, Hooijer A, Vernimmen R, Liew S C and Page S E 2017 From carbon sink to carbon source: extensive peat oxidation in insular Southeast Asia since 1990 *Environ. Res. Lett.* **12** no 2 p 024014 doi: 10.1088/1748-9326/aa5b6f

Focal Mechanism Analysis of the Earthquakes Beneath the Sunda-Banda Arc Transition, Indonesia, Using the BMKG Data

Pepen Supendi[1,2*], Nicholas Rawlinson[1], Andri Dian Nugraha[3], Sri Widiyantoro[3,4], Chalid Idham Abdullah[5], Daryono[2], Bambang S. Prayitno[2], Muhammad Sadly[2], Dwikorita Karnawati[2]

[1] Department of Earth Sciences – Bullard Labs, University of Cambridge, Cambridge CB30EZ, United Kingdom
[2] Agency for Meteorology, Climatology, and Geophysics (BMKG), Jakarta 10720, Indonesia
[3] Global Geophysics Research Group, Faculty of Mining and Petroleum Engineering, Institut Teknologi Bandung, Bandung 40132, Indonesia
[4] Faculty of Engineering, Maranatha Christian University, Bandung 40164, Indonesia
[5] Geodynamic and Sedimentology Research Group, Faculty of Earth Sciences and Technology, Institut Teknologi Bandung, Bandung 40132, Indonesia

*E-mail: pepen_geophysics@yahoo.com

Abstract. Structural complexity in the Sunda-Banda arc transition is a topic of much debate amongst Earth scientists. We have processed focal mechanism study using moment tensor inversion for 20 events in the region using the Agency for Meteorology, Climatology, and Geophysics (BMKG) data from 2014-2016 for earthquakes of magnitude Mw ≥ 5.0. Our result shows different solutions that depend on the source region of the earthquakes that include subduction zones and collision zones, which host active faults and the back-arc thrusts. Earthquakes that occurred in the subduction zone appear to rupture on thrust faults for shallow and intermediate events, while the deep events have normal fault mechanisms. The shallow events in the collision zone occur on thrust faults with differing strike directions, but generally, there are largely parallel to the Timor trough. We also found normal fault mechanisms at play for deep events below the collision zone. The occurrence of deep earthquakes in this area is consistent with remnant slab activity that persists to the present day. Focal mechanism solutions for shallow events in the north of Sumbawa island indicate a thrust fault with a strike direction that is almost parallel to the back-arc thrust in this area. We also found evidence of strike-slip motion along local-scale active faults in the area.

Keywords: Focal mechanism, earthquakes, Sunda-Banda Arc.

1. Introduction

The Sunda-Banda arc transition lies at the confluence of the eastern end of the Australian plate subduction margin, defined by the Sunda arc, and the western section of the collision between the Australian plate and the Banda arc. Based on the Agency for Meteorology, Climatology, and Geophysics (BMKG) earthquakes catalog from 1975 to 2018 that spans the Sunda-Banda arc transition, there have been 40 destructive earthquakes that resulted in fatalities, three of which were followed by a tsunami:

IOP Conf. Series: Earth and Environmental Science 1031 (2022) 012012 doi:10.1088/1755-1315/1031/1/012012

The 1977 Mw 8.3 Sumba earthquake and tsunami, the 1992 Mw 7.5 Flores earthquake and tsunami, and the 1995 Mw 6.5 Timor earthquake and tsunami [1]. Earthquakes in the region are a consequence of subduction, collision, back-arc thrust faults, and faults located on land. The structural complexity of the region and its importance in understanding fundamental plate tectonic processes has resulted in intensive discussions and disagreements amongst scientists, and more high-quality data is needed to achieve consensus. Since 9-10 Ma, the Australian continent has been effectively colliding with the western section of the Banda arc [2], along the islands of Timor, Sumba, and Flores [3]. Recent seismic tomography and earthquake relocation results suggest that the Indo-Australian slab is dipping at 10-30° from trench to the arc, before dipping more steeply down to the transition zone [4]. The slab dip along the Sunda arc is ~55–60°, while along the Banda arc it is ~75–80° [5]. In this study, we have conducted focal mechanism analysis to reveal fault types of shallow, intermediate, and deep earthquakes using moment tensor inversion for 20 events (magnitude Mw ≥ 5.0) in the region using the BMKG station network from 2014-2016.

2. Data and Method

We used waveforms from 21 BMKG stations (Fig. 2) from March 2014 to March 2016. We used the ISOLA package [6] to perform moment tensor inversion. We used displacement data from at least three BMKG stations at a narrow epicentral location to invert for full moment tensors. The observed data were pre-processed using a low pass filter with a corner frequency of 0.01 Hz to 0.05 Hz and we used the Ak135 seismic velocity model in the inversion [7]. The ground motion recorded by a seismometer can be viewed as a convolution of three processes: The earthquake source, the response of the Earth's crust, and the seismometer response. The source is defined by a tensor (**m**) and the Green's function describes the impulse response of the medium through which the waves propagate. The best-fitting between synthetic and observed seismograms (example in Fig. 1) is essential for assessing the achievement of the inversion.

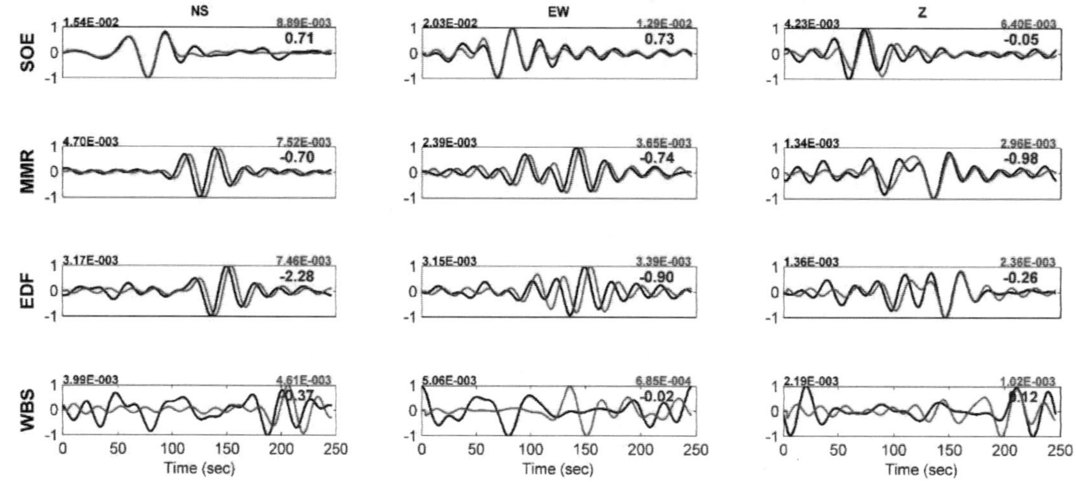

Figure 1. Example of waveform fitting for event number 17 (19 Nov 2014). Black and red waveforms are observation and synthetic waveforms, respectively. Blue numbers depict variance reduction (VR) values for each component.

3. Results and Discussion

We have calculated focal mechanism solutions of 20 events from the subduction zone, collision zone in the south of the islands, and the back-arc thrust fault in the north of islands (Table 1 and Fig. 2). The similarity between the observed and synthetic waveforms can be expressed as a variance reduction (VR), where a VR value close to 1 indicates that our model predictions are close to the actual data predictions. The average VR value for the 20 events is 0.3 (Fig.3).

Table 1. List of the focal mechanism solutions along the Sunda-Banda arc transition.

Date	Time (UTC)	Lat (°)	Long (°)	Depth (km)	Mw	Nodal Plane 1			Nodal Plane 2			Event ID
						strike	dip	rake	strike	dip	rake	
30/05/14	00:56:21	-8.54	119.47	180	5.1	119	52	43	360	58	133	13
07/07/14	09:15:29	-7.27	123.6	561	5.4	150	27	-171	53	86	-63	6
06/08/14	11:45:29	-6.89	127.72	19	6.2	314	70	157	53	68	22	11
19/11/14	02:56:27	-10.47	123.92	27	5.1	71	60	84	263	31	100	17
31/01/15	17:30:10	-7.31	126.51	352	5.5	289	48	-37	46	63	-132	10
27/02/15	13:45:09	-7.85	122.61	483	7.0	41	22	-77	207	69	-95	4
10/06/15	22:01:34	-10.35	120.99	61	5.1	341	86	-177	251	87	-4	15
15/06/15	17:41:00	-9.67	125.13	23	5.8	170	20	36	46	78	106	19
15/06/15	22:09:50	-8.03	118.16	21	5.0	304	69	100	97	23	65	1
19/06/15	09:14:00	-9.03	127.32	49	5.0	127	48	137	249	59	51	20
02/07/15	07:04:31	-7.25	120.18	430	5.2	56	49	-18	158	77	-138	3
31/07/15	22:26:09	-7.68	118.24	23	5.0	77	59	74	286	35	115	2
17/08/15	23:39:32	-7.88	123.08	19	5.0	100	42	123	239	56	64	5
31/10/15	04:43:57	-9.13	124.12	65	5.0	287	53	99	92	38	78	18
03/11/15	21:25:04	-7.89	125.28	26	5.6	55	83	21	322	70	172	8
04/11/15	03:44:21	-8.2	124.95	12	6.5	47	62	-16	145	76	-151	7
15/12/15	19:01:47	-7.13	127.8	246	5.0	204	19	-147	82	80	-74	12
03/02/16	20:27:31	-9.99	123.46	20	5.1	70	80	121	176	32	18	16
12/02/16	10:02:30	-9.87	119.35	38	6.2	63	50	56	288	51	123	14
12/03/16	04:26:00	-7.66	125.87	16	5.2	32	63	62	261	38	132	9

The focal mechanisms of several events in the Sunda-Banda arc transition reveal different solutions that depend on the source of the earthquakes, which include subduction zones, collision zones, active faults on land, and back-arc thrust faults. Event 13 (180 km depth) and 14 (38 km depth) occurred in the subduction zone and are thrust faults with different strikes. The strike direction for event 14, located in the south of Sumba island is roughly trench parallel, while event 13 is more trench-perpendicular. The northern boundary of the Scott Plateau now lies below the Sumba Ridge, this condition has resulted in the young thrusting north of Savu and Rote [8]. A deep earthquake that occurred beneath this area at 430 km depth (event 3) is a normal fault earthquake. This event may have been caused by a gravitational slab pull force that was more dominant than the buoyancy force at this depth. Event 15 is a shallow earthquake that occurred in the southeast of Sumba island on a strike-slip fault.

Event numbers 16, 17, 18, 19, and 20 are close to the collision zone between the Australian plate and the Banda arc and occur at depths of less than 100 km. They are all thrust fault mechanisms but with various strikes, though they are all roughly, parallel to the Timor trough. Event 4 (483 km depth) occurs on a normal fault with a strike almost parallel to the collision zone, and may be caused by the ongoing sinking of remnant slab material from the Australian plate that followed the collision millions of years ago. The presence of remnant slabs in the area has previously been shown by [4]. Furthermore, event 6 (561 km depth), event 10 (352 km depth) and event 12 (246 km depth) all have normal fault mechanisms that strike parallel to the collision zone.

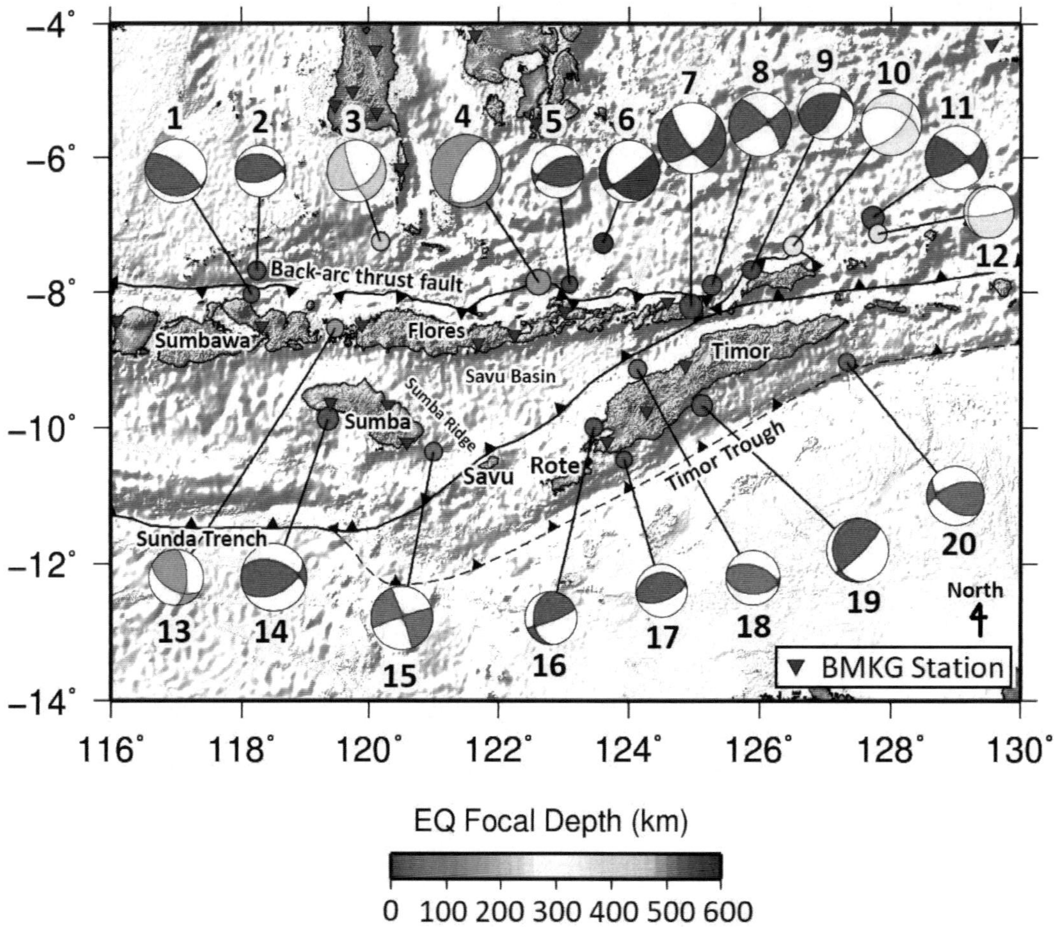

Figure 2. Map view of focal mechanism solutions along the Sunda-Banda arc transition. The black-arc thrust fault locations are taken from [9], while trench (solid line) and trough (dashed line) data are taken from [10]. The epicenters of earthquakes are represented by coloured circles as a feature of focal depth.

The occurrence of deep events in this area indicates that the remnant slab material is still capable of producing earthquakes in the present day. Events located along the back-arc thrust fault, i.e., event 1 (21 km depth), event 2 (23 km depth), and event 9 (16 km depth) indicate a thrust fault with a strike that is almost parallel to the back-arc thrust. The continental collision is the primary driving force behind the back-arc thrust in the eastern Sunda Arc [11]. The difference between back-arc thrusting mechanisms has a considerable influence on the development of the convergent margin [12]. Event 7 (12 km depth) and event 8 (26 km depth) are located in the north of Alor island and occurred on a strike-slip fault. Both of these earthquakes were likely generated on a terrestrial transform fault that trends northeast-southwest.

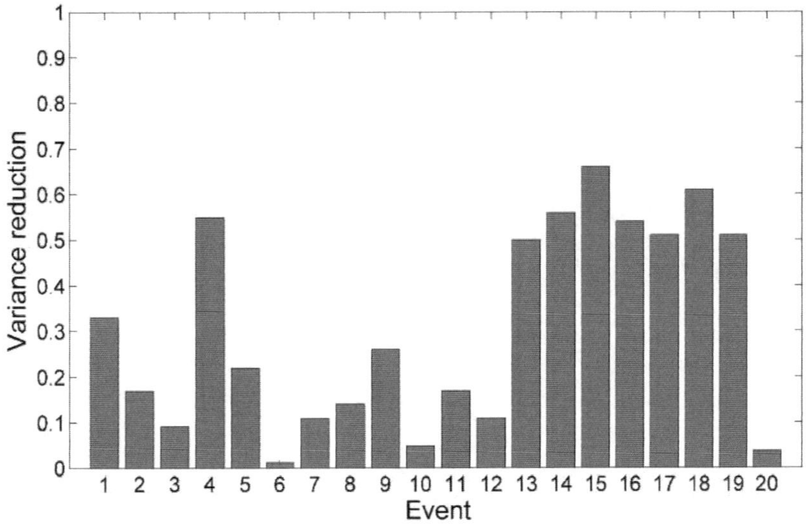

Figure 3. Waveform fitting variance reduction for the 20 events used in this study.

4. Conclusion

The focal mechanisms of shallow earthquakes in the subduction zone, collision zone, and back-arc thrust in the Sulu-Banda arc transition reveal thrust faulting, but the deep earthquakes reveal normal faulting, which may be related to the ongoing sinking of post-collisional slab remnants. Shallow earthquakes are also found to occur on active strike-slip faults with different orientations to the plate margins, and likely accommodate different movements caused by slip along a relatively complex configuration of thrust faults in the region.

Acknowledgments

We thank BMKG for the earthquake data. This research funded by Program Disertasi Doktor (DIKTI) 2019-2020 and Hibah PDUPT (DIKTI) 2019-2020-2021 awarded to A.D.N. This research is also funded by the University of Cambridge through a Herchel Smith Research Fellowship awarded to P.S.

References

[1] Supendi P, Nugraha A D, Widiyantoro S, Pesicek J D, Thurber C H, Abdullah C I, Daryono D, Wiyono S H, Shiddiqi H A and Rosalia S 2020 Relocated aftershocks and background seismicity in eastern Indonesia shed light on the 2018 Lombok and Palu earthquake sequences *Geophysical Journal International* **221** 1845–55

[2]. Keep M, Haig DW 2010 Deformation and exhumation in Timor: distinct stages of a young orogeny *Tectonophysics*. **483** 93–111

[3]. Hall R 1997 Cenozoic plate tectonic reconstructions of SE Asia, Petroleum Geology of Southeast Asia; *Geol. Soc. London*, Spec. Publ. **126** 11–23.

[4]. Widiyantoro S, Pesicek JD, Thurber CH 2011 Subducting slab structure the eastern Sunda arc inferred from non-linear seismic tomographic imaging *Geol. Soc. London, Spec. Publ.* **355** 139-155.

[5]. Royden LH, Husson L 2009 Subduction with variations in slab buoyancy: Models and application to the banda and Apennine systems. In: Lallemand S., Funiciello F. (eds) Subduction Zone Geodynamics *Front. Earth Sci*. Springer, Berlin, Heidelberg.

[6]. Sokos E, Zahradník J 2008 ISOLA a Fortran code and a Matlab GUI to perform multiple-point source inversion of seismic data *Comput. & Geosci.* **34** 967–977.

[7]. Kennet BLN, Engdahl ER, Bulland R 1995 Constraints on seismic velocities in the earth from traveltimes *Geophys J. Int* **122** 108 – 124.

[8]. Rigg JWD and Hall R 2011 Structural and stratigraphic evolution of the Savu Basin Indonesia *the Geol. Soc. London*, 0305-8719/11.

[9]. Irsyam M, Widiyantoro S, Natawidjaja DH, Meilano I, Rudyanto A, Hidayati S, Triyoso W, Hanifa NR, Djarwadi D, Faizal L, Sunarjito 2017 Earthquake sources and hazard map of Indonesia. Research and Development Center, Minister for Public Works and Human Settlements (in Indonesian).

[10]. Barber P, Carter P, Fraser T, Baillie P, and Myers K 2003 Paleozoic and Mesozoic petroleum systems in the Timor and Arafura seas, eastern Indonesia in *Proc. Indo. Petrol. Assoc., 29th Ann. Conv*. (Indonesian Petroleum Association).

[11]. Silver EA, Reed DR, McCaffrey R, Joyodiwiryo Y 1983 Back-arc thrusting in the eastern Sunda Arc, Indonesia: A consequence of arc-continent collision *J. Geophys. Res*. **88**,7429-7448

[12]. McCaffrey R, Nabelek J 1984 The geometry of back arc thrusting along the Eastern Sunda Arc, Indonesia: Constraints from earthquake and gravity data *J. Geophys. Res*. **89**, 6171-6180.

IOP Conf. Series: Earth and Environmental Science 1031 (2022) 012013 doi:10.1088/1755-1315/1031/1/012013

Integrated Design and Analysis of Corrosion Rate Model on Tubular Pipe of Oil and Gas Production Well

B T H Marbun[1*], B Z Fadholi[1], S Z Sinaga[1], and B A Purbantanu[1]

[1] Petroleum Engineering Study Program, Bandung Institute of Technology (ITB), Jalan Ganesha 10, Bandung 40132, Indonesia

*Email: bonar.marbun@tm.itb.ac.id

Abstract. Corrosion occurrence in the oil and gas production system is an example of an unavoidable problem. This problem is caused by the presence of water and corrosive material fraction along with the produced hydrocarbon. On the way from the reservoir to the surface, the mixture between water fraction and the corrosive substance might adhere to the wall of conduit steel pipe (e.g. tubing, casing, and flowline) and can lead to corrosion on the wall. The corrosion problem can have a major impact on the economics of the production because the cost of production equipment is relatively expensive and workover operation is very costly. Therefore, planning and mitigation for a better production system need to be performed, so that the rate of corrosion can be controlled, and its economic impacts can be minimized. The purpose of this study was to create an integrated model of the corrosion rate calculation on the tubing by the Petroleum Engineering Study Program of Bandung Institute of Technology (ITB). Several terms related to the production process were considered in this model: the characteristics of the produced fluid, flow regimes, pressure and temperature conditions along the tubing, the trajectory of the well, the type of tubing material used, and the use of inhibitors. The model was developed from the model developed by C. de Waard, Liane Smith, and Mike Billingham. The corrosion rate calculation model was then validated using field data and the results were compared with the result from industrial commercial software, Electronic Corrosion Engineer (ECE®). The corrosion rate calculation model were able to predict the rate of corrosion in oil and gas wells accurately. A software-based system was created based on this integrated and accurate proven model. By using this software, the production system can be simulated and tested for various conditions, and then the result can be compared with each other to obtain the optimum fluid production rate with minimum corrosion rate. In addition, this software is capable to determine the most suitable tubing material which will be used in a given production environment and condition.

1. Introduction

1.1. Background

During the production life of an oil and gas well, tubing corrosion and deterioration cannot be avoided because of the presence of the corrosive material in hydrocarbon production [1] [2]. The corrosion problem leads to several failures such as tubing cracking, tubing leaks, or at worst tubing part-off. Activities that require high expenditure such as well intervention and workover, are required to deal with these kinds of failures.

An accurate corrosion rate prediction model is required to control the corrosion that considers not only the composition of the production fluid, but also other factors such as fluid properties, environmental condition, production rate profile, well deviation, tubing characteristics, and the use of inhibitor for the corrosion protection [1] [2] [3] [4].

1.2. Objectives

The objectives of this study were to:

1. develop a model of corrosion rate prediction (called tubular corrosion desktop or TCD) on tubing based on the model developed by C de Waard, Lianne Smith, and Mike Billingham [5] [3] [6] [4] [7] [1];
2. create a software interface for ease of use of the model. In principle, the TCD model considered the material selection for the downhole production tubing as one of the key capital expenditure items in oil and gas field development [8].

1.3. Study Methodology

The methodology used in this study was to develop the basic concept of the corrosion rate prediction model (TCD model) from the model developed by De Waard, Smith, and Billingham (Intetech's corrosion model) [9] [10] [11]. Then, a software interface was created for the ease of use of the corrosion model. The corrosion model result of self-created software was called Tubular Corrosion Desktop (TCD). This corrosion rate prediction model was then compared with the commercial software, Electronic Corrosion Engineer (ECE®) that was based on Intetech's Corrosion Model [11]. Finally, several improvements were made to overcome the drawback of Intetech's model, and several recommendations were provided for the future development of the corrosion rate prediction model and the TCD software.

2. Result and Discussion

The corrosion model in ECE® software was based on de Waard-Lotz corrosion equations and further model developed by C. de Waard, Liane Smith, and Mike Billingham [11] [3] [4] [9] [10] [5] [7] [1] [9]. Table 1 shows the list of the input data for corrosion modeling.

Table 1. Input data for corrosion modeling [1] [5] [10]

Parameter	Data	Unit
Fluids Properties	CO_2 content	%mole
	N_2 content	%mole
	H_2S content	%mole
	Dissolved Bicarbonate	ppmw
	Gas Gravity	
	Equivalent NaCl	ppmw
	Water Density	kg/m^3
	Liquid Density	kg/m^3
	Oil API Gravity	^oAPI
Environmental Condition	Wellhead temperature	^oC

Parameter	Data	Unit
	Bottomhole temperature	°C
	Wellhead pressure	bar
	Bottomhole pressure	bar
	Standard temperature	°F
	Standard pressure	psia
Production Profile	Gas flowrate at wellhead	MMSm³/d
	Oil flowrate at wellhead	m³/d
	Water flowrate at wellhead	m³/d
	Liquid holdup change	%
Tubing Deviation	Deviation angle	degree
Tubing Properties	Tubing length	ft
	Outside diameter	m
	Inside diameter	m
	Wall thickness	mm
	Chromium content	%
Inhibitor	Inhibitor Availability	%
	Inhibitor Efficiency	%

Further, the corrosion rate model was developed in this study based on the following equation [1] [3] [4] [7] [5] [9] [12] [13] [10]:

$$Corrosion\ rate = V_{cor} x F_{scale} x F_{H_2S} x F_{cond} x F_{oil} x F_{inhibit} x F_{glyc} \quad (1)$$

The above equations were adjusted for the presence of protective scale, H_2S, crude oil or condensate, glycol, and inhibitor using multiplier factors for the basis of CO_2 corrosion rate. However, in this study, F_{H2S}, F_{cond} and F_{glyc} were not considered in the equation. Therefore, the corrected corrosion rate equation model used in this study was as follows:

$$Corrosion\ rate = V_{cor} x F_{scale} x F_{oil} x F_{inhibit} \quad (2)$$

where V_{cor}, $F_{inhibit}$, F_{oil}, F_{scale} are the corrosion rate accounting the combination between mass-transfer controlled part and reaction controlled part (mm/year), the corrosion rate for mass-transfer controlled part (mm/year), the corrosion rate for reaction controlled part (mm/year), carbon compositional correction factor, chrome compositional correction factor, the correction factor for the presence of inhibitor, the correction factor for the presence of oil, and correction factor for the presence of protective scale, respectively.

Following factors were considered in TCD model development [14] [15] [6] [4] [16] [17] [18] [19] [20] [21] [22] [23] [24] [25] [26] [27] [28]:

1. Pressure and temperature
2. Fugacity of CO2
3. pH calculation

4. Gas compressibility factor (Z-factor)
5. Condensation of water
6. Flow profile and pattern
7. Mixture density
8. Erosion-corrosion
9. Influence of carbonate scales
10. Influence of crude oil
11. Influence of inhibitor
12. Compositional correction factor

The TCD software was created for the ease of use of the corrosion prediction model using C++ programming language and Qt and Qwt Wrapper. The algorithms were established based on the corrosion rate model for software development.

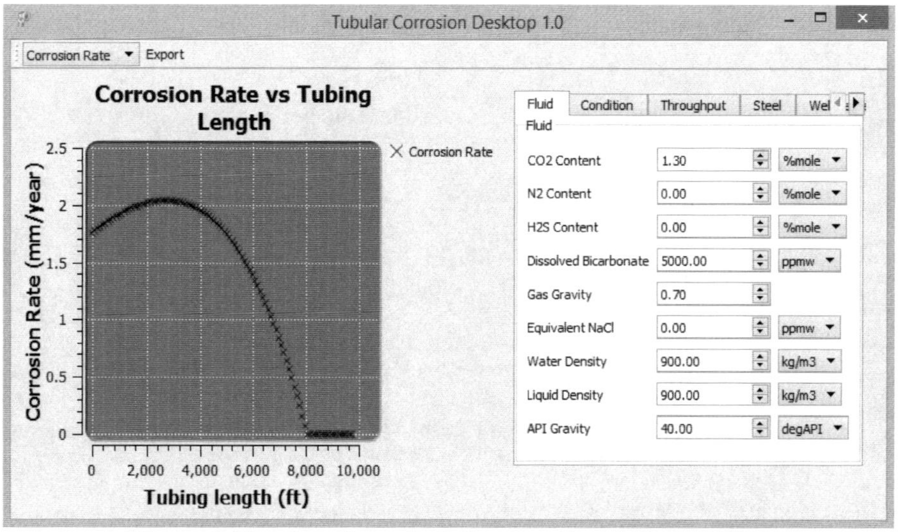

Figure 1. User interface of tubular corrosion desktop (TCD)

The TCD software accommodated more input data for a more comprehensive analysis and more accurate calculation of corrosion rate. Besides, there were additional parameters that were considered in the software, such as N_2 gas fraction, standard temperature and pressure used by the user, and dissolved NaCl. In TCD software, the Z-factor was calculated based on the condition at each tubing segment, providing better and more accurate gas effect and corrosion prediction calculation [20] [16] [19]. The Z-factor was calculated using Hall-Yarborough Correlation.

3. Conclusion

1. TCD model was accurate for carbon/low alloy steel.
2. TCD model accommodated additional data parameter.
3. The model only considered corrosion rate prediction calculation for quenched and tempered steel, but it did not consider normalized steel.
4. The TCD software accommodated user for both metric unit input data and field unit input data.

4. Further Study

Future development of the TCD model should include and consider: oil condensate formation correction factor, the compositional correction factor for chrome and carbon content, glycol correction factor, normalized steel, squeeze inhibition system, more accurate density model, multiphase fluid flow in the tubing, more accurate pH calculation, dissolved Fe, H_2S and pitting, H_2S correction factor, acetic acid, gas gravity, more flow pattern profiles, smoothness of pipe in erosion calculation, top-line corrosion, high deviated and horizontal wellbore, high pressure and high temperature (HPHT).

5. Nomenclature and List of Abbreviations

API:	Gravity of the oil in °API
ECE®:	Electronic Corrosion Engineer
F_{cond}:	Correction factor for the presence of condensate (the value between 0 – 1)
F_{glyc}:	Correction factor for the presence of glycol (the value between 0 – 1)
F_{H_2S}:	Correction factor for the presence of H_2S (the value between 0 – 1)
$F_{inhibit}$:	Correction factor for the presence of inhibitor (the value between 0 – 1)
F_{oil}:	Correction factor for the presence of oil (the value between 0 – 1)
F_{scale}:	Correction factor for the presence of protective scale (the value between 0 – 1)
HPHT:	High pressure and high temperature
RP:	Recommended practice
V_{cor}:	Corrosion rate accounting the combination between mass-transfer controlled part and reaction-controlled part (mm/year)
TCD:	Tubular corrosion desktop
Z-factor:	Gas compressibility factor

Acknowledgements

The authors would like to acknowledge funding from the Institute for Research and Community Services (Lembaga Penelitian dan Pengabdian kepada Masyarakat) Institut Teknologi Bandung (ITB) through the 2020 Research, Community Service, and Innovation (P3MI) Program.

References

[1] Bellarby J 2009 *Well Completion Design* (Amsterdam: Elsevier Science)
[2] American Petroleum Institute (API) 2000 *API RP 14E: Recommended Practice for Design and Installation of Offshore Production Platform Piping Systems* [Internet] (Washington D.C.: American Petroleum Institute) Available from: https://www.api.org/
[3] Waard C de, Lotz U 1994 Prediction of CO2 Corrosion of Carbon Steel. Working Party Report on Predicting CO2 Corrosion in the Oil and Gas Industry *European Federation of Corrosion*
[4] de Waard C, Lotz U, Milliams DE 1991 Predictive Model for CO2 Corrosion Engineering in Wet Natural Gas Pipelines *Corrosion* **47**(12) 976–85
[5] Billingham MA, King B, Murray A 2012 Maximising the Life of Corroding Tubing by Combining Accurate MultiFinger Caliper Data with Corrosion Modelling. In: Abu Dhabi International Petroleum Conference and Exhibition. Abu Dhabi: SPE
[6] Waard C de, Lotz U, Dugstad A 1995 Influence of Liquid Flow Velocity on CO2 Corrosion: A Semi-Empirical Model. In: Corrosion 95. NACE
[7] Nyborg R 2010 CO2 Corrosion Models For Oil And Gas Production Systems. In: Corrosion [Internet]. San Antonio, Texas: NACE Available from: https://onepetro.org/NACECORR/proceedings-abstract/CORR10/All-CORR10/NACE-10371/127015
[8] Tenaris University 2009 The Effect of Corrosion on Material Selection-Sweet Corrosion

Environments

[9] Smith L and De Waard C 2005 Corrosion Prediction and Materials Selection for Oil and Gas Producing Environments. In: Corrosion. Houston, Texas: NACE

[10] Smith L, Bartlett P, Cunningham H 2001 Modelling Corrosion Rates in Oil Production Tubing. In: Eurocorr

[11] Wood Group Intetech Ltd. 2019 Electronic Corrosion Engineer, ECE®

[12] Smith L, Craig B, Waard K de 2003 The Influence of Crude Oils on Well Tubing Corrosion Rates. In: Corrosion [Internet]. California: NACE Available from: https://onepetro.org/NACECORR/proceedings-abstract/CORR03/All-CORR03/NACE-03629/114463

[13] Van Hunnik E, Pots BFM, Hendriksen ELJA 1996 The Formation of Protective FeCO3 Corrosion Product Layers in CO2 Corrosion. In: Corrosion. NACE

[14] Brennen CE 2005 *Fundamentals of Multiphase Flow* (New York: Cambridge University Press)

[15] Craig B 1998 Predicting the Conductivity of Water-in-Oil Solutions as a Means to Estimate Corrosiveness *Corrosion* **54**(8) 657–62

[16] Whitson CH, Brule MR 2000 Gas Gravity and Brine Correction. In: SPE Phase Behavior Monograph. Richardson: SPE

[17] Khaled AAF 2007 *A Prediction of Water Content in Sour Natural Gas* (Riyadh: King Saud University)

[18] IHS Markit. Pressure Loss Calculations [Internet]. Fekete Associates Inc. Available from: http://www.fekete.com/SAN/TheoryAndEquations/HarmonyTheoryEquations/Content/HTML_Files/Reference_Material/Reference_Materials.htm

[19] Gas Processors Suppliers Association (GPSA) 2004 *GPSA Engineering Data Book, FPS Version* [Internet] (Tulsa: Gas Processors Suppliers Association (GPSA)) Available from: www.gasprocessors.com

[20] Guo B 2005 *Ghalambor A. Natural Gas Engineering Handbook* (Houston: Gulf Publishing Company)

[21] Pots BFM, Hendriksen ELJA 2000 CO2 Corrosion Under Scaling Conditions – The Special Case of Top-of-Line Corrosion in Wet Gas Pipelines. In: Corrosion. NACE

[22] Russell R, Nguyen H, Sun K 2011 Choosing Better API RP 14E C Factors for Practical Oilfield Implementation. In: Corrosion. NACE

[23] Society of Petroleum Engineers 2012 Petrowiki [Internet]. Society of Petroleum Engineers (SPE) Available from: https://petrowiki.spe.org/Temperature-depth_profiles

[24] Cai J, Nesic S, C. De Waard 2004 Modeling of Water Wetting in Oil-Water Pipe Flow. In New Orleans, Louisiana: NACE Available from: https://onepetro.org/NACECORR/proceedings-abstract/CORR04/All-CORR04/NACE-04663/115011

[25] Mcketta JJ, Wehe AH 1958 Use This Chart for Water Content of Natural Gases. Petroleum Refiner p 153

[26] George E. King Engineering. George King [Internet]. GEK Engineering. Available from: https://www.gekengineering.com/

[27] Sagar R, Doty DR, Schmldt Z 1991 Predicting Temperature Profiles in a Flowing Well *SPE Prod Eng* [Internet] **6**(04) 441–8. Available from: https://onepetro.org/PO/article-abstract/6/04/441/53475/Predicting-Temperature-Profiles-in-a-Flowing-Well?redirectedFrom=fulltext

[28] Petalas N, Aziz K 2000 A Mechanistic Model for Multiphase Flow in Pipes *J Can Pet Technol* [Internet] **39**(06) Available from: https://onepetro.org/JCPT/article-abstract/doi/10.2118/00-06-04/30038/A-Mechanistic-Model-for-Multiphase-Flow-in-Pipes?redirectedFrom=fulltext

Iterative 2D Gravity Modelling for Basement Depth Estimation

Faradilla Thelma Ariesti and Hendra Grandis*

Geophysical Engineering Program, Faculty of Mining and Petroleum Engineering, Institut Teknologi Bandung (ITB), Bandung, Indonesia.

*e-mail: grandis@geoph.itb.ac.id

Abstract. In gravity modelling, iterative refinement of 2D basement geometry is a classical problem where the density contrast has to be determined in advance. We present a relatively simple method to estimate 2D basement depth along with its density contrast. Known basement depths at several points and their associated gravity anomalies determine the gravity - depth relationship that is used to obtain the first estimate of the overall basement depth. The initial model response is calculated with a unitary density. The linear regression between calculated model response and actual gravity data results in the density contrast estimation. The misfit or residuals and the gravity - depth relationship are used to update the model iteratively. The proposed technique is relatively stable and results in a recovered model quite similar to the synthetic model. We also explore the possibility to include estimation of the constant the regional anomaly.

1. Introduction

The well-known method proposed by Bott in 60s allows 2D basement modeling by iterative algorithm [1]. The Bouguer correction formula is be used to estimate the basement depth from the gravity anomaly at each point, i.e. Bouguer plate or 1D approximation. The horizontal position and depth points constitute a polygon defining the cross-section of an initial 2D model. Talwani's method [2] is used to calculate the gravity response of the initial model. The misfit between calculated and observed gravity anomaly determines the correction factor for basement depth, also by using the Bouguer plate approximation. The process is done iteratively until convergence. As an extension of the iterative 2D modeling, similar method has also been proposed for 3D basement geometry [3]. The initial model and its iterative refinement are similar to the method for 2D modelling, while the 3D gravity forward modelling is calculated by using vertical prism as the building block of the 3D basement model [2].

The classical Bott's method needs a relatively accurate estimate of the density contrast which is critical in gravity modelling. Slightly different density contrast may lead to significant differences in model geometry. In addition, this approach is inherently unstable. However, a regularization or damping can be included to stabilize the iteration. The latter is done by formulating the algorithm in the formal Gauss-Newton method for non-linear inverse problems [4]. Recently, other type of stabilization was implemented for 2D and 3D iterative gravity modelling, i.e. using maximum differences of a factor in reducing the residual [5,6]. In this paper, we adopt the algorithm proposed by Florio [7,8] originally for 3D basement modeling for 2D gravity modeling. We also evaluate the validity of the method in the presence of noise. Furthermore, we propose to include the estimation of the constant regional gravity anomaly in the iterative process. In all cases, we used synthetic gravity data associated with a relatively realistic basement model.

2. Method

The Bouguer plate approximation is the simplest form of linear relationship between gravity and its anomalous source's geometry, i.e. thickness. In basement modelling, the thickness of the sediment is in fact the depth of the basement topography. A more realistic gravity - depth linear relationship can be deduced from known basement depth at several points with their actual associated gravity anomalies [7-9]. Such gravity - depth relationship allows the first estimate of the basement depth without assuming the density contrast value in advance. This avoids one of the difficulties in Bott's and other similar methods.

We consider gravity data distributed along a profile line perpendicular to the strike of a 2D model. The cross-section of the 2D model is represented by a polygon defined by its corners (x_k, z_k), $k = 1, 2, \ldots, N$ with N is the number of vertices. Overall, the method can be summarized in the following algorithm:

1. A linear regression is done between known basement depths at several points (d, in km) with their associated observed gravity anomalies (g, in mGal). With the observed gravity as horizontal axis (x) and the known basement depths as vertical axis (y), we have,

   $$d = a\,g + b \qquad (1)$$

   where a is the gradient or slope of the regression line and b is the constant. Equation (1) is used to estimate the basement depth, i.e. the initial model, from the gravity anomaly at every point.
2. The theoretical gravity response of the initial model is calculated by using Talwani's method [2] with a unit density contrast, i.e. 1 gr/cm^3 or 1000 kg/m^3 in SI.
3. A linear regression is performed between observed gravity (y-axis) and calculated gravity with a unit density contrast (x-axis). The gradient of the straight-line curve gives the estimated density contrast that will be used for all subsequent forward modeling calculations.
4. The forward modeling is done for the initial model with the estimated density contrast to obtain the difference between observed and calculated gravity (E) at each point.
5. The gravity difference (E) associated with the initial model is used to obtain the depth modification at every point by using the following formula,

   $$d_{new} = d_{old} + E\,a \qquad (2)$$

   where a is the gradient or slope in Equation (1).
6. The steps number (4) and number (5) are repeated until a convergence is achieved or other stopping criteria are reached.

The criteria for convergence are either of the following: the total misfit falls below the target misfit (e_1), or the total misfit difference at two consecutive iterations becomes very small, i.e. less than a threshold value (e_2). The maximum number of iterations is set to avoid infinite loop if the convergence cannot be achieved.

3. Application to Synthetic Data

The iterative 2D gravity modeling algorithm was applied to invert synthetic gravity data associated with a relatively complex basement model. The synthetic model and data without noise (density contrast -0.5 g/cm^3) are shown in Figure 1. Since we deal with synthetic data, the linear regression was performed by using depths from the vertices of the synthetic model and calculated gravity anomalies associated with them (Equation 1). The term "observed" gravity will be used for the synthetic gravity data. The results of the regression are: $a = -0.0890$ and $b = -1.8764$ (Figure 2) which are then used for gravity to depth conversion to obtain the initial model. In Figure 2, the slope of the regression line does not represent a negative value since the depth is set to positive downwards.

IOP Conf. Series: Earth and Environmental Science 1031 (2022) 012014 doi:10.1088/1755-1315/1031/1/012014

The regression between the initial model response with a unit density (calculated) and actual gravity anomaly (observed) has led to estimated density contrast of -0.4651 g/cm^3 with a constant term of -1.2744 (Figure 3). Note that the regression line reflects the correlation between the negative values of both observed and calculated gravity. Figure 4 shows the initial basement model and its calculated gravity response. The gravity to depth conversion results in a very approximative geometry of the basement model. However, the model response is quite similar to the observed anomaly except for the details in the spatial high frequency variations.

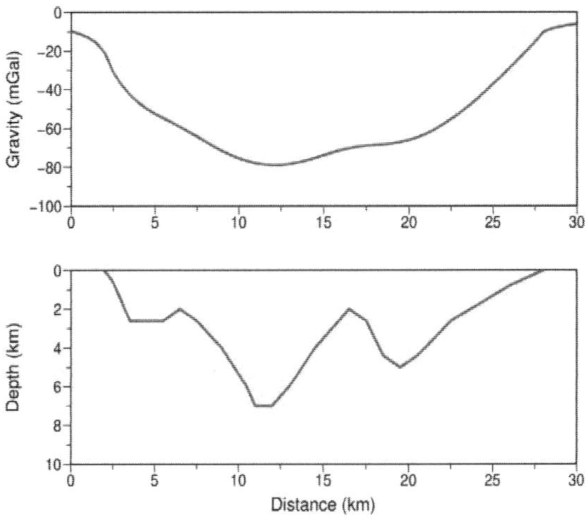

Figure 1. Synthetic gravity data without noise (top) associated with the synthetic basement model with the density contrast of -0.5 g/cm^3 (bottom).

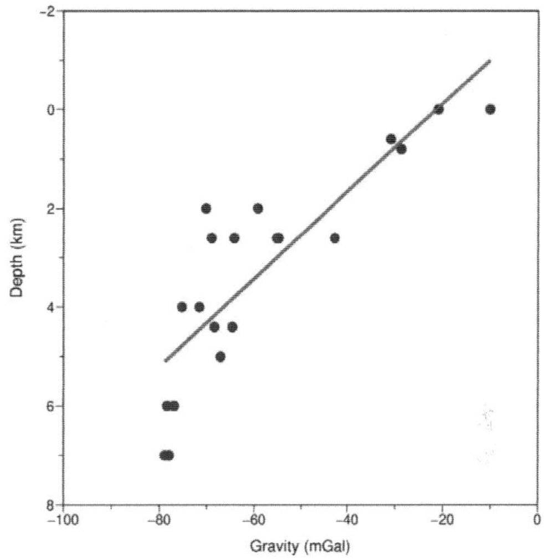

Figure 2. Gravity - depth relationship from linear regression between basement depth with synthetic gravity data at the same points.

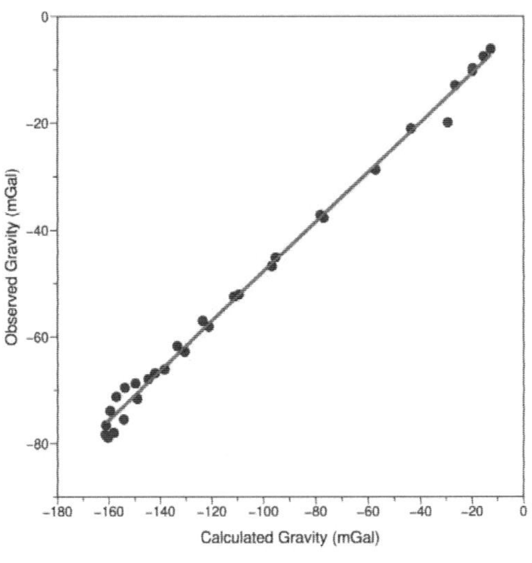

Figure 3. Regression between observed and calculated gravity of the initial model with unit density to estimate the density contrast, a = -0.4651, b = -1.2744 (see step 3 in the text).

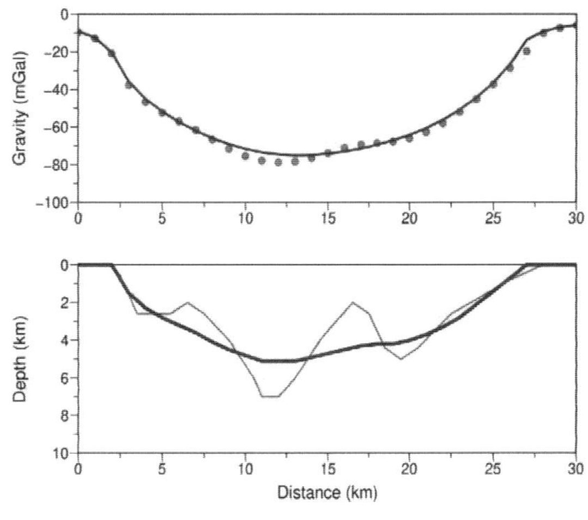

Figure 4. Synthetic (red dots) and calculated data (blue) of the initial model with estimated density contrast (top) and synthetic (red) and initial (blue) models (bottom).

IOP Conf. Series: Earth and Environmental Science 1031 (2022) 012014 doi:10.1088/1755-1315/1031/1/012014

Iterative refinements of the initial model led to the model presented in Figure 5 with overall or average misfit of 0.044 mGal after 31 iterations. Although the synthetic data do not contain noise, the iterations cannot be done further to obtain smaller misfit. In this case, e_1 was set to 0.05 mGal. Similar exercise was also performed with synthetic data containing noise of 0.5 mGal with Gaussian distribution. The final results are presented in Figure 6. The misfit of 0.294 mGal was obtained after 15 iterations. The final misfit around or slightly smaller than the noise content in the data is in general considered adequate, i.e. e_1 was set to 0.3 mGal. Inverse models from synthetic data with and without noise recovered the synthetic model satisfactorily. Higher noise in the data (e.g. 1 mGal or more) would result in more oscillations of the final model. In such case, a smoothing process, either during or after inversion would be needed to obtain more meaningful results.

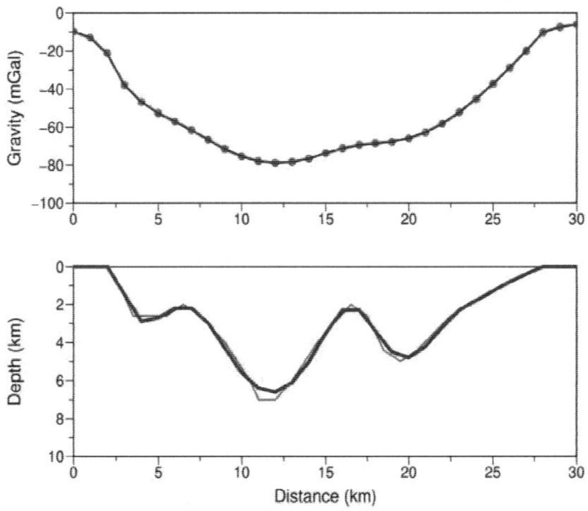

Figure 5. Modeling of synthetic data without noise: observed and calculated data of the final basement model (top) and resulting model along with the synthetic model (bottom).

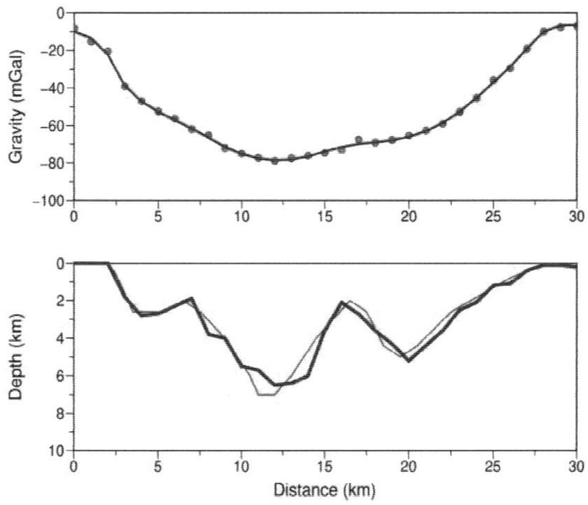

Figure 6. Modeling of synthetic data with additional Gaussian noise of 0.5 mGal: observed and calculated data of the final model (top) and resulting model along with the synthetic model (bottom).

4. Estimation of the Regional Anomaly

In the case of a linear gravity - depth relationship, a constant addition to the gravity anomaly would lead to similar feature of the basement model. Therefore, we investigate the possibility to model gravity data containing the simplest, i.e. constant, regional anomaly. For this preliminary attempt, we consider only the synthetic data without noise to simplify the problem. We add to the synthetic gravity data a +6 mGal constant regional anomaly and proceed with the same iterative algorithm as before.

The regression between depths and associated gravity anomalies gives a = -0.0890 and b = -1.3423 that are used to obtain the initial model. The gradient is the same as before, while the constant term differs due to the presence of the additional regional anomaly. The regression between the initial model response calculated with a unit density and the observed gravity anomaly results in the same density contrast of -0.4651 g/cm^3 with a different constant term of 4.7226. The initial model and its gravity response calculated with the estimated density is shown in Figure 7. With the same initial model and density contrast as before, the discrepancy between calculated and observed gravity is certainly due to the regional anomaly contained in the gravity data.

The constant factor of the second regression (to obtain the density contrast) is proportional to the added regional gravity anomaly. Thus, this constant can be considered as a representative of the regional anomaly, although its value is not exactly the same. Furthermore, an approximation to the regional anomaly led to "correct" basement model. In this case, we add the estimated regional anomaly to the calculated anomaly associated with the basement model estimates. In such case, the observed gravity anomaly shown in Figure 7 is not changed in the iteration process. The result is presented in Figure 8, with misfit of 0.148 mGal after 50 iterations. The basement model recovers well the synthetic model and it does not differ significantly with the basement model obtained from gravity data without the regional anomaly added (see Figure 5).

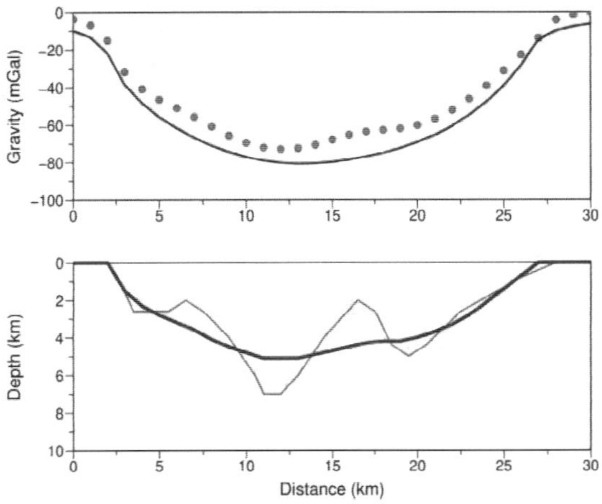

Figure 7. Synthetic data with regional anomaly added and calculated gravity data of the initial model with estimated density contrast (top), synthetic and initial models (bottom).

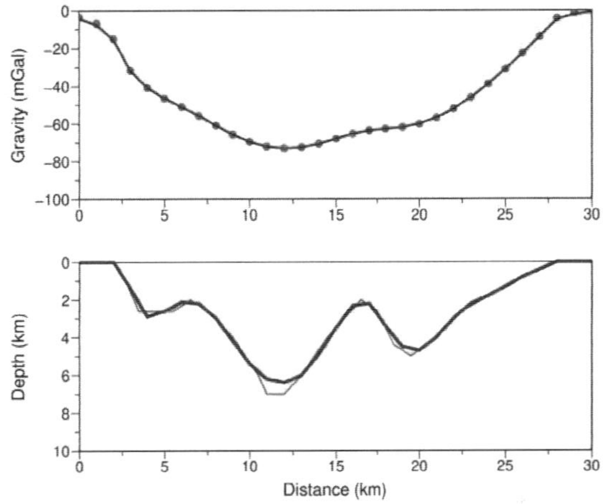

Figure 8. Modeling of synthetic data with regional anomaly, observed and calculated data of the final basement model (top), synthetic and inverse models (bottom).

5. Conclusion

We have presented a relatively simple iterative 2D gravity modelling similar to the iterative method that relies on the gravity - depth or thickness relationship. It does not involve complex matrix multiplications nor inversions. The proposed technique is stable for synthetic data without and with 0.5 mGal Gaussian noise. In general, the final models agree well with the synthetic model. We also improve the algorithm to include the estimation of the constant component of the regional anomaly with satisfactory results. The algorithm can be extended to involve more complex density contrast that varies with depth. In such case, the regression would not be linear but follows the density variation with depth, e.g. exponential.

Acknowledgment

This research was supported by ITB research grant through P3MI 2020.

References

[1] Reynolds JM 2011 An introduction to applied and environmental geophysics, Wiley.
[2] Cordell L, Henderson RG 1968 *Geophysics* **33** 596-601.
[3] Blakely RJ 1996 Potential theory in gravity and magnetic applications, Cambridge.
[4] Silva JBC, Santos DF, Gomes KP 2014 *Geophysics* **79** G79-G91.
[5] Zhou X 2013 *Geophysical Prospecting* **61** 220-234.
[6] Handyarso A, Grandis H 2017 *J. Engineering and Technological Sciences* **49** 358-372.
[7] Florio G 2018a SEG Technical Program Expanded Abstracts 2018 1415-1419.
[8] Florio G 2018b *Journal of Geophysical Research: Solid Earth* **123** 9101-9120.
[9] Leao JWD, Menezes PTL, Beltrao JF, Silva JBC 1996 *Geophysics* **61** 1702-1714.

IOP Conf. Series: Earth and Environmental Science 1031 (2022) 012015 doi:10.1088/1755-1315/1031/1/012015

Resources Optimization in Underground Gold Mining Case Study the Kubang Cicau Gold Vein in Pongkor Area, West Java, Indonesia

M. Akmal[1], L.E. Widodo[2*], M.N. Heriawan[2], F.A. Rosyid[3], A.R. Trilaksana[4]

[1]Mining Engeering Graduate Program, Bandung Institute of Technology
[2]Earth Resources Exploration Research Group, Bandung Institute of Technology
[3]Mining Engineering Research Group, Bandung Institute of Technology
[4]Geomin, PT Aneka Tambang, Indonesia

*Corresponding author: lew@mining.itb.ac.id

Abstract. Mining activities are typically controlled by the smallest mining units (SMU). These SMUs comprise mining blocks with dimensions that are generally adjusted to the mining equipment used as well as various geotechnical factors. In addition to SMUs, the smallest mining decision unit (SMDU), which is a collection of several SMU blocks, is usually established. In this paper, the SMDU is represented by an underground (UG) mining panel. This paper presents the results of a study on the optimization of gold resource in SMU blocks within SMDU and cut-off grade (CoG) variations. The study focuses on how to design a panel as an SMDU that is based on a weekly production quota that defines an optimum weekly rate of production that meets processing and milling capacities. The UG mining scheme in this paper is realized using a level, a panel, and SMU blocks. The level represents the progress of the mining in a vertical direction, while the panel represents the progress of the mining in a horizontal direction. Acting as the objective functions are the ore tonnage and gold mass. In relation to the gold mass, the goal is to reach the optimum level within the SMDU and fulfill the constraint function through the number of SMU blocks within the SMDU and the range of CoG variations. The optimization process was conducted using a genetic algorithm scheme. Kubang Cicau Vein B at Pongkor Gold Mine Business Unit was selected for the case study.

1. Introduction

In underground (UG) gold mining, resources are often optimized due to mining constraints and modification factors related to the aspects of technical, social, environmental, geometry, costs, metal prices and gold content limitations usually referred to as cut-off grades (CoGs). Resource optimization is a process used to produce optimal reserves, and thus resulting in optimal benefits, which is typically measured by the net present value (NPV). This paper focuses on the optimization of gold resources in the smallest mining unit (SMU) blocks and CoG variations. More specifically, the paper explores how to determine the smallest mining decision unit (SMDU) that contains SMU blocks under certain CoG variations to obtain the optimal level of gold resources in the SMDU.

Selective mining is predominantly based on CoGs, meaning that not all materials are extracted – only those in the zone of high grade. Conversely, the geometry of the SMDU also depends on the CoG. The optimization of the CoG is essential to maximize the NPV [1], which was initially developed by [2] and [3], who limited every operation that affected the profits made from mining, processing and refining. Lane's theory in [2] and [3] focused on the CoG optimization of mining, not on the optimization of mining against SMDU. Thus far, Lane's theory in [2], [3] has been predominantly applied to open-pit mining. However, the application of Lane's method to UG mining was conducted by [1], [4] and [5], who applied the method to CoG-based sublevel stope mine optimization.

The objective of this paper is to explore the development of an optimized model of gold resources to accommodate long-term planning based on dynamic CoGs, which comprise the preliminary method for optimizing gold resources in this study. The case study was conducted at Pongkor Gold Mine Business Unit (UBPE Pongkor) in the Kubang Cicau gold vein mineralization zone. The optimization of gold resources was carried out in the presence of mining constraints using the cut and fill mining method and variations in the number of SMU blocks and CoGs.

2. Methods

Gold resources within orebodies are defined as the ore tonnage or quantity of gold mass in blocks that are limited by the grade of the gold as well as physical boundaries of geology, hydrogeology, hydrology, geomorphology, etc. The blocks were designed according to the SMU using geometry that is limited by the size of the mining equipment used, and the geometry of the orebody and various geotechnical constraints. In this paper, gold resources within the orebody were optimized subject to these constraints to obtain optimal ore tonnage and, or gold mass. The optimal number of blocks in the SMDU were selected according to the CoGs within the range stated. To support a long-term UG plan, the UG process was optimized using the cut and fill technique by proceeding downwards using CoG variations. The data and the method used will now be explained.

2.1. Smallest Mining Unit (SMU)

The modeling of spatial gold grade distribution was conducted using ordinary kriging for block sizes that were designed as SMUs. Geometrical variables were used in the direction of dip with adjustments for the geometry of the orebody, and geotechnical factors as well as the mining equipment used. The number of blocks needed to be limited based on the zone / body of the mineralization domain and according to the mining target. This limitation resulted in variations in the dimensions of the block. However, reducing the dimensions of the SMU blocks affected the resources, as each block had a different value. The criteria of the resources that can be converted into reserves were measured and indicated, based on the Australian Code for Reporting of Exploration Results, Mineral Resources and Ore Reserves (JORC), which comprised the base model used in this study.

2.2. Panel and Level

The spatial distribution of the SMU blocks was then arranged according to the short-term plan, which was based on the weekly production capacity referred to as the panel. The panel is a cumulation of SMU blocks along the strike of the orebody, while the level accommodates those panels at each elevation interval that fit the height of the SMU blocks. The dimension of the panel used in this study indirectly provided the quantity and gold grade per panel. Thus, the panel can be considered the most short-term SMDU in terms of weekly production. These determinants were closely related to the operating costs. Its size in each level varied due to variations in the configuration and dimension of the SMU blocks.

2.3. Optimization and Decision Model

Corcenning the main objective of this study, the number of SMU blocks and the variations in the CoG represented the parameters to be determined. These parameters are expressed by chromosomes in the optimization procedure using a genetic algorithm (GA) scheme. The panel is considered as the basis of optimization, i.e., the mining decision unit. The goal was to optimize the quantity of ore tonnage and, or the quantity of gold mass in the total number of considered levels. This quantity was then formulated into the maximization of the objective function, which is expressed as follows:

$$Zt = \sum_{i}^{i=nL} \sum_{j}^{j=nP} \sum_{k}^{k=nB} \left(B_x^{ijk}.B_y.B_z\right).(RD) \tag{1}$$

$$Zg = \sum_{i}^{i=nL} \sum_{j}^{j=nP} \sum_{k}^{k=nB} \left(B_x^{ijk}.B_y.B_z\right).(RD).\left(\gamma^{ijk}\right) \tag{2}$$

$$K = \frac{Zt}{Zg} \tag{3}$$

where *i, j, k* are the summation indexes; *nL, nP, nB* are the number of levels, the number of panels, and the number of SMU blocks within each panel, respectively; B_x, B_y, B_z, *RD*, γ are the width of the SMU blocks in the dip direction, the length of the SMU blocks in the strike direction, the height of the SMU blocks, and the rock density and variable CoG, respectively; *Zt, Zg, K* are the ore production (tonnage), quantity of gold mass, and average quantity of gold mass in the considered number of levels, respectively. The objective function expressed by Eq. (1) was calculated based on the assumption that there were no geological losses. Eq. (1) defines the equation of ore tonnage, while Eq. (2) quantifies the gold mass in the considered number of levels.

Eq. (1) and (2) are limited by two constraints – namely, the range of SMU blocks within the panel and the range of CoG variations. Concerning the summary of the decision model given above, some constraints were noted and are presented below:

a. The number of SMU blocks was set within the range of nB_{min} = 2 blocks to nB_{max} = 8 blocks. The constraint related to the range of SMU blocks can be expressed as:
$nB = integer\left[nB_{min}, nB_{max}\right].$ (4)
b. The CoG variation was set within the range of γ_{min} = 2.00 g/ton to γ_{max} = 12.58 g/ton. The constraint related to the range of CoG variation can be expressed as:
$\gamma = \left[\gamma_{min}, \gamma_{max}\right].$ (5)

The minimum CoG for the resources model (i.e., the economic threshold that may be converted into reserves) was placed at 2 g/ton of gold grade in each block following [6]. This reduce the weight, as 33.5% of blocks contained a grade lower than 2.00 g/ton of gold content. The maximum gold grade used in this paper for the optimization of gold resources was set at 12.58 g/ton. The dimension of the SMU blocks in the direction of the strike (B_y), the height of the SMU blocks (B_z), and the rock density comprised the constants. Concerning the number of SMU blocks, the integer constraint was used to facilitate the simplified optimization process.

2.4. Optimization Process for the Optimization of Gold Resources

The following steps explain the optimization process, which used the GA scheme.

- **Step 1**: The CoG value was determined within the range of CoG variations. The determination of said value was based on an economic analysis that served as the basis for the optimization calculations.
- **Step 2**: The dimension of the SMU blocks, the rock density, the range of CoG variations, and the number of levels. Panels were given and SMU blocks were then calculated.

- **Step 3**: The ore tonnage and gold mass were calculated by completing the objective functions in Eq. (1) and (2) for the number of SMU blocks within the range of SMU blocks ($k = nB_{min}$ .... nB_{max}), for the number of panels on a certain level within the range of panels ($j = 1$, ... nP) and for the number of levels ($i = 1$, ... nL).
- **Step 4**: The ore tonnage and gold mass were optimized using the GA scheme by selecting the maximum ore tonnage and maximum gold mass obtained on the particular number of SMU blocks and CoG values.
- **Step 5**: Step 1 was repeated for the other CoG values within the range of CoG variations.

Optimization using the GA scheme shows its effectiveness and efficiency for a wide range of SMU blocks in large-scale mining production. For small-scale weekly production, the range of number of SMU blocks is not so large, so optimization can be conducted manually.

3. Case Study

UG mining in the Kubang Cicau gold vein mineralization zone was chosen as the case study, as some of the requirements for the implementation of the proposed method were met, including: (1) having a wide-range grade distribution and (2) having a medium-sized average grade.

3.1. Kubang Cicau Vein B

According to [7], Kubang Cicau is one of the four main veins of the Pongkor deposit. The three other veins are Pasir Jawa, Ciguha, and Ciurug. It is between 950–1,500m long and has an average thickness of 1.6–7.8m. The Kubang Cicau vein has a similar facies assemblage and mineral zone as the Ciguha vein, with dominant BOQ facies. The pitch of the ore shoots is steep to the south, meaning that it is parallel to the mineral zoning of the vein. The southern extension of the Kubang Cicau vein has an anastomosing calcitic stockwork bounded by a rim of chlorite and disseminated sulfides. The study will focus on the Kubang Cicau Vein B, which is 500 masl, between 375 and 475 masl, and possesses the following characteristics:

- X_{local} : 10,714–10,812 m
- Y_{local} : 7,914–8,136 m
- Z_{local} : 353–475 masl
- Strike : N 341^0 E
- Dip direction : N 79^0 E
- Dip : 72^0

The local coordinates were set by the UBPE Pongkor, which is the primary mining operator in the area. Based on a statistical analysis of 1,474 drill samples, the following data were obtained: the gold grade averaged at 17.75 ppm, with a standard deviation of 9.77 ppm and a top cut of 90% probability at 29.47 ppm. The spherical semivariogram shows the correlation between data with a range of 15 m in the direction of 90°N, 70°E/ (N–S), and a range of 18 m in the direction of N0°E (E–W). The estimation of the block models was conducted using ordinary kriging with a configuration of a block size of 15m x 3m x 5m against the XYZ axis and then cut using sedimentary geometrical constraints. The dimension of the X-axis was variable, while the dimensions on the Y and Z axis were constant. The gold resources can be categorized as having been measured or indicated based on the kriging variance. The geometry of the block models used in this study was limited by the following boundary: 10,642–10,816 m on the local X-axis, 7,857–8,647 m on the local Y-axis and 350–475 masl on the Z-axis.

4. Results and Discussion

The resource models that may be converted to reserves based on long-term CoG planning comprise the blocks that have a CoG greater than 2.00 ppm and that are within the classification of the measured and indicated reserves that are economically mined from 475 masl to 350 masl along a strike length of 800m. The number of SMU blocks used for resource estimation in this study was 6,600 blocks. The information was then processed using MATLAB software.

Table 1 presents the results of probable gold reserve optimization for the panel with variations in the number of SMU blocks as well as CoG variations. As mentioned, when using the terminology of the reserve, no classification is needed. Conversely, if the terminology of resources is to be used, then the average total resource is used to combine several resource classifications. At a CoG lower than 6 g/ton, the panel, as the smallest decision unit, will contain four SMU blocks that produce an optimum probable reserve. At a CoG of 6 g/ton, the panel will contain two SMU blocks that provide an optimum probable reserve.

Table 1. Probable reserves in various panel sizes and CoG values

No	Number of SMU blocks	CoG (g/ton)	Ore (ton)	Av. grade (g/ton)
1	2		449,677	4.37
	3		452,285	4.35
	4	2.00	462,234	4.29
	6		462,957	4.28
	8		459,292	4.28
2	2		321.989	5.11
	3		322.303	5.11
	4	3.11	324.244	5.09
	6		322.025	5.08
	8		324.281	5.05
3	2		228,580	5.72
	3		229,970	5.70
	4	4.05	233,800	5.67
	6		229,890	5.67
	8		235,210	5.60
4	2		63.907	7.97
	3		61.848	8.02
	4	6.00	63.366	7.94
	6		59.286	8.01
	8		57.605	8.00

The results of the optimization of gold resources, as stated in Table 1, are necessary for panels as the SMDU, which is a collection of several SMU blocks. The panel is based on short-term planning, which, in this paper, was a weekly production level of between 7,000–8,000 tons. Based on specific CoG criteria, the number of SMU blocks can be established in order to achieve optimum production capacity. Because panels are SMDUs, different treatments can be applied in terms of the mining process.

Resource estimation using CoG provides a greater amount compared to resources with independent CoG. This is because the use of panels involving individual blocks that have not been economically mean-graded, but the distribution is close to high-grade blocks, meaning that when they are combined, they provide an average grade according to the expected CoG. Based on spatial distribution modeling, the use of decision units is more applicable.

5. Conclusion

The computation of probable reserves using a panel that considers the short-term mining process contributes to more gold recovery than the use of a single SMU block, and it is practically advantageous in the case of a high variation of block grades in close range. The optimization of resources for short-term production capacity will ensure the stability of supply to processing / milling units.

Acknowledgments

The authors would like to thank the Research Institution and Community Services of Bandung Institute of Technology (LPPM-ITB) for Riset ITB 2018 and P3MI 2019 research funding. The authors would also like to thank the management at PT Aneka Tambang for the valuable information they provided and for the permission to publish the findings of the study.

References

[1] King B and Newman A 2018 Optimizing the Cutoff Grade for an Operational Underground Mine *Interfaces* pp 1–15

[2] Lane K F 1964 *Choosing the Optimum Cut-off Grade* (Quarterly of the Colorado School of Mines, USA)

[3] Lane K F 1988 *The Economic Definition of Ore* (Brisbane: Comet Strategy Pty Ltd)

[4] Gu X W, Wang Q, Chu D Z and Zhang B 2010 Dynamic Optimization of Cutoff Grade in Underground Metal Mining *Journal of Central South University of Technology* **17** pp 492-497

[5] Bootsma M T 2013 *Cut-off Grade Based Sublevel Stope Mine Optimization* (Report, Section for Resources Engineering, Department of Geoscience & Engineering, Delft University of Technology, the Netherlands)

[6] Hedenquist J W, Arribas A and Urien E G 2000 Exploration for Epithermal Gold Deposits *SEG Reviews* **13**(2000) pp 245-277

[7] Milesi J P, Marcoux E, Sitorus T, Simandjuntak M, Leroy J and Bailly L 1999 Pongkor (West Java, Indonesia): A Pliocene Supergene-enriched Epithermal Au-Ag-(Mn) Deposit *Mineralium Deposita* **34** pp 131-149

Petrophysical Analysis and Acoustic Impedance Inversion to Determine the Storage Target of Carbon Dioxide in Gundih Block, Central Java

Tara Annisa[1], Mohammad Rachmat Sule[1], Fatkhan[1]

Geophysical Engineering Major, Faculty of Mining and Petroleum Engineering, Bandung Institute of Technology

tara.annisa@gmail.com

Abstract. A pilot project of Carbon Capture and Storage in Indonesia is being prepared in the Gundih field, Central Java. This project acquired Kujung formation as their target area. However, an alternative to that formation is being investigated with Kujung formation's data as a benchmark to determine the total capacity of storage in Gundih Block. To determine the other carbon dioxide storage target, several tests are needed, one of which is petrophysical analysis. The results of this analysis are porosity, permeability, and thickness of a reservoir. After determining the porosity cut off from the Well data, we use the acoustic impedance inversion approach to assess the spread of the reservoir laterally. In this paper, we estimated that the porosity ranges from 7-12% with an average permeability of 10 mD. After examining the lateral distribution of the reservoir through the inversion, we can roughly estimate the distribution of good porosity reservoir. With the same method and logs derived from Kujung formation, we estimate that the upside potential from Ngrayong formation has an average porosity of 30% and average permeability of 16 mD. which are good enough to become an area of interest. With these results, we conclude that determining the area of interest from other target formation through petrophysical analysis and acoustic impedance inversion is sufficient to determine reservoir areas in a formation, though further injection feasibility studies are needed to determine the feasibility of injection of CO2 in Ngrayong Formation.

1. Introduction

Gundih Block is a target for the pilot project of Carbon Capture and Storage (CCS) development in Indonesia. Since December 2013, Gundih block, owned by Pertamina EP actively produces gas as much as 70 mmscfd with a CO2 content of 20%. This field is projected to keep in production for 12 years. In this block, a central processing plant was established to separate 80% of methane gas which are sold to generate electricity in Tambak Lorok, Central java and 20% of CO2 which are disposed to the air. The purpose of this pilot project is to utilize 20% of CO2 to reduce the effect of global warming [2].

To store CO2, there are certain criteria to find a feasible storage reservoir. A CO2 reservoir needs to have a caprock and a good porosity and permeability [4]. To find these characteristics, we use petrophysical analysis by looking into lithology, porosity, water saturation, and permeability. After determining the reservoir characteristics, we incorporate seismic data to find the continuation of the reservoir so we can roughly estimate the volume capacity of storage in that reservoir.

The target formation of this project is in Kujung Formation. However, an alternative to that formation is being investigated with Kujung formation's data as a benchmark to determine the total capacity of storage in Gundih Block. Even though the data in Kujung formation is enough to do petrophysical analysis, the data in the upside potential chosen, which is in Ngrayong Formation is insufficient for petrophysical analysis. To do so, some logs are needed to be generated with Kujung formation as a benchmark. This research studies if petrophysical analysis and acoustic impedance inversion are able to roughly estimate the reservoir characteristics in both Kujung formation and Ngrayong formation properly. The targeted output of the research is the estimated volume of the reservoir adequate to store CO2.

2. Methods

The data that were used in this research are well logs and seismic data. The wells that are being used are KTB-1 and RBT-2. 7 types of logs were used, such as GR, Caliper, RHOB, DT, RT, NPHI, PEF, and Check shot. However, there are only two logs that are available in Ngrayong formation, which are GR log and DT log. A density log is needed to make an Acoustic Impedance log, but since in Ngrayong formation that data is unavailable, therefore synthetic logs are generated to counter this setback. The synthetic logs are generated based on the same formation but in a different well nearby, which is JEPON-1 well. The seismic data that was used is 3D Post stack time migration data of the Gundih block.

In this research, two types of analyses were done: petrophysics analysis and Seismic inversion analysis. The inversion method that was used is model-based Acoustic Impedance Inversion. We start with finding out the porosity, permeability, and thickness of the reservoir based on the parameter of reservoir cutoff of porosity, permeability, thickness, water saturation, and volume of shale. After getting the porosity parameter, we determine the relationship between acoustic impedance and porosity on the well. After that, we did model-based acoustic impedance inversion to apply the relationship of acoustic impedance and porosity from well to the seismic [6]. After getting the rough estimate of the area of the reservoir. We estimate the volume of the reservoir by using a volumetric equation.

3. Results and discussion

Table 1 shows the porosity, permeability, Vclay and water saturation for well KTB-1 and RBT-2 in Kujung formation. In this table, we can infer that the clay content supports the report that Kujung formation consists of carbonate and shale. From these results, we can also infer that Kujung formation is mostly porous and are saturated by water.

Table 1. Vclay, Porosity, Permeability and SW for Kujung Formation.

Well	Vclay	Porosity	Permeability	Sw
	%	%	mD	%
KTB-1	9,0-80	7,0-20	0.0006 – 96	20-100
RBT-2	0,5-35	0,01-20	0,0001- 600	10-100

With the reservoir cut off shown in Table 2, we can estimate the net pay or estimated thickness of the CO2 reservoir. The reservoir area for Kujung formation is shown in Table 3. Table 3 shows that the reservoir areas of Kujung formation have a thickness of 17.3 meters in KTB-1 and 17.9 in RBT-2. With that thickness, the reservoir in Kujung Formation is also thick enough to store CO2.

Table 2. Vclay, Porosity, Permeability and Thickness parameter for CO_2 Storage

Parameter	Porosity	Permeability	Vcl	Thickness
Value	>10%	>10 mD	25%	>5 m

Table 3. Target zone for Kujung Formation

Well	Top	Bottom	Gross	Net	N/G	Av Phi	Av Sw	Av Vcl	Av Perm
KTB-1	2887.5	2921.8	34.3	17.3	0.504	0.135	0.404	0.198	10.68
RBT-2	2992.5	3065.07	72.54	17.98	0.248	0.165	0.058	0.085	8752.823

With the target area of the reservoir estimated, the distribution of the reservoir can be estimated to figure out the location of the reservoir on seismic data. The seismic inversion utilizes 2 wells, KTB-1 and RBT-2. In Kujung formation we averaged the value of the log's Acoustic Impedance to minimize the error of the Inversion. The result of the inversion is shown in Figure 1.

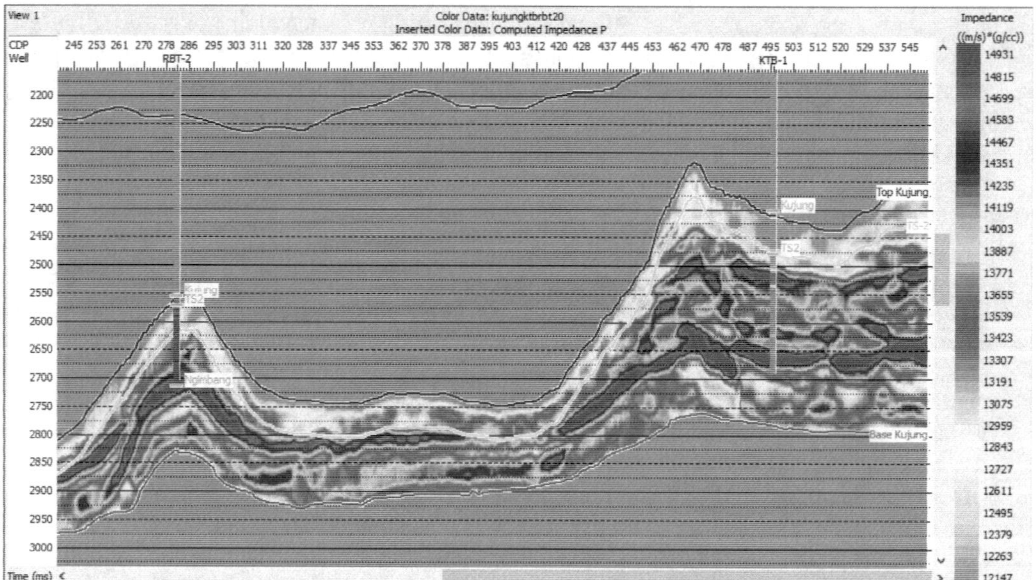

Figure 1. The acoustic impedance inversion result of Kujung formation

After acquiring the value of inverted Acoustic Impedance from the seismic data, we applied the relationship between Acoustic Impedance from well and porosity from well, demonstrated in Figure 2, to the seismic AI. With that relationship, we can make a porosity map displayed in Figure 3.

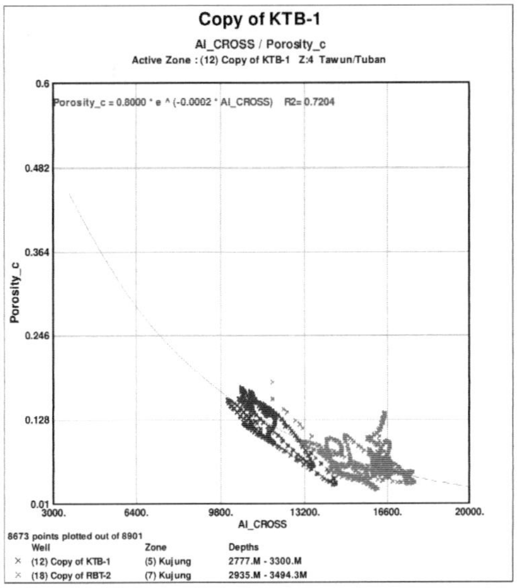

Figure 2. Relationship of acoustic impedance log and porosity log.

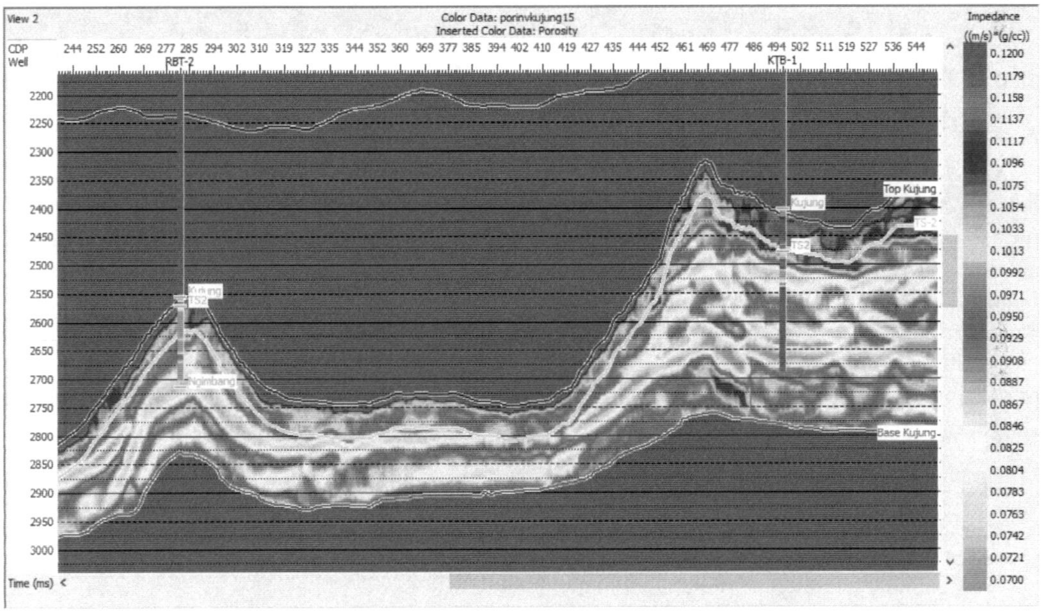

Figure 3. Porosity distribution in Kujung formation.

Figure 3 shows that the porosity that is considered good porosity ranges from 2700-2900 ms. This supports the result from the petrophysical analyses that show the reservoir is at 2887.70 m to 2921.40 m in KTB-1 and 2999.2 m – 3007 m. in RBT-2. From the results taken, we made a porosity distribution map with time slice varied from 0, 30, 60 and 90 ms with a centered window of 30 ms to map how deep is the good porosity distribution. The time structure map is shown in Figure 4.

IOP Conf. Series: Earth and Environmental Science 1031 (2022) 012016 doi:10.1088/1755-1315/1031/1/012016

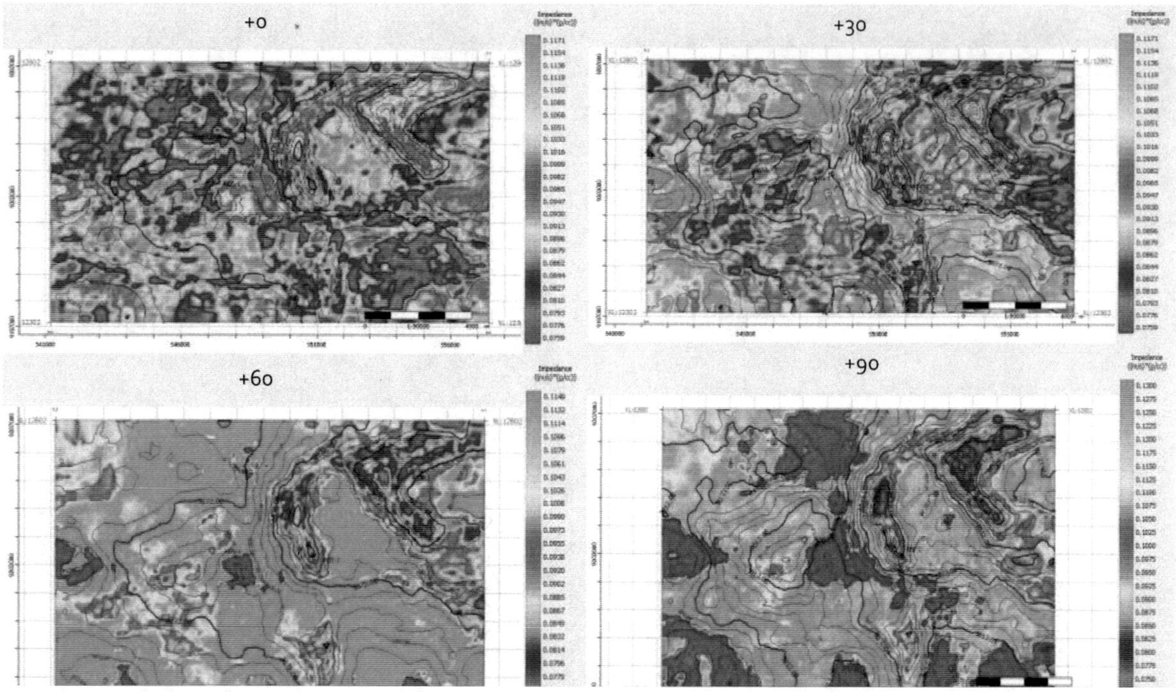

Figure 4. Slice of the Porosity distribution in Kujung formation from top Kujung (a), +30ms (b), +60ms(c), and + 90ms(d).

In Ngrayong formation, there is an additional step to do before the seismic inversion process, which is density modeling. The modeling of density was done by using Gardner's equation, which implies the relationship between Density and Velocity (VP). We used JEPON-1 well as a benchmark, which also has Ngrayong formation in-depth 1000-1500m. Since Ngrayong formation consists of sand and shale, two equations are needed to determine the density value of Ngrayong formation in KTB-1 and RBT-2. The types of lithologies are separated with GR index with the cut off of 50 API. The result of Gardner equation shown in Figure 5. After acquiring the synthetic density log, the same petrophysical analysis steps are done in Ngrayong formation also. The results of the target area of the reservoir are shown in Table 4.

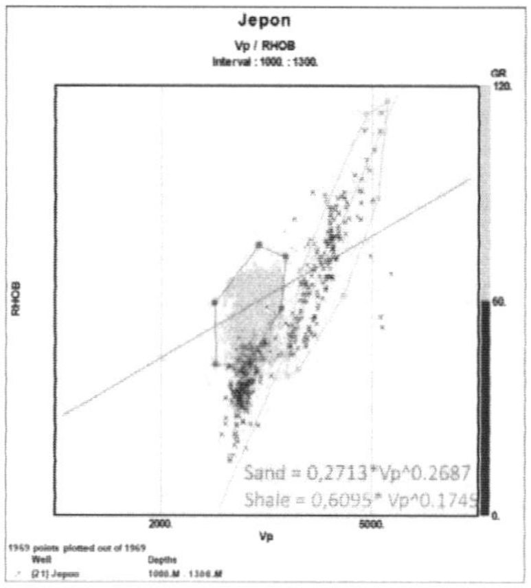

Figure 5. Gardner estimation for JEPON 1.

Table 4. Target zone for Ngrayong Formation

Well	Top	Bottom	Gross	Net	N/G	Av Phi	Av Sw	Av Vcl	Av Perm
KTB-1	1465.7	1560.1	94.4	8.5	0.09	0.315	0.477	0.156	76.07
RBT-2	879.96	925.53	45.57	26.75	0.587	0.39	0.456	0.121	23.979

With the target area of the reservoir estimated, the distribution of the reservoir can be estimated to figure out the location of the reservoir on seismic data. The seismic inversion utilizes 2 wells, KTB-1 and RBT-2. In Ngrayong formation we averaged the value of the log's Acoustic Impedance to minimize the error of the Inversion. However. The relationship between porosity and AI is hard to distinguish, therefore we estimated the reservoir distribution by using the difference in lithologies. The lithology separation and reservoir distribution are displayed in Figure 6 and Figure 7.

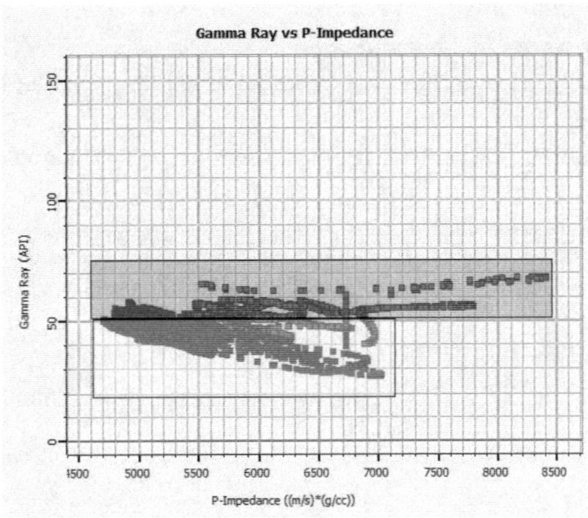

Figure 6. Relationship of Gamma ray log and Acoustic Impedance from KTB-1(Blue) and RBT-1 (Red).

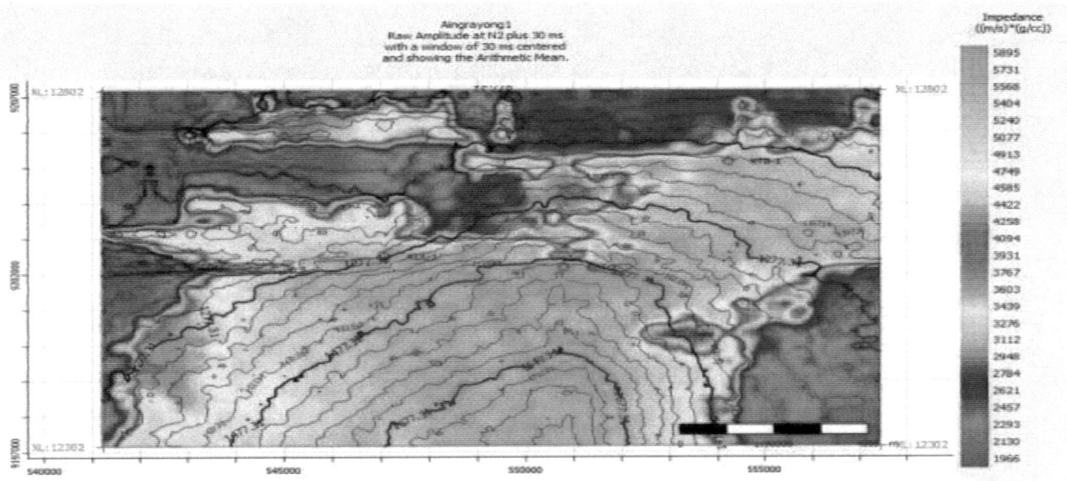

Figure 7. Sandstone thickness distribution in Ngrayong formation.

IOP Conf. Series: Earth and Environmental Science 1031 (2022) 012016 doi:10.1088/1755-1315/1031/1/012016

4. Conclusion

According to the analysis given. Petrophysical analysis and Acoustic Impedance inversion are enough to determine the target area of the reservoir in Kujung formation. Even though the petrophysical analysis in Ngrayong formation shows that Ngrayong formation is a good fit as an upside potential for CO2 Injections, the relationship between AI and porosity cannot be established due to factors such as inadequate log availability. Further data collection and coherent investigations are needed to establish a more accurate and complete knowledge of the potential in Ngrayong formation.

References

[1] BP, C. (Juni 2011): *Petrophysical Analysis to characterize field "X,"* Universitas Indonesia.
[2] Center of Excellence CCS ITB (2015): *The Planned CCS Pilot Project in Gundih Area, Blora District, Central Java Province, Indonesia.*
[3] Mavko, G., Mukerji, T., dan Dvorkin, J. (2009): *The Rock Physics Handbook (2 ed.)*, Cambridge University Press, Cambridge, 102.
[4] Nordbotten, J. M., dan Celia, M. A. (2012): *Geological Storage of CO2 : Modeling Approaches for Large-Scale Simulation.*, John Wiley & Sons, Hoboken, 11.
[5] Purwaningsih, M. E. M., Satyana, A. H., Noeradi, D., Budiyani, S., dan Malik, N. M. (2002)*: The Late Oligocene Kujung Reef Complex In The Western East Cepu High, East Java Basin : Seismic Sequence Stratigraphic Study*, 18.
[6] Russell, B. H. (1988): *Introduction to Seismic Inversion Methods*, Society of Exploration Geophysicists.
[7] Tiab, D., dan Donaldson, E. C. (2004): *Petrophysics (2 ed.)*, Elsevier, Oxford, 87– 312.
[8] Tsuji, T., et al. (2014): *Reservoir Characterization for site Selection in the Gundih CCS Project, Indonesia*, Energy Procedia, **63**, 6335– 6343

IOP Conf. Series: Earth and Environmental Science 1031 (2022) 012017 doi:10.1088/1755-1315/1031/1/012017

Application of Geo-electrical Method in Preliminary Study for Earth Electrode Site Selection

Prihadi Sumintadireja

Faculty of Earth Science and Technology, Institut Teknologi Bandung (ITB), Bandung, Indonesia

Abstract. Electrical power transmission is usually done as high voltage transmission for efficiency. For that purpose, effective grounding is necessary especially at the location of the power transmission control center. To obtain a good connection to the ground, very conductive or low resistivity soil should be present up to an appropriate depth at the earth electrode site. In this paper we present the results of the geoelectrical survey performed in a candidate site in the southern part of South Sumatra province, Indonesia. The objective of this preliminary study is to delineate the most appropriate location for the grounding electrodes. Data acquisition for 2D resistivity tomography was done along several lines crossing the survey area. Low resistivity layer of 10 Ohm.m or less up to 50 m depth was found at the north-western sector of the area.

1. Introduction

In a high voltage system for transmission of bulk electrical power over long distances, a proper earth electrode system is required. A significant amount of current is injected into the ground, especially in case of system failure. The electrical conductivity of the host medium for an earth electrode is important for electrode functionality and also for minimizing environmental impact due to the injected current. High resistivity medium around the electrode will create high potentials and strong electric fields leading to an effect on infrastructure. For a proper design, the site selected for the ground electrode should allow current penetration relatively deep into the earth. Therefore, it is important to obtain information about the resistivity structure of the subsurface close to the electrode and also on a regional scale [1,2].

The most common choice for shallow to intermediate depth investigation (50 to 100 m) of the subsurface conductivity structure is the geoelectrical survey. Conventionally, direct current is injected into the ground and its effect, i.e. potential, is measured using a standard electrode configuration. The 2D resistivity tomography consists of systematic measurement with varying the electrode array separation along a traverse to obtain the resistivity variation both laterally and vertically. The latter is obtained from the application of the inversion algorithm where we seek for a plausible resistivity model that fits the observed data [3,4]. In this paper, we focused only on the preliminary study for earth electrode location in a pre-selected area in the southern part of South Sumatra province, Indonesia.

2. Geoelectrical Survey and 2D Modelling

In the geoelectrical imaging technique, a large number of electrodes is placed on the ground along a profile and connected with the multi-core cable to an automatically controlled switching and data acquisition system. Then, each electrode can serve as either current (C_1 or C_2) or potential (P_1 or P_2) electrodes in a 4-electrode resistivity measurement array. We used ARES G-5 resistivity-meter from GF-Instruments Inc. powered by a 12 Volt DC car battery. The current source can deliver up to 1000

IOP Conf. Series: Earth and Environmental Science 1031 (2022) 012017 doi:10.1088/1755-1315/1031/1/012017

Volt DC and 5 Ampere of current into the ground [5]. The number of channels or electrodes connected to take-outs of the multi-core cable is 48 with the unit electrode spacing is 12.5 m. Therefore, the measurement at a traverse line corresponds approximately to a 600 m profile.

A combination of Wenner and Schlumberger electrode configurations (Figure 1) was used to optimize both horizontal and vertical resolution that are characteristics of each configuration [4]. The generalized formula for calculating the apparent resistivity (ρ_a, in Ohm.m) from measured values of current (I, in Ampere) and potential difference (ΔV, in Volt) is expressed by:

$$\rho_a = K\,\frac{\Delta V}{I} \quad ; \quad K = 2\pi\left(\frac{1}{C_1P_1} - \frac{1}{C_2P_1} - \frac{1}{C_1P_2} + \frac{1}{C_2P_2}\right)^{-1} \tag{1}$$

where C_1P_1 denotes the distance (in meter) between current electrode C_1 and potential electrode P_1 and so on, while K is the geometrical factor of the electrode configuration. With automatic data acquisition system, the measured parameters are directly resistance values or $R = \Delta V/I$, in Ohm. For Wenner electrode configuration, $K = a$, where a is the electrode spacing (Figure 1) which is an integer or n multiple of the unit electrode spacing. For Schlumberger configuration, K is function of a nd b defined in Figure 1. In more general cases of geoelectrical imaging technique, the distance between electrodes is an integer or n multiple of the unit electrode spacing that determines the investigation depth and the position for the data plotting for the construction of pseudosection.

A sequence of all possible measurements in a traverse consisting of 48 electrodes following a configuration of electrodes results in apparent resistivity data that can be plotted as pseudosection. In the latter, the apparent resistivity data are plotted horizontally on the traverse line at mid-points of the current and potential electrode pairs, while their vertical position is a scaled value of the electrode spacing that depends on the configuration used. The colour-contoured values show variations of the measured data with position and with effective investigation depth, rather than with the true depth. It is used for preliminary qualitative interpretation, before more quantitative interpretation using 2D inversion modelling.

The geoelectrical survey was conducted only in the vicinity of the proposed earth electrode site at 6 traverse lines N, W, E, SE, NW and NE, according to their direction or azimuth (Figure 2). We used Earth-Imager 2D resistivity modeling software from Advanced Geo-sciences Inc. (AGI) with apparent resistivity as function of electrode spacing as the input data to obtain 2D resistivity model representing resistivity distribution of the subsurface [6].

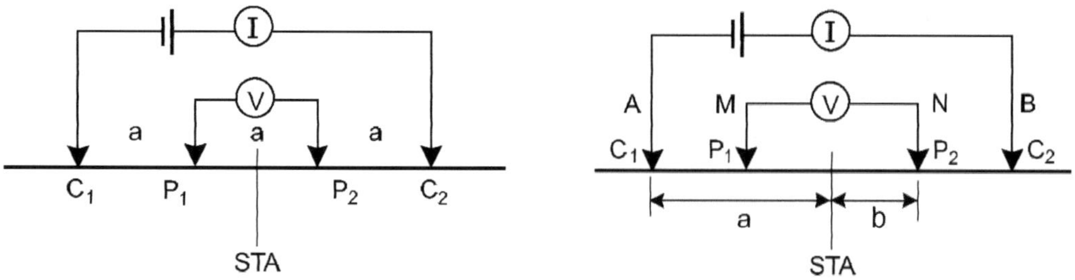

Figure 1. Wenner electrode configuration with the electrode spacing a (left) and Schlumberger electrode configuration with a and b are half-distances between current electrodes (C_1 C_2) and between potential electrodes (P_1 P_2) respectively (right). STA is the measured position.

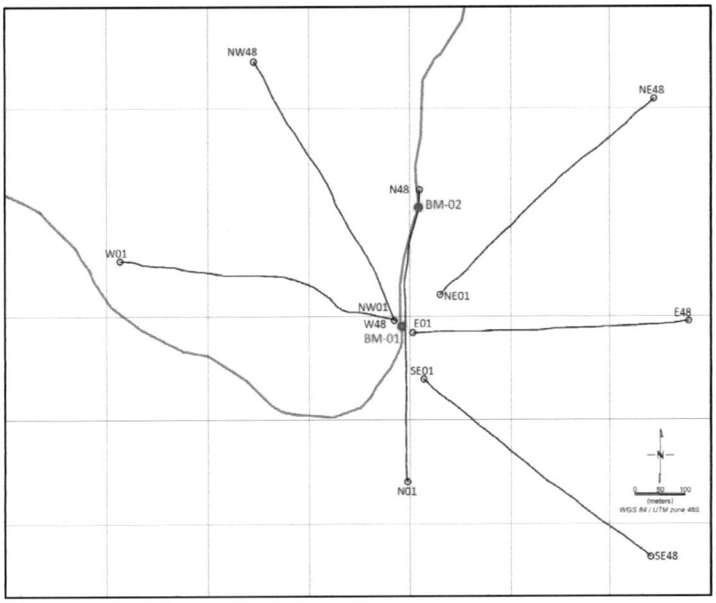

Figure 2. Outline of the geoelectrical survey profiles at the survey location. The lines are named according their orientation, i.e. N, W, E, SE, NW and NE, with numbers are the electrode number.

3. Results and Discussion

We present only representative results with significant evidence on the existence or non-existence of very conductive layer at the surface up about 50 m depth as required for the in-situ conductive medium for the earth electrodes. The result for line N extending from South to North at the central part of the study area is presented in Figure 3. The figure contains apparent resistivity pseudo-section (data), theoretical response of the inverted model and 2D resistivity section (model) resulting from inversion modelling. The model is limited to a maximum depth of approximately 125 m due to limitation in the unit electrode spacing (12.5 m) and the number of electrodes for each traverse, i.e. 48 electrodes. Note that the vertical axis of the 2D resistivity model is presented as elevation above the mean sea level (MSL). More detailed 2D models for line W, NW and SE are presented in Figure 4 to 6 with vertical exaggeration showing also the topography.

The 2D resistivity model for line N exhibits an almost horizontal and continuous layer with resistivity less than 10 Ohm.m at the extremity of the profile (southernmost and northernmost). The superficial conductive layer has approximately 20 m thick. The central part of the profile is dominated by a heterogeneous zone with more resistive medium up to 100 Ohm.m (see Figure 3, bottom). The subsurface resistivity distribution to the western part of the study area shows more promising conductive zones that extends almost all along the profile W, except at a small part at the eastern end of the profile. The conductive zones are relatively thick with the most homogeneous and horizontal western part is approximately 20-25 m thick (Figure 4). The subsurface resistivity distribution of line NW shows low resistivity layer extending from the center up to south-eastern part of the profile with a limited extent of heterogeneity towaards the end of the profile (Figure 5).

As an example of the non-existence of near-surface conductive layer is shown in Figure 6 for 2D model from line SE that extends from North-West to South-East. The superficial layer is dominated by resistive heterogeneity, while more conductive layer is at deeper part of the subsurface. Line E and NE (not shown) exhibit almost similar characteristics of the subsurface. Therefore, the most appropriate area for the earth electrode is to the West and North-West of the central part of the survey area indicated by the position of the bench-mark BM-01 (see Figure 2).

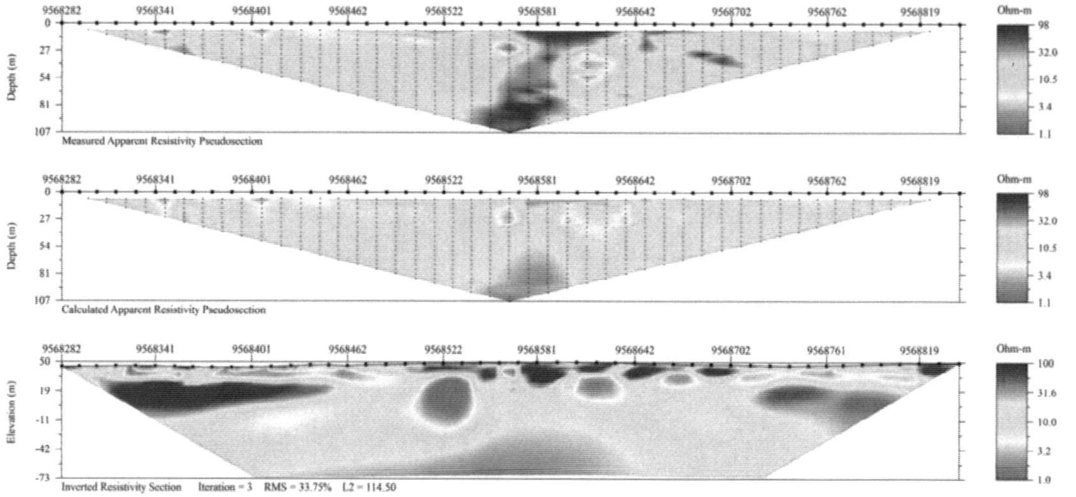

Figure 3. Resistivity section of line N with measured apparent resistivity pseudosection (top), calculated apparent resistivity pseudosection (middle) and 2D resistivity model (bottom).

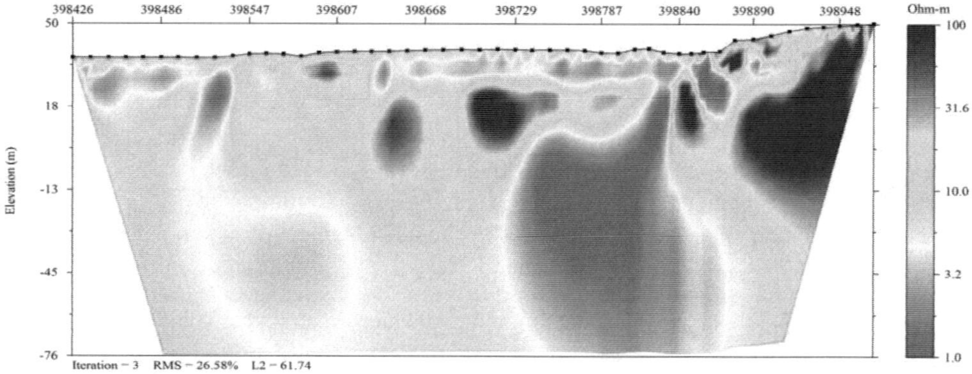

Figure 4. 2D resistivity model of line W from West to East with vertical exaggeration and topography.

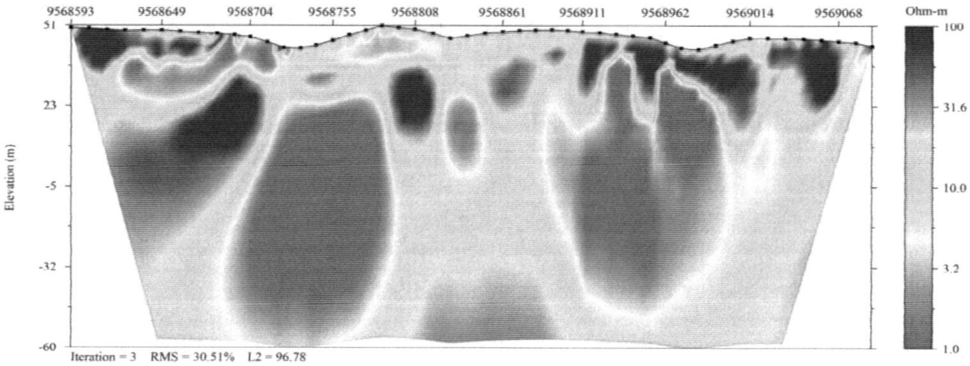

Figure 5. 2D resistivity model of line NW from South-East to North-West with vertical exaggeration and topography.

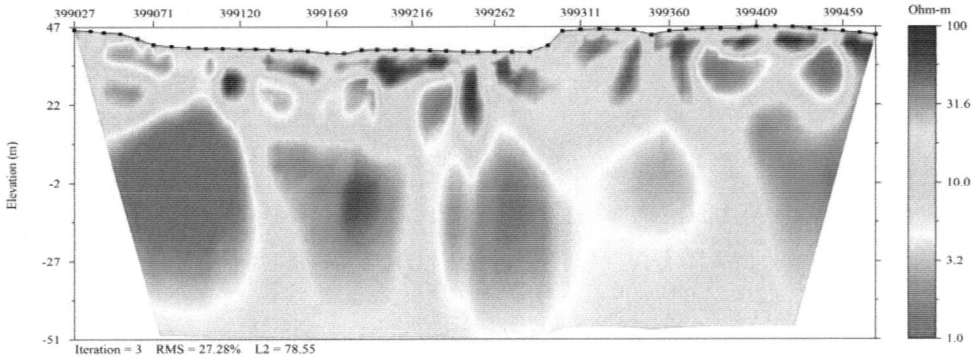

Figure 6. 2D resistivity model of line SE extending from North-West to South-East with vertical exaggeration and topography. Superficial heterogenities dominated by resistive layer is more pronounced than those from other profiles shown in Figure 3 to Figure 5.

4. Conclusion

We have presented the application of geoelectrical method for earth electrode site selection for high voltage electrical power transmission. The tomography technique was performed both in data acquisition and inversion modelling resulting in 2D resistivity models that fit the observed data. The RMS error of all inversion modelling is in the order of 20% to 30% which is quite high due to relatively high level of noise in the survey area. Heterogeneous superficial layer also contributes to noisy data since the contact resistance of the electrodes was also relatively high (more than 1 kOhm). Nevertheless, our results have been successful in delineating the most promising and appropriate zone for the earth electrode site. The location with dominating low resistivity superficial layer up to 20 to 30 m depth is located to the West and North-West of the central part of the survey area.

References

[1] Thunehed H, Åström U and Westman B 2007 *IEEE PES PowerAfrica*.
[2] Villas JET and Portela CM 2003 *IEEE Transaction on Power Delivery* **8**(3).
[3] Reynolds JM 2011 An Introduction to Applied and Environmental Geophysics, Wiley.
[4] Loke MH 2012 Tutorial: 2-D and 3-D electrical imaging surveys, www.geoelectrical.com.
[5] GF-Instruments 2010 ARES G-5 resistivity-meter Users' Manual.
[6] Advanced Geosciences Inc. 2005 EarthImager 2D Instruction Manual Version 1.9.0.

IOP Conf. Series: Earth and Environmental Science 1031 (2022) 012018 doi:10.1088/1755-1315/1031/1/012018

A New Approach for Storage Capacity and Allowable Injection Time Calculation in Carbon Capture and Storage (CCS) Reservoir

Anthony Kurnia Jaya[1], Zuher Syihab[1]

[1] Bandung Institute of Technology, Bandung, Indonesia
[1] anthonykurniaj@gmail.com, [1] zuher.syihab@tm.itb.ac.id

Abstract. In reducing the CO_2 emissions in the atmosphere, some energy industry start to implement the Carbon Capture Storage (CCS). However, there still less restriction and considerations in calculating the storage capacity and allowable injection time of CCS reservoir which might lead into leakage problem. In overcoming this problem, there is a study conducted to make a new calculation with higher accuracy. In this study, the reservoir model is using CMG software with base model of East Natuna Field. Then, this model is used to make a sensitivity analysis for 10 CCS fields with a support of CMG-CMOST and Eureqa software. The result is an equation with an R-Square of 96.6% for storage capacity and 95.4% for allowable injection time. The considered parameter in the new equation are permeability, porosity, fracture pressure, thickness, area, reservoir temperature, and reservoir pressure. To validate the new equation, a case study is made for 3 CCS fields. The result shows that the new equation shows up to 21% smaller storage capacity and 7 years reduction in allowable injection time. The result is reasonable due to additional restrictions. Therefore, it is believed that the generated equation is valid and more beneficial compared to the conventional equation.

1. Introduction

The oil and gas sector still holds the majority energy source in the world with almost 63% in 2017 and it could not be replaced immediately [1]. The extraction of hydrocarbon will make a carbon footprint which will damage the environment and global warming. With the Paris Agreement that has a goal to reduce Greenhouse Gas (GHG) to achieve the global temperature increase below 1.5°C. As one of GHG type, CO_2 is the most GHG that being released into the atmosphere which takes 82% of total GHG in the atmosphere [2].

With the purpose to reduce the emission of CO_2, the CCS technology in the oil and gas industry start to be implemented. This component could be captured and compressed into reservoir. The critical factor in evaluating the storage is the reservoir capacity and allowable injection time. Currently, the calculation of storage capacity is using the Bachu equation that considers the parameter of effective porosity, irreducible water saturation, bulk volume, capacity coefficient and density of CO_2 [3]. For the allowable injection time, it use equation from Szulczewski which require a pre-calculation [4].

In order to reduce the complexity of calculation, engineers that assess the CCS storage will assume that the equation with less restriction could have an accurate result. This less restriction condition could make an over injection or under injection in the reservoir due to inaccurate calculation. These inaccuracy will reduce the interest from company to execute a CCS project. Therefore, these concrete problems show urgency to make a new approach in calculating storage capacity and allowable injection time.

2. Basic Theory

2.1. Reservoir Modelling
Numerical modeling is commonly used to calculate an injection process as it is accurate in prediction and optimization. The prediction will be used to know the flow path of the injected CO_2 and the optimum well location will need to be assessed. One of the simulations processes that have been applied to analyze the storage capacity in a saline aquifers reservoir is the GEM simulator generated by the Computer Modeling Group [5]. In CCS reservoir model, there are some parameters being used: (1) Grid specification, (2) Initial water saturation, (3) Regions, (4) Permeability, (5) Porosity, (6) Transmissibility, (7) Aquifer specification, (8) Basic reservoir component, (9) Rock fluid specification, (10) Injection well position, (11) Depth, and (12) Thickness.

2.2. CMG-CMOST Analysis
There are some parameters selected to be processed in this analysis which will be based on the relation with storage capacity and allowable injection time. Each parameter is analyzed in a specific range of value obtained by doing an analysis on other CCS field. The experimental case generator in this analysis is categorized into 3 types, Classic Design, Latin Hypercube Design, and User Defined Design. The result of this analysis is a proxy model generated by a regression process. Following the proxy model, there also a result comparison with the simulated result.

2.3. Eureqa Analysis
To find out the best equation that has a higher R-Square and simplest form, another sensitivity analysis was made using Eureqa. The concept of this sensitivity process is to classify the important parameter relation which will have a big effect on the equation and make the regression result. Data input for this analysis comes from the experimental case from CMG-CMOST. For some parameters that have a low impact on the objective function, it will be eliminated automatically by Eureqa. This process differentiates Eureqa with the CMG-CMOST analysis that will try to make a correlation between every parameter with the objective function. This process also helps to generate a simpler form equation without reducing the confidence level. The generated equation in Eureqa sensitivity analysis needs to have a high maturity and stability to be a valid equation. The cut off for the maturity and stability needs to be higher than 50%.

2.4. Storage Capacity Calculation
The storage capacity of a CCS reservoir could be calculated by a volumetric method generated by Bachu which is shown in Equation (1) [3]. The concept of this equation is by calculating the volume of a reservoir through the thickness and area which is represented by the integral in 3 directions of x, y, and z.

$$M_{co_2} = Cc \int \int \int \o_{eff}(1 - SWirr)\rho_{co_2} dxdydz \quad \text{.....(1)}$$

2.5. Allowable Injection Time Calculation
Besides calculating the storage capacity, the other important parameter is the allowable injection time. The definition of allowable injection time is the total duration from the injection start until the rate is too small and couldn't be injected again. However, the current calculation of this parameter need a pre calculation to determine the maximum injection rate first as shown in the equation (2) below. This equation was generated by Szulczewski which consider the parameter of density from CO_2, the reservoir dimension, reservoir characteristic and the pressure [4]. Besides the needs of pre calculation, the determination for dimensionless pressure parameter will need a specific study from the reservoir.

$$t = \frac{k}{\mu\left(\frac{Qmax*\dot{p}}{2\rho*H*A*FP}\right)^2} \quad \text{..........(2)}$$

IOP Conf. Series: Earth and Environmental Science 1031 (2022) 012018 doi:10.1088/1755-1315/1031/1/012018

3. Methodology

3.1. Assumption
In building the model, there are some consideration and assumption that are used: (1) Seal for each reservoir layer is assumed to have no leakage, (2) Reservoir is assumed to be filled perfectly with saline aquifer, (3) Each layer of the reservoir have a homogeneous parameter, (4) The generated equation have a specific range of value for each parameter, (5) Mineral in the reservoir is assumed neglectable.

3.2. Sensitivity Analysis
10 CCS fields in the world that is used to determine the possible affecting parameters. Also, these fields are used to determine the specific range of value for each parameters. These fields are East Natuna, Indonesia; Sleipner, Norway; Snohvit, Norway; In Salah, Algeria; Gundih, Indonesia; Farnsworth, United States; Southeast Queensland, Australia, Lacq, France, SPE-196691-MS, and SPE-194629-MS [6, 7, 8, 9, 10, 11, 12, 13, 14, 15].

After analysing the possible affecting parameter, some parameters were chosen to be used in this process are CO2 content, Fracture pressure, Permeability, Porosity, Reservoir pressure, Specific Gravity, Thickness, Reservoir temperature, Area, Project's lifetime and Irreducible water saturation.

4. Case Study

In this study, a reservoir model with the base model of East Natuna Basin data will be shown in Figure *1*. This model was generated by a total of 1125 grid blocks (15x15x5) which represent 24000 acres area of this field. Properties from the same layer were assumed to have the same value. The composition for reservoir fluid was modeled by CMG-Winprop where the composition was 98% H_2O and 2% CO_2 as the targeted reservoir to store the CO_2 in Deep Saline Aquifer.

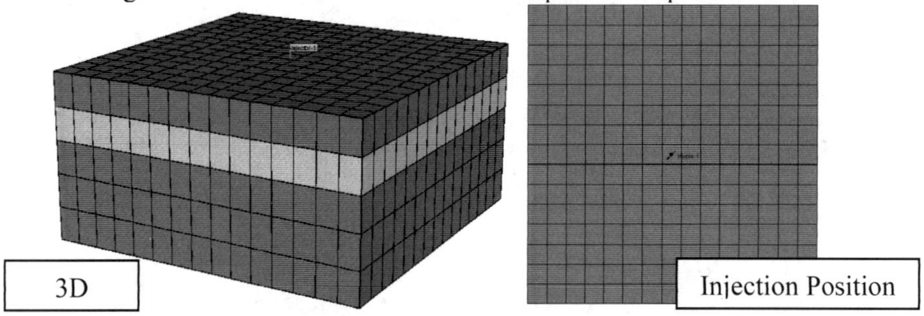

Figure 1 Reservoir Model

These generated equations from CMG-CMOST and Eureqa was used to calculate the base model first. This process aims to determine whether the simpler form of the equation could have a similar result or not. Besides the R-Square, the simplicity of the equation will be an additional concern in this study. The chosen equations were used to calculate the storage capacity and allowable injection time of three CCS field: East Natuna, Snohvit, and Southeast Queensland. Result and Discussion

4.1. CMG-CMOST Equation
The equation generated in CMG-CMOST will be based on the experimental cases with each parameter have a different range of values. The range of value from each parameter will be shown in Table 1.

Table 1 Parameter Range

No.	Parameter	Lower Limit	Upper Limit	Unit
1	CO_2 content	0.6275	1	%
2	Fracture pressure	1042	8000	psia
3	Permeability	20	400	mD

4	Porosity	0.1	0.528	frac.
5	Reservoir pressure	1091	8000	psia
6	Specific gravity	0.5	2	frac
7	Thickness	40	250	ft
8	Reservoir temperature	80	400	F
9	Area	40	60000	Acres
10	Injection duration/ project's lifetime	20	40	years
11	Irreducible water saturation	0.15	0.4	frac.

The generated equation is shown below for both Storage Capacity and Allowable Injection Time.

$$Storage\ Capacity\ (lb) = Coeff_1 * Term_1 + Coeff_2 * Term_2 \dots\dots + Coeff_n * Term_n \dots\dots\dots\dots\dots (3)$$

With the tabulation for each coefficient and term is shown in Table 2.

Table 2 Term and Coefficient Storage Capacity

Term	Coefficient	Term	Coefficient
Intercept	2.1477E+11	Por*Por	-4.25632E+11
CO2Percentage	-1.0547E+11	Por*Pres	-9.21601E+07
Fracpres	-4.34664E+07	Por*Tres	4.21992E+08
Perm	-2.62375E+08	Por*ACRE	867216
Por	-1.6809E+11	Pres*Pres	1340.65
Pres	4.03745E+07	Pres*Thickness	-222813
Thickness	-3.95385E+08	Pres*Tres	53261
Tres	-1.78007E+08	Pres*ACRE	-1052.88
ACRE	-1.94064E+06	Pres*SWirr	-2.26283E+07
SWirr	4.11982E+09	Thickness*Thickness	-1.90654E+06
CO2Percentage*Tres	3.24801E+08	Thickness*ACRE	9425.35
Fracpres*Fracpres	-1613.64	Thickness*SWirr	4.73789E+08
Fracpres*Por	9.50941E+07	Tres*Tres	-830098
Fracpres*Thickness	249657	Tres*ACRE	1723.77
Fracpres*ACRE	961.434	ACRE*ACRE	-23.1227
Perm*Perm	-283645	Perm*Thickness	913964
Perm*Por	9.43819E+08	Perm*ACRE	5847.75
Perm*Pres	-20498.5		

$$Allowable\ Injection\ Time(month) = Coeff_1 * Term_1 + Coeff_2 * Term_2 \dots\dots + Coeff_n * Term_n \dots (4)$$

With the tabulation for each coefficient and term is shown below in Table 3.

Table 3 Term and Coefficient Allowable Injection Time

Term	Coefficient	Term	Coefficient
Intercept	238.156	Fracpres*Tres	6.01997E-05
CO2Percentage	-100.775	Fracpres*ACRE	-5.20313E-07
Fracpres	-0.034797	Perm*Perm	0.00182654
Perm	-1.22066	Perm*Por	-0.44554
Por	585.354	Perm*ACRE	-9.8006E-06
Pres	-0.00895164	Por*Tres	-0.9363
Specific_Gravity	25.4938	Por*ACRE	0.00639336
Thickness	0.243962	Por*DURATION	-6.71426
Tres	-0.0797618	Pres*ACRE	7.90001E-08
ACRE	0.00627766	CO2Percentage*DURATION	3.23769
DURATION	-1.83037	Fracpres*Perm	7.60457E-05

4.2. Eureqa Equation

Below is the generated equation from Eureqa for storage capacity:

$$SC' = 1.55 * 10^8 * k * ø + \frac{346*H*Por^3*A^2}{k} - [ø * A * H * (29.2 * Pres - 27.8 * FP)] \text{.........}(5)$$

$$Storage\ Capacity\ (lb) = 6.46 * 10^9 + \frac{(1.49*SC'*k)-(26.4*SC')}{k+(0.401*Pres*ø^2)} - (5.97 * 10^7 * k) \text{..............}(6)$$

The calculation is used two step equation, where the equation (5) is the first step equation and is used as the coefficient in the second step equation (6).

The generated equation from Eureqa for allowable injection time is shown in equation (7) shown below:

$$Maximum\ Injection\ Time\ (month) = \frac{(8*10^5)+171*ø*A-13.4*k*Tres}{9.46*Tres+[FP*(1+0.0145*k])} \text{.......................................}(7)$$

4.3. Case Study Result

Both of the generated equation from CMG-CMOST and Eureqa are used to calculate the verification data generated in CMG-CMOST. The result of this comparison is shown in Table 4. It is shown that the generated equation from Eureqa have a higher R-Square and lower error. Therefore, the Eureqa-generated equation is used for case study.

Table 4 Result Comparison between Eureqa and CMG-CMOST

	Tool: Eureqa		Tool: CMG-CMOST	
	Storage Capacity	**Allowable Injection Time**	**Storage Capacity**	**Allowable Injection Time**
Maximum Error (%)	51.27	53.2	53.5	65.72
Minimum Error (%)	5.35	5.68	7.48	5.79
Average Error (%)	9.88	10.27	10.07	10.73
R-Square (%)	93.23	92.8	89.21	89.77

The generated equation from Eureqa was used to calculate CCS Field in East Natuna, Snohvit, and Southeast Queensland as the case study. The storage capacity result is shown in Table 5 and allowable injection time is shown in Table 6.

Table 5 Case Study Storage Capacity Calculation Result

	East Natuna Field			Snohvit Field			Southeast Queensland Field		
Lyr.	**Predictive (mmTon)**	**Conv. (mmTon)**	**Diff (%)**	**Predictive (mmTon)**	**Conv. (mmTon)**	**Diff (%)**	**Predictive (mmTon)**	**Conv. (mmTon)**	**Diff (%)**
1	6.96	7.85	11.3	0.69	0.70	1.2	0.27	0.35	21.9
2	63.19	78.14	19.1	0.88	1.03	14.8	0.49	0.50	2.8
3	72.06	80.67	10.7	1.07	1.09	1.9			
4	73.57	82.34	10.7						
5	74.31	83.77	11.3						

Table 6 Case Study Allowable Injection Time Calculation Result

Layer	East Natuna Field (Years)	Snohvit Field (Years)	Southeast Queensland Field (Years)
1	18.78	3.44	4.14
2	17.25	4.02	5.77
3	16.95	4.88	
4	16.58		
5	16.39		

5. Conclusion

The result of this study shows that the equation generated by Eureqa software for both storage capacity and allowable injection time has a more accurate result if compared with the conventional equation. The utilization of this equation is simpler if compared with the CMG-CMOST generated equation. The R-Square of this equation reaches 96.6% and 95.4% for storage capacity and allowable injection time. The final equation for both equations only considers seven parameters, permeability, porosity, fracture pressure, reservoir pressure, reservoir temperature, thickness, and area.

The result from calculating the three CCS fields shows that the generated equation from Eureqa could give a difference ranging from 1-21% less than the current result due to the additional restriction and consideration. For the allowable injection time, the result shows reduction for up to 7 years.

References

[1] Smil V 2017 *Energy and Civilization* (Massachusetts: MIT Press)

[2] EPA U 2015 *2014 Annual Performance Report* (Washington D.C.: US Environmental Protection Agency)

[3] Bachu S 2015 Review of CO2 Storage Efficiency in Deep Saline Aquifers *International Journal of Greenhouse Gas Control* 188-202.

[4] Szulczewski M 2009 *Storage Capacity and Injection Rate Estimates for CO2 Sequestration in Deep Saline Aquifers in the Conterminous United States* (Massachusetts: Massachusetts Institute of Technology, Libraries Archives)

[5] Ajayi T, Gomes J S and Bera A 2019 A Review of CO2 Storage in Geological Formations Emphasizing Modeling, Monitoring and Capacity Estimation Approaches *Petroleum Science*, 1028-1063

[6] Cherdasa J R 2018 Formation Evaluation and Contingent Storage Capacity Estimation for Cabrbon Capture Storage and Utilization: A Case Study from East Natuna *Modern Applied Science* Vol. **12** No 4

[7] Chadwick R 2004 Geological Reservoir Characterization of a CO2 Storage Site: The Utsira S and, Sleipner, Northern North Sea *Energy Journal* vol. **29** no. 9-10 1371–1381.

[8] Buscheck T 2016 Managing Geologic CO2 Storage with Pre-Injection Brine Production: a Strategy Evaluated with a Model of CO2 Injection at Snøhvit *Energy & Environmental Science* vol **9** no 4 1504–1512.

[9] Rutqvist J 2010 Coupled Reservoir-Geomechanical Analysis of CO2 Injection and Ground Deformations at In Salah, Algeria *International Journal of Greenhouse Gas Control* vol **4** no 2 225-230.

[10] Sapiie B 2015 Geological Characteristic and Fault Stability of the Gundih CCS Pilot Project at Central Java, Indonesia *12th SEGJ International Symposium* Tokyo

[11] Balch R and Mcpherson B 2016 Integrating Enhanced Oil Recovery and Carbon Capture and Storage Projects: A Case Study at Farnsworth Field, Texas *SPE Western Regional Meeting* pp 23-26

[12] Cinar Y 2007 Geo-Engineering and Economic Assessment of a Potential Carbon Capture and Storage Site in Southeast Queensland *SPE Asia Pacific Oil and Gas Conference and Exhibition* Australia
[13] Monne J 2012 The Lacq CCS Pilot, a First *International Conference on Health, Safety and Environment in Oil and Gas Exploration and Production.*
[14] Rizzato P 2019 Dynamic Modeling of a High Temperature CO2-Rich Giant Gas Field with a Carbon Capture and Storage Strategy *SPE Reservoir Characterisation and Simulation Conference and Exhibition*
[15] Mishra G and Kumar 2019 Planning India's First CO2-EOR Project as Carbon Capture Utilization Storage: A Step Towards Sustainable Growth *SPE Oil and Gas India Conference and Exhibition.* India

IOP Conf. Series: Earth and Environmental Science 1031 (2022) 012019 doi:10.1088/1755-1315/1031/1/012019

Identification of Carbonate Reservoir Prospective Zones Using Rock Physics Approach and Extended Elastic Impedance (EEI) in "BAP" Field, South Sumatera Basin

B A Prayoga[1], S Winardhi[2]
[1]Geophysical Engineering Master Program, Faculty of Mining and Petroleum Engineering, Bandung Institute of Technology, Ganesa Number 10, Bandung, 40132
[2]Reservoir Geophysics Research Group, Faculty of Mining and Petroleum Engineering, Bandung Institute of Technology, Ganesa Number 10, Bandung, 40132

Email: bimaamarta04@gmail.com

Abstract. One type of reservoir that has a large enough potential is the carbonate reservoir. Carbonate reservoir has heterogeneity in relation to pore shape, so it is necessary to apply a specific rock physics approach and seismic inversion which takes into account the elastic parameters of the rock in order to know the prospect zone of the carbonate reservoir. The rock physics approach taken is to estimate the modulus of the matrix and the estimation of the aspect ratio shows that the prospect carbonate has an aspect ratio range of 0.10 - 0.30 dominated by the pore shape of the Stiff Pore. Seismic inversion shows using extended elastic impedance shows carbonate porous (Mu-Rho < 33 GPa*g/cc) and saturated hydrocarbons (Lambda-Rho < 40 GPa*g/cc) are in the middle area (around the well) where the lateral spread is getting thinner with a thickness of up to 10 ms. The prospect zone analysis from rock physics modeling and extended elastic impedance shows that reef Baturaja Formation carbonate is a zone that has a good reservoir quality dominated by a pore shape that develops is stiff pore.

1. Introduction

The identification of prospect areas in a carbonate reservoir is not very easy because the carbonate reservoir has heterogeneous properties related to various pore shapes (Mavko et al., 2009). Therefore, it is necessary to do rock physics modeling with these various approaches so that the modeling of the elastic properties of the rock in the P-wave and S-wave data closer to the actual geological conditions and used as a reference in identifying the heterogeneity of carbonate rocks (Dvorkin, 2007). An interpretation stage is carried out on the seismic data using seismic inversion model-based by carrying out the process in extended elastic impedance order to strengthen the results of the identification of the carbonate reservoir distribution pattern in the "BAP" field. In the case of the "BAP" field. The Telisa Group is located in the Baturaja Formation, which is a formation formed from the transgression phase where the composition of the Baturaja Formation consists of Limestone (Limestone) or platforms and reef (De Coster, 1974). The thickness of the lower part of these formations varies, but averages 200-250 feet (about 60-75 m). With the age range in this formation it is equivalent to the fauna (foraminifera planktonic) in the N5-N6 (Early Miocene) age range formation as shown in Figure 1.

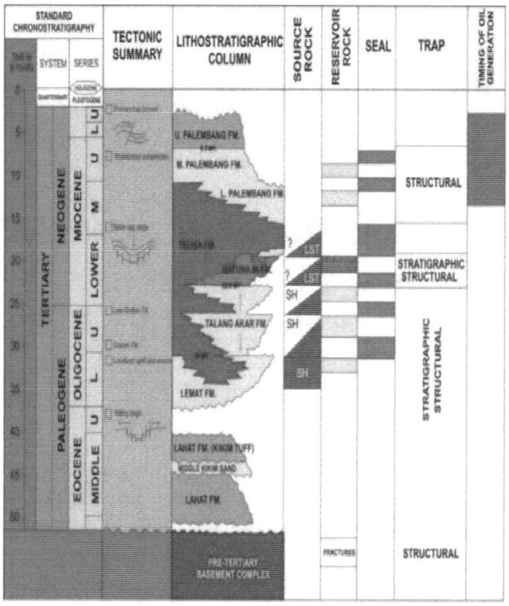

Figure 1. Stratigraphy of the South Sumatera Basin

2. Research Methodology

2.1. Data

In the field research "BAP" the identification process was carried out on the "BAP" well. The field consists of wells BAP-1, BAP-2, BAP-3, and BAP-4. The process of identifying carbonate reservoirs is carried out in two stages, the first stage is identification with the rock physics approach, which requires sufficient data, namely the BAP-1 well. Then at the identification stage by inversion is carried out on wells BAP-2, BAP-3, and BAP-4 which are covered by seismic data. The seismic data used in the "BAP" field are three-dimensional (3D) seismic data partial angle-stack which is limited by in-line 1100-1500 and x-line 1001-2000. With information partial stack with angles near (4-140), mid (14-220), far (22-380). Seismic data in the "BAP" field has 3 horizons, namely Baturaja, Pendopo, and Basement horizon. Research area in the field "BAP" is shown in Figure 2.

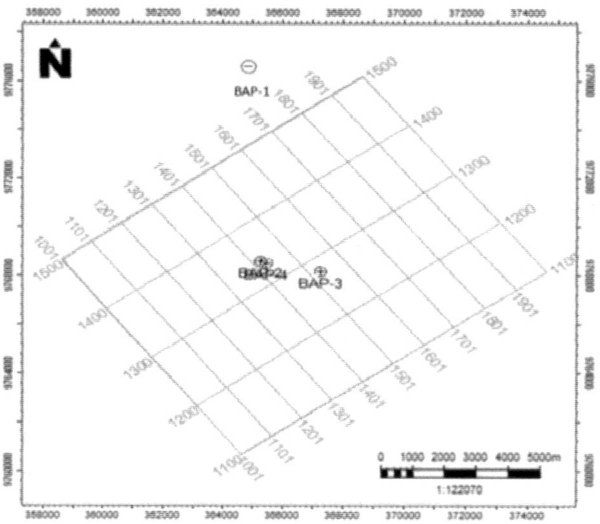

Figure 2. Basemap of research area

IOP Conf. Series: Earth and Environmental Science 1031 (2022) 012019 doi:10.1088/1755-1315/1031/1/012019

2.2. *Method*

Data processing carried out in the "BAP" field uses two approaches, namely the approach with rock physics modeling in the BAP-1 well which aims to determine the heterogeneity of carbonate rocks to suit the actual geological conditions. Then the approach with seismic inversion aims to identify the validated prospective zone of the carbonate reservoir from the modeling results. The data processing flow is shown in Figure 3.

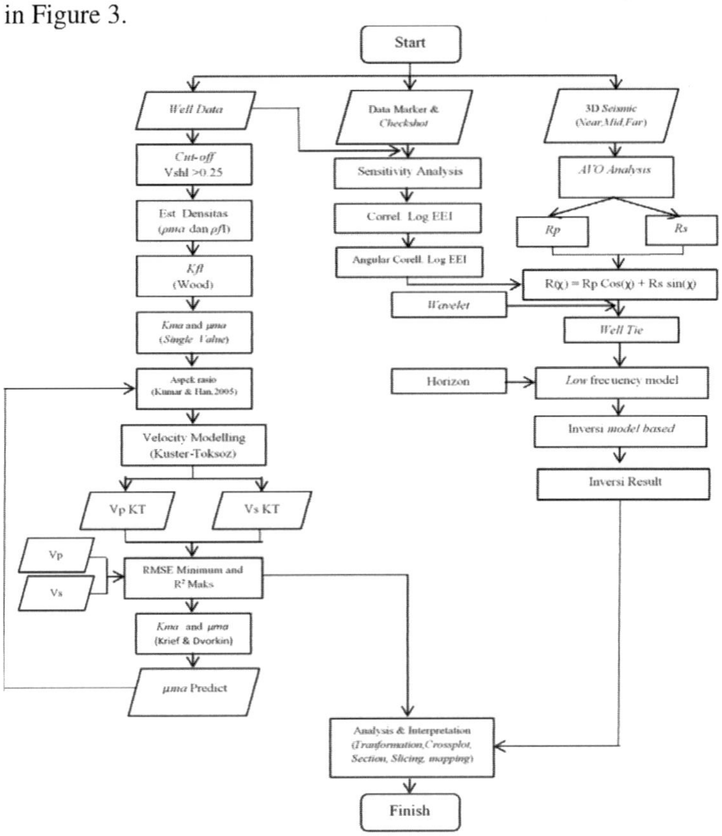

Figure 3. Research Flowchart

2.2.1. Cut Off V-shale more than 0.25. The data is sorted with a cut off V-shale > 0.25. This is done so that this modeling is actually carried out on carbonate rocks and a void the presence of claystone lithology inserts in it.

2.2.2. . Density Estimation. Determination of density matrix and fluid density obtained from linear line equations is very important to determine the elastic modulus of solid rock. The equation y = -1.8448x + 2.7044 is obtained where y is the value bulk density and x is the porosity data. Then the matrix density value obtained is 2.7044 *g/cc* in Figure 4.

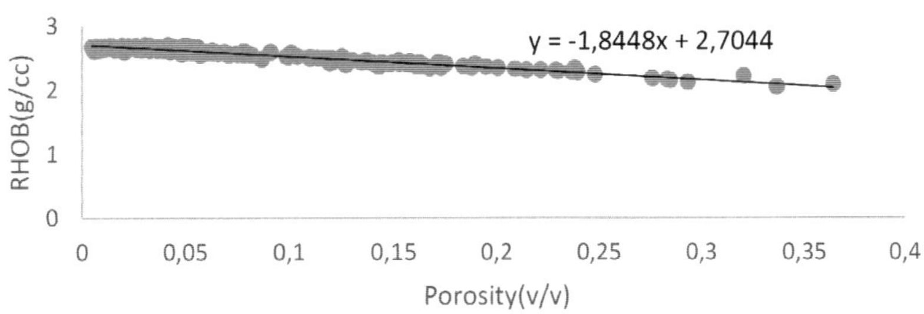

Figure 4. Density Estimation

IOP Conf. Series: Earth and Environmental Science 1031 (2022) 012019 doi:10.1088/1755-1315/1031/1/012019

2.2.3. Estimation of Modulus of Elastic Fluid. Estimation is performed using the Wood equation (1955). In the absence of pressure, temperature, and salinity data, the assumptions used in determining the value of fluid elastic parameters are K-brine = 2.2 *GPa*, K-oil = 1.1 *GPa*, ρ-brine = 1 *g/cc*, and ρ-oil = 0.8 *g/cc*.

2.2.4. Estimation of Elastic Modulus Solid Rock. Absence of XRD data and the elastic modulus of solid rock for each depth, an estimate of the elastic modulus of carried out solid rock is using the approach single value and Krief (1990) and Dvorkin (2007). The results of the validated equations are used to estimate the modulus value for shear solid rock at each depth point and furthermore, it can estimate the aspect ratio value of each depth point.

2.2.5. Determination of Aspect Ratio Value. Determination of the aspect ratio value is obtained based on the experimental modification of Kumar and Han (2005). This experiment aims to obtain the aspect ratio value of each depth point and can be used to identify the heterogeneity of carbonate rocks in the target zone. The model for determining the aspect ratio value based on the upper limit (Hashin Shtrikman Upper Bound), middle limit (Wyllie), and lower limit (Hashin Shtrikman Lower Bound) is shown in Figure 5.

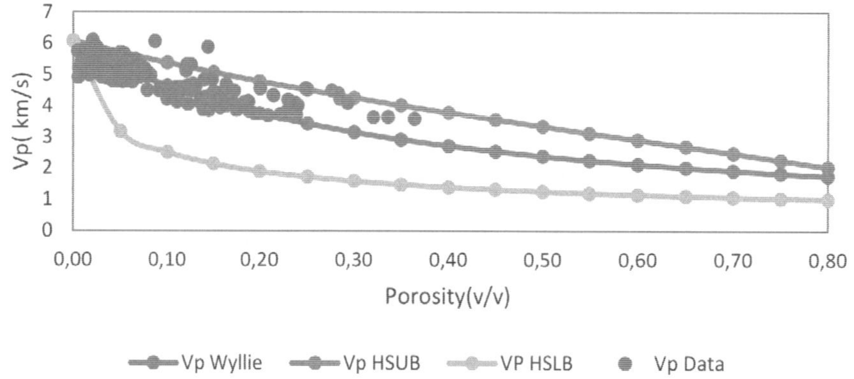

Figure 5. The relationship between velocity-p and the pore shape of the rock

2.2.6. Kuster-Tokoz Velocity Validation. After obtaining the value aspect ratio based on the experimental modification Kumar and Han (2005). The next step is to validate by modeling the P and S velocities of saturated rock using the Kuster-Toksoz model. The results of the validation of the maximum correlation and error were minimum used in the analysis stage of the pore properties of the carbonate rock. Correlation and results were obtained error in the initial experiment using the elastic modulus of solid rock single value and the final experiment using the modulus shear solid rock model Krief (1990) is shown in Table 1.

Table 1. Error and correlation coefficient

No	Zone	Information	Initial Experiment		Final Experiment	
			RMSE	R^2	RMSE	R^2
1	Stiff	Vp	0.0492	0.9930	0.0264	0.9938
		Vs	0.1180	0.8974	0.0938	0.8938
2	Soft	Vp	0.0689	0.9508	0.0512	0.9499
		Vs	0.0749	0.9050	0.0563	0.8958
3	All Zone	Vp	-	-	0.0424	0.9694
		Vs	-	-	0.0747	0.8945

2.2.7. Analysis Tunning Thickness. In the seismic data processing, process analysis is necessary tuning thickness to determine the minimum thickness of a layer so that it can be resolved by seismic data. The obtained average P-wave velocity (Vp) around the target formation is 4872,091 m/s. With a dominant seismic frequency of 18 Hz. Thus, with a value tuning thickness around the target area of 67,667 m while the thickness of the target reservoir is around 77.75 m and above the value tuning thickness, the seismic data can resolve the target reservoir.

2.2.8. Logging Derivative. In this process, the production of derived logs is obtained from existing well data. Such as Vp, Vs, and density data. Logs derivatives such as P-Impedance, S-Impedance, Lambda-rho, Mu-Rho, Lambda/Mu, Vp/Vs. In this case, the log data that can be processed with the derivative log is the well data "BAP-4". Meanwhile, for data from wells "BAP-2" and "BAP-3", the log derivatives were carried out to obtain the P-impedance value which was used as inversion validation for the well "BAP-4".

2.2.9. Sensitivity Analysis. Sensitivity analysis is carried out to find elastic parameters that are sensitive to the presence of changes in lithology and fluid in the target area carried out by the inversion process.

2.2.10. Extended Elastic Impedance. This stage aims to determine the volume of extracted seismic reflectivity based on the reflectivity of the P wave and the reflectivity of the S wave which represents certain elastic parameters according to the correlation angle χ with the elastic parameters of the well data carried out by the EEI process.

2.2.11. Well Seismic Tie. This process aims to tie seismic data and well data. Considering the difference in the data domain between time domain seismic data (domain-two-way time) and depth-well data.

2.2.12. Seismic Inversion. An inversion process is carried out using a model based, which begins with the process of creating a model at low frequency controlled by well data. Furthermore, pre-inversion analysis is carried out in order to determine the inversion parameter which gives the optimal inversion result. The optimal inversion results can be identified from the inversion results that approach log data and seismic data with a value error minimum. The inversion process stage is carried out for each EEI reflectivity which corresponds to a certain elastic parameter.

3. Result and Discussion

3.1. Target Zone Analysis

The target zone in the field well "BAP" is in the Baturaja formation. Where this target zone is identified by lithology of carbonate types saturated by hydrocarbons. From a reading volume-shale low, the result of cross-over between low neutron porosity and density with a low change in value water saturation.

3.2 Prospect Zone Analysis

For the hydrocarbon prospect zone analysis based on petro-physical parameter data and the results of rock physics modeling. Figure 6 shows the prospect zone analysis carried out on BAP-1 well which is divided into 3 zones, namely zone A, zone B, and Zone C. So that the three zones are in one Baturaja Formation, for zone A and zone B shows the existence of a good type of reservoir quality as indicated by the high values of porosity and hydrocarbon saturation. And if it is compared with the results of the related rock physics model, the result of the aspect ratio value which is relatively large compared to the elastic parameter shows that the best reservoir quality is in zone B because it has a developed rock pore type, namely types vuggy and mouldic porosity (stiff pore). This pore type produces small Lambda-Rho and K/μ dry values when porosity > 0.20 and large Lambda-Rho and K/μ dry when porosity < 0.20. In accordance with the geological review of the Baturaja Formation, it is developing due to the existence of a platform which is a relatively zone tight with low porosity and hydrocarbon saturation values. in accordance with the results of the modelling shown by zone C is dominated by rock types soft pore. And the part of reef the Baturaja Formation which is a relatively zone porous indicated by the porosity and high hydrocarbon saturation values in Zone A and Zone B are zones that have good reservoir quality with pore types that develop in the zone dominated by types reference pore + stiff pore.

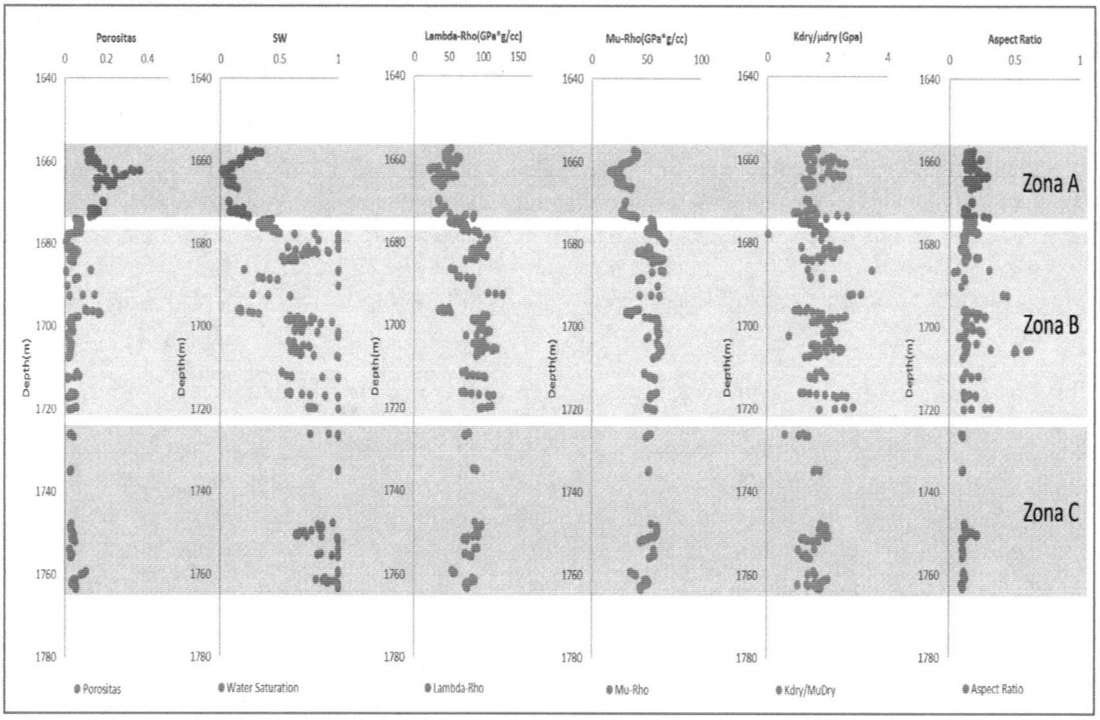

Figure 6. Prospect zone based on log data and aspect ratio value.

3.3 *Sensitivity Analysis*

From the results of the cross-plot preliminary on the well "BAP-1", it was found that the target formation area between baturaja and pendopo horizon found that the majority of rock lithology types were carbonate rock types. And in determining the cut-off between the carbonate porous and tight the value was Mu-Rho cut-off 33 *GPa* g/cc*. Where the carbonate with high porosity is below Mu-Rho 33 *GPa*g/cc* while the carbonate with low porosity is above Mu-Rho 33 *GPa*g/cc*. Then by using the fluid sensitive parameter, namely lambda-rho, from the results of the analysis cross-plot, the cut-off value for carbonate saturated with hydrocarbons or low SW is at a value Lambda-Rho small of 40 *GPa*g/cc*. Meanwhile, the carbonate that is saturated with water or high SW is at a Lambda-Rho value of 40 *GPa*g/cc* which is shown in Figure 7.

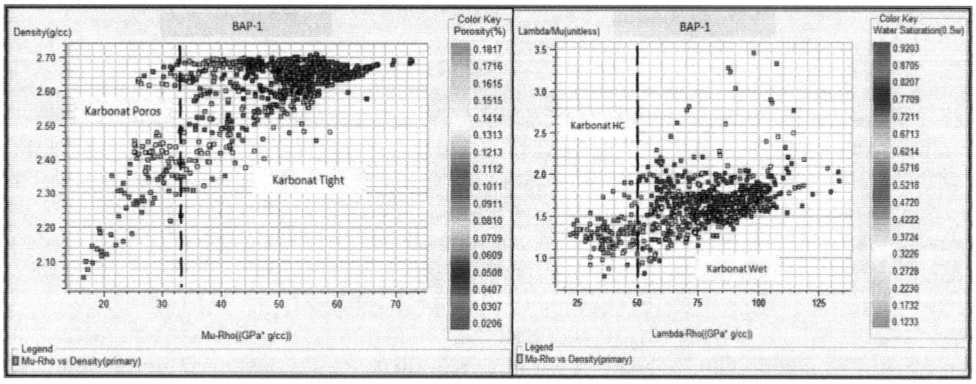

Figure 7. Sensitivity Analysis Baturaja Formation

IOP Conf. Series: Earth and Environmental Science 1031 (2022) 012019 doi:10.1088/1755-1315/1031/1/012019

3.4 Extended Elastic Impedance

The optimal correlation on sensitive log parameters is lambda-rho corresponds to log EEI (-30^0) and mu-rho corresponds to log EEI (71^0). The result of reflectivity Mu-Rho on the horizon of the Baturaja horizon is dominated by a positive color (red), this is related to changes in lithology. The results of EEI analysis and changes in reflectivity, which lambda-rho are dominated by negative colors (blue) in the Baturaja horizon, are associated with changes in fluid content as shown in Figure 8 and Figure 9.

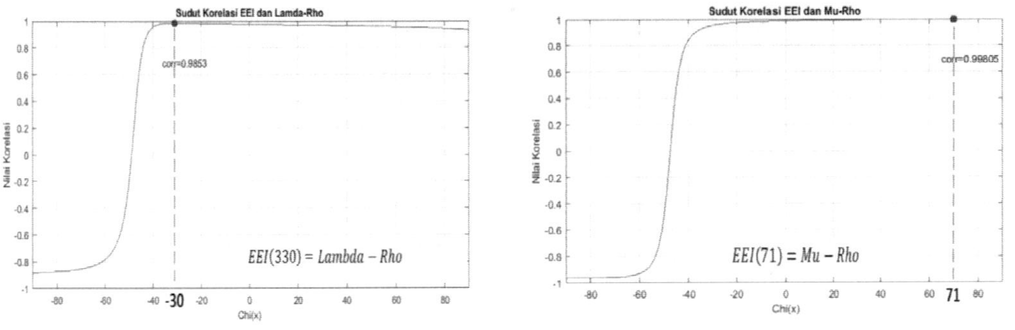

Figure 8. Correlation analysis and error EEI log on rock elastic parameters

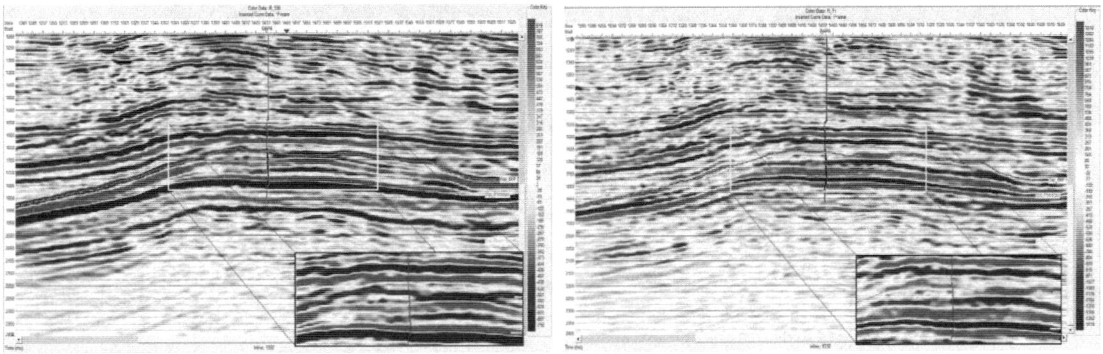

(a) (b)

Figure 9. (a) EEI reflectance volume (330^0), (b) EEI reflectance volume (71^0)

3.5 Inversion Results.

In the initial process of inversion, well seismic tie was carried out. Parameters that were carried out by well seismic tie were sensitive parameters in the reservoir identification process, namely Lambda-Rho and Mu-Rho. The time-depth relationship is obtained in the correlation process around the target formation so that it can be matched between the well data and the seismic data in Table 2.

Table 2. Well Tie parameters AI, Lambda-Rho, and Mu-Rho well "BAP-4"

No	Sensitivity Parameter	Wavelet	Correlation
1	Lambda-Rho	Ricker 16 Hz -15^0	0.680
2	Mu-Rho	Ricker 16 Hz +15^0	0.762

The results of the inversion Mu-Rho of the Baturaja formation are in the yellow-red color zone with a range of values in Mu-Rho between the values < 33 *GPa*g/cc* around the horizon of the Baturaja formation which is estimated to be carbonate lithology with high porosity and Lambda-Rho < 40 *GPa*g/cc* is a type of carbonate saturated with hydrocarbons as shown in Figure 10. Therefore, from the inversion analysis in the "BAP" field around the horizon of the Baturaja formation, it is a type of carbonate rock. reef build-up relatively porous saturated with hydrocarbons. It is also in line with the geological review that the identification of carbonates in the baturaja formation is widely distributed, consisting of platform relatively tight carbonates. Carbonates build-up and and reef relatively porous under the Baturaja horizon.

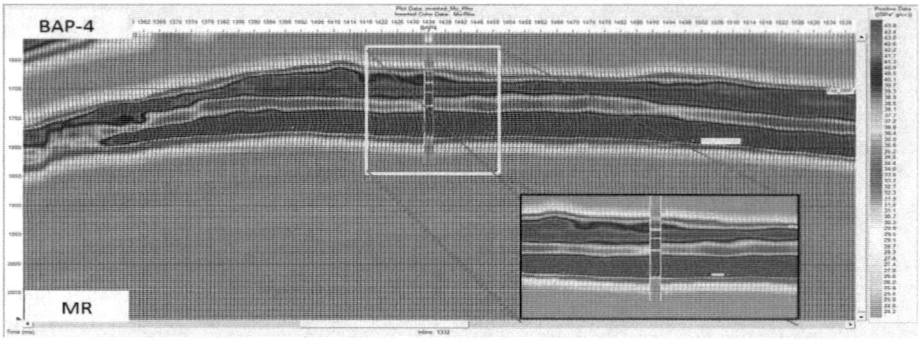

(a)

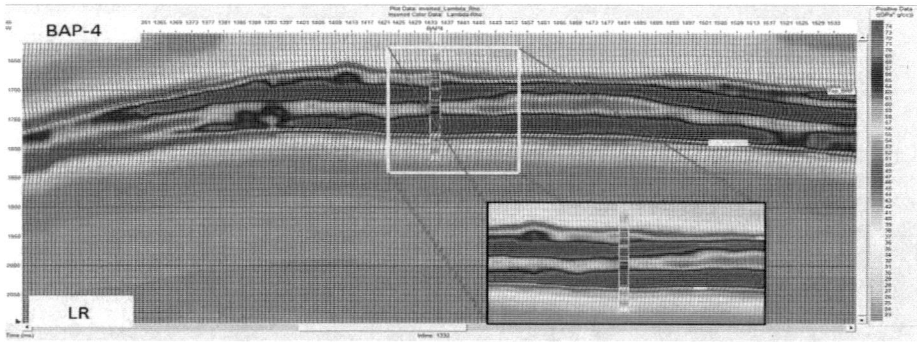

(b)

Figure 10. Inversion results (a) *Mu-rho*, (b) *Lambda-Rho*

3.6 Identification of Reservoir Prospective

From the results of slicing every 5 ms down on the horizon of the Baturaja Formation, it was identified that the lateral distribution of porous carbonate in the Baturaja Formation has a thickness of up to 10 ms (down) with a Mu-Rho value < 33 *GPa*g/cc* and saturated carbonate distribution The hydrocarbons laterally in the Baturaja Formation also have a thickness of up to 10 ms (down) with a Lambda-Rho value < 40 *GPa*g/cc* is shown in Figure 11. Where the more the porous and saturated carbonate expanse of hydrocarbons the thinner the orientation of the northwest-southeast area research. The thickening of the carbonate in the middle (the area around the well) occurs because the carbonate grows in the marine environment which has been exposed and dissolves, which is a characteristic of vuggy and mouldic porosity (stiff pore) carbonate growth. Therefore, the carbonate in the reef of the Baturaja formation has the potential to be of good reservoir quality by having high porosity and hydrocarbon saturated as evidenced by the results of rock physics modeling and EEI inversion which takes into account the elastic parameters of the rock.

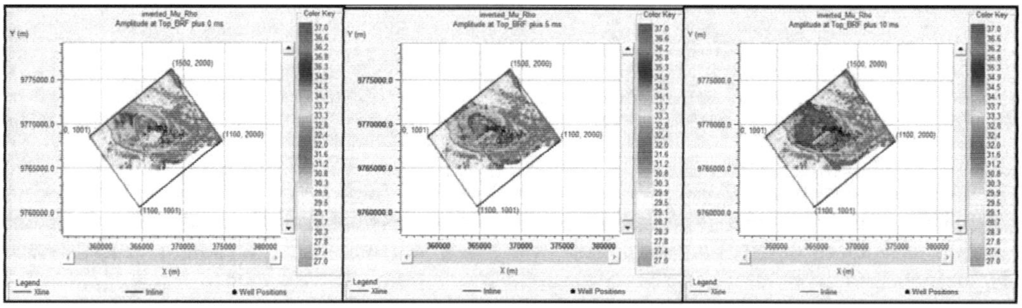

(a)

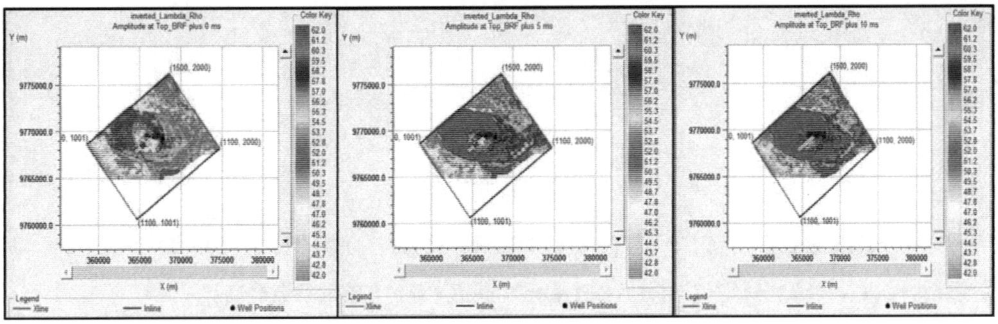

(b)

Figure 11. Slicing horizon volume (a) Mu-Rho, (b) Lambda-Rho every 5 ms and down

4. Conclusion

According to the discussion result above, can be concluded that:

1. The identification of the quality of the carbonate reservoir in the field "BAP" from the results of rock physics modeling and EEI inversion obtained a spread of porous carbonate (Mu-Rho < 33 *GPa*g/cc*) and accumulated HC (Lambda-Rho < 40 *GPa*g/cc*) with a dominant value aspect ratio 0.10 - 0.30 (reference + stiff pore) is found on the reef part of the Baturaja formation.
2. Where in the reef part of the Baturaja formation there is a thickening of the carbonate prospects in the middle (the area around the well), the more the carbonate spread is getting thinner.
3. This shows that the carbonate grew in a marine environment that was exposed and dissolved, which is a characteristic of vuggy and mouldic porosity (stiff pore) carbonate growth which is oriented towards the northwest-southeast of the study area.

5. Recommendation

1. Pore type analysis is required in all research wells to ensure a better quality reservoir distribution (reference + stiff pore) in the spatial prospect zone.
2. It is necessary to develop research by finding the value of effective porosity in the target zone to be able to as certain the development area or location of other prospect wells that can be produced.

References

[1] De Coster G L 1974 The Geology of the central and south sumatera basin *Proceeding's 3rd annual convention IPA June 1974 Jakarta*

[2] Dvorkin, J P 2007 Yet another Vs equation *Geophysics* Vol **73** No 2 E35-E39

[3] Krief M, Garat J, Stellingwerff J and Ventre J 1990 A petrophysical interpretation using the velocities of P and S waves (full-waveform sonic) *Log Analyst* Vol **31** 355–369

[4] Kumar M and D Han 2005 Pore shape effect on elastic properties of carbonate rocks *SEG/Houston 2005 Annual Meeting* page 1477-1480

[5] Kuster G T and Toksoz M N 1974 Velocity and attenuation of seismic waves in twophase media, Part I: theoretical formulations *Geophysics* **39**(5) 587– 606

[6] Mavko G, Mukerji T and Dvorkin J 2009 *The Rock Physics Handbook* (New York: Cambridge University Press)

[7] Thomas M, Ball V, Blangy J P and Tenorio L 2016 Rock-physics relationships between inverted elastic reflectivities *The Leading Edge* **35**(5) 438-444

[8] Ursenbach C P and Stewart R R 2008 Two-term AVO Inversion: Equivalences and New Methods *Geophysics* **73** no 6 C31–C38

[9] Whitcombe D N, Connolly P A, Reagan R L and Redshaw T C 2002 Extended elastic Impedance for fluid and lithology prediction Geophysics **67** 63–67

[10] Wyllie M R J, Gardner G H F and Gregory A R 1963 Studies of elastic wave attenuation in porous media *Geophys.* **27** 569-589

IOP Conf. Series: Earth and Environmental Science 1031 (2022) 012020 doi:10.1088/1755-1315/1031/1/012020

Study of terrain correction computation using square pattern and sloped triangle methods in Karangsambung

R A Salam[1], **E J Wahyudi**[1], **and Setianingsih**[1]

[1] Department of Geophysical Engineering, Bandung Institute of Technology, Bandung, West Java, Indonesia.

rafifabdussalam2@gmail.com

Abstract. Conventional assessments of terrain correction are carried out by laying transparent paper containing the Hammer chart on topographic maps, then estimating the elevation for each of Hammer chart's compartment. This procedure has disadvantages, the number of compartments is too small, and there is a subjectivity from the observer in estimating the compartment's height, especially in an area with varied topography, such as Karangsambung. These research aims are to study the effect of the topography around Karangsambung on terrain correction value and to overcome problems in conventional terrain correction estimation. Estimation of terrain correction carried out using square pattern and sloped triangle method, where the area around measurement points divided into square and triangle compartment. Based on the results, South Seraju Ranges, with an altitude of 1000 m at a distance of 20 – 30 km from Karangsambung gives an effect of 0.05 mGal on terrain correction, while the Quaternary Volcano with height of 3000 m at range of 30 – 40 km gives the effect of 0.1 mGal. The results of applying the program at the gravity data show that the use of the square pattern and sloped triangle method can correct errors from conventional estimation up to 3 mGal.

1. Introduction

Conventional assessment of terrain correction performed by spreading transparent paper containing a Hammer chart on a topographic map, then the average height of each compartment is estimated. This procedure has weaknesses, namely the relatively small number of compartments and the observers is subjective in determining the average height of compartments, especially in an area with varied topography [1]. A practical example of this method is applied at the Geophysical Engineering ITB field study in Karangsambung. The division of zones on the Hammer chart is simplified into four concentric ring compartments with an inner radius of 2 meters and an outer radius of 100 meters. The height difference of the compartment to the height of the station is estimated visually in the field.

Terrain correction estimation method is continuously developing. In 1962, Kane presented a study of terrain correction calculation using computer, by dividing the area around the station into grids [2]. Nagy (1966) developed the application of the prism rock body approach for terrain correction [3]. The study of elevation satellite data for calculation of terrain correction [4]. Whereas the authors' aims in this research are to see the effect of the topography around Karangsambung on terrain correction value and to overcome problems in conventional terrain correction estimation. In this study, the topography data was taken from DEM (Digital Elevation Model), and the calculation of terrain correction was performed using Matlab software.

2. Data and methodology

2.1. Data

The data used in this study are topographic and gravity data. Topographic data are taken from the 2012 SRTM (Shuttle Radar Topography Mission) elevation model, with a resolution of 3 arcseconds or around 90 m, covers 103 km x 103 km, with the center being the Karangsambung area. Meanwhile the gravity data are from the result of a field study of Geophysical Engineering ITB in 2018. The tool used was La Coste Romberg.

2.2. Methodology

In the calculation of terrain correction using the square pattern and sloped triangle methods, the area is divided into the near zone and intermediate zone, as shown in figure 1. The intermediate zone will be approached using a quadrilateral prism rock body published by Plouff [5], while the near zone approached using a sloped triangle rock body from Kane [2]. The illustration of the rock body approach can be seen in figures 2 and 3.

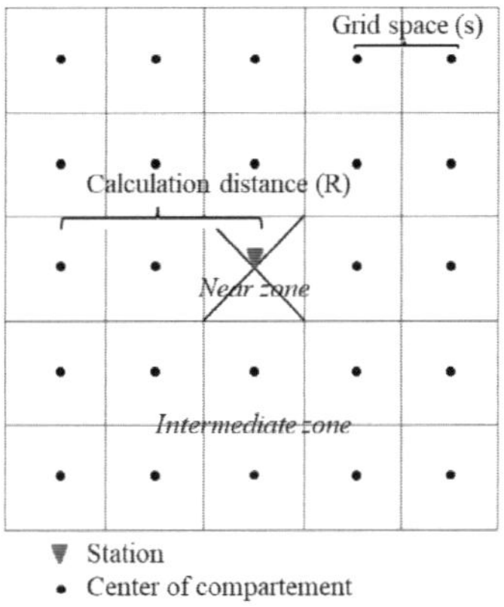

Figure 1. Division of the compartments.

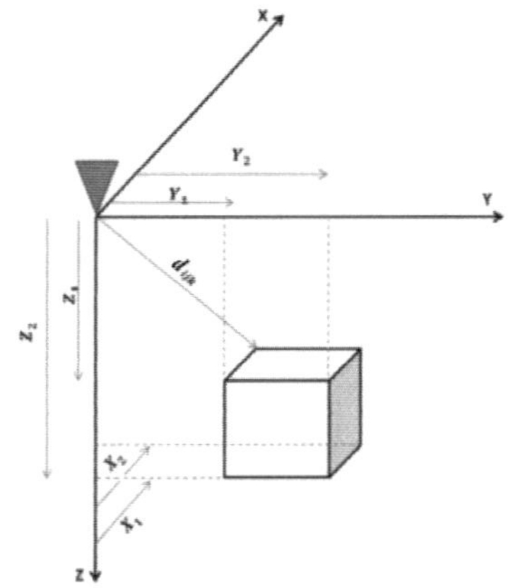

Figure 2. Quadrilateral prism rock body [5]

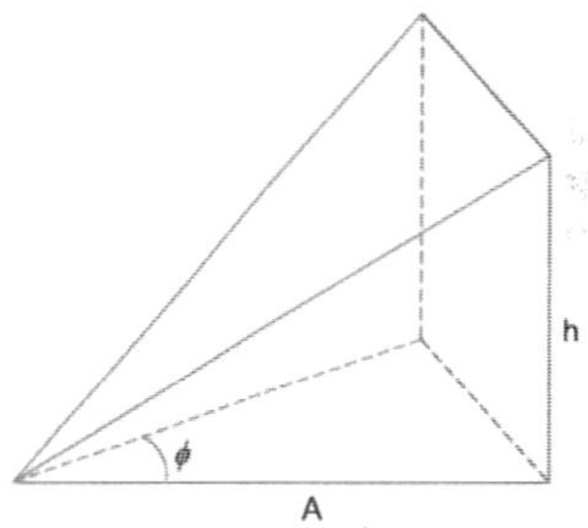

Figure 3. Sloped triangle rock body [2]

The formulation of the vertical direction gravity for the quadrilateral prism rock body approach is as follows [5]:

$$g = G\rho \sum_{i=1}^{2} \sum_{j=1}^{2} \sum_{k=1}^{2} \mu_{ijk} \left[z_k \arctan\frac{x_i y_j}{z_k d_{ijk}} - x_i \ln\left(d_{ijk} + y_j\right) - y_j \ln\left(d_{ijk} + x_i\right) \right]$$

$$d_{ijk} = \sqrt{x_i^2 + y_j^2 + z_k^2},$$

$$\mu_{ijk} = (-1)^i (-1)^j (-1)^k \tag{1}$$

while calculation of the sloped triangle rock body based on the results of Kane's research [2] is as follows:

$$g = G\rho\phi \left(A - \sqrt{A^2 + h^2} + \frac{h^2}{\sqrt{A^2 + h^2}} \right) \tag{2}$$

IOP Conf. Series: Earth and Environmental Science 1031 (2022) 012020 doi:10.1088/1755-1315/1031/1/012020

3. Result and discussion

3.1. Study of the topographic area around Karangsambung

The topographic map of Karangsambung and surrounding areas can be seen in figure 4. In Hammer chart, the radius of terrain correction calculation is about 21.9 km, in case the topography around the station is not varied too much [6]. In Karangsambung, there are South Serayu Mountains range and the Quaternary Volcano that have the potential to provide significant terrain correction values even though the distance is above 21.9 km. To find this out, six trial points were taken, namely points A, B, C, D, E, and F, which had elevation variations around 40-500 m. The illustration can be seen in figure 5.

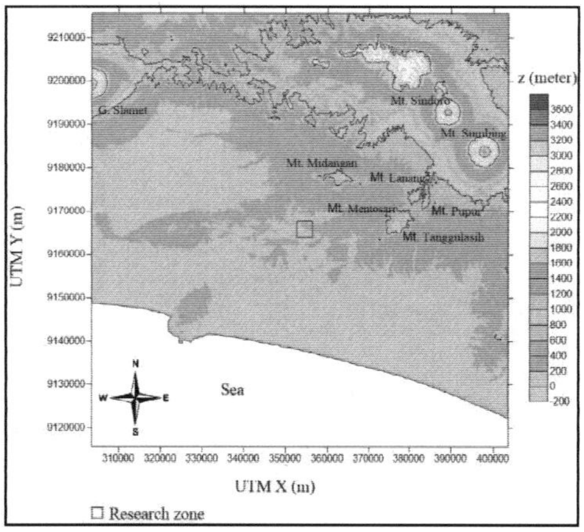

Figure 4. Topographic map of Karangsambung and the surrounding area.

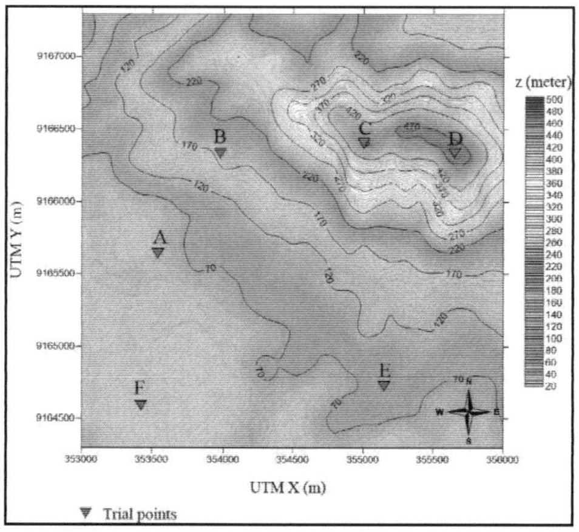

Figure 5. Distribution of trial points in the study area.

Each point's terrain correction is calculated with the variation of calculation distance (R), start from 0 to 50 km. The parameters used in the calculation are 100 m grid, and rock density 2400 kg m^{-3}. These value is taken from the clay rock density, which is the basic matrix of the Karangsambung. The results are shown in figure 6. The plot results show after a distance of 21.9 km, there is still a rise in the value of terrain correction. To see the increase more clearly, a plot composed of distance range and increased value with a spacing of 5 km was made. The results are presented in figure 7.

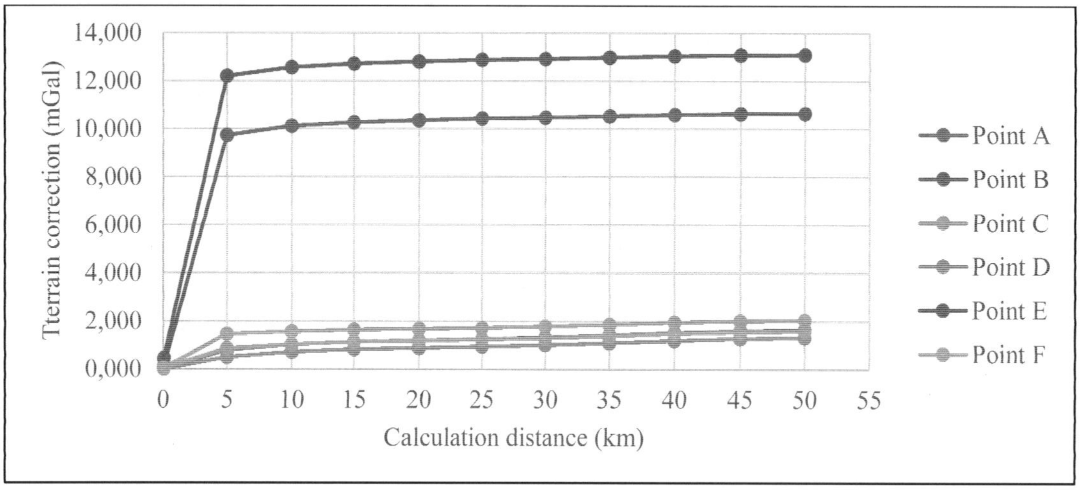

Figure 6. Terrain correction of trials points in variated calculation distance (R).

IOP Conf. Series: Earth and Environmental Science 1031 (2022) 012020 doi:10.1088/1755-1315/1031/1/012020

Based on the results, South Seraju Ranges, with an altitude of 1000 m at a distance of 20 – 30 km from Karangsambung gives an effect of 0.05 mGal on terrain correction, while the Quaternary Volcano with height of 3000 m at range of 30 – 40 km gives the effect of 0.1 mGal. These additions can be taken into consideration when determining the farthest distance in the calculation of terrain correction.

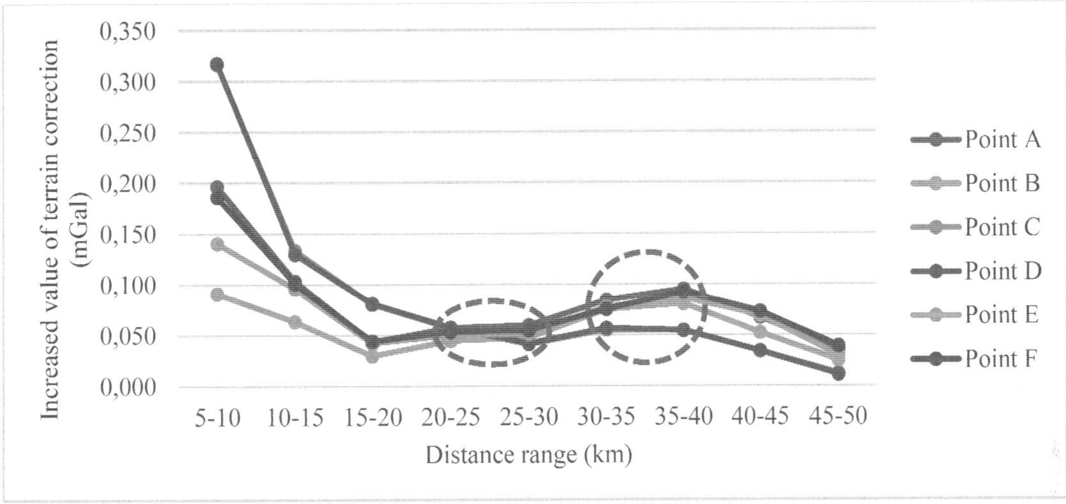

Figure 7. Increased value of terrain correction for every 5 km distance.

3.2. Comparison of the program with the conventional method

The program was applied to the 2018 Karangsambung Geophysical Engineering ITB field study data. Figure 8 shows the distribution of stations.

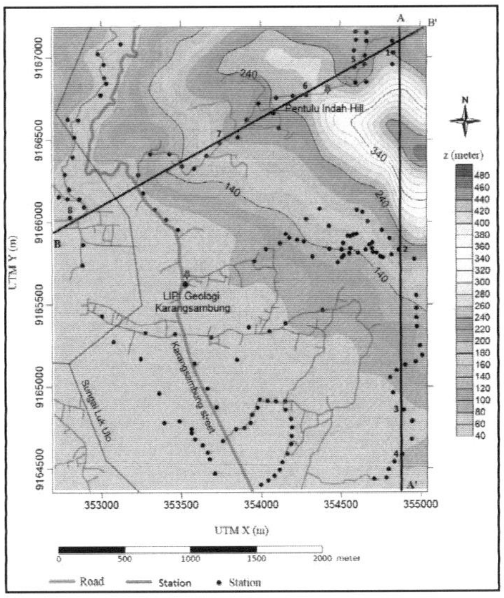

Figure 8. 175 data are collected within an area of 2.5 x 2.5 km, with an elevation variation of 40 – 270 m above mean sea level. The tool used for acquisition is La Coste Romberg. The topographic data used is derived from the SRTM (Shuttle Radar Topography Mission) elevation model released in 2012.

The terrain correction value is estimated using two methods, the first method is based on visual observations in the field using a simplified Hammer chart method (from now on referred as TC 1). In contrast, the second method estimates terrain correction based on the SRTM elevation model using the square pattern and sloped triangle methods (referred as TC 2). The difference between the two is analyzed (TC 2 - TC 1). The results obtained show that the value of TC 2 is relatively higher than TC 1.

IOP Conf. Series: Earth and Environmental Science 1031 (2022) 012020 doi:10.1088/1755-1315/1031/1/012020

From 175 data taken, there is only 1 data that have less value. The difference values vary from -0,518 to 3,176. The histogram of the data can be seen in figure 9. Based on this, calculations using this method can get the value of terrain correction better than the simplification of the Hammer chart method up to about 3 mGal.

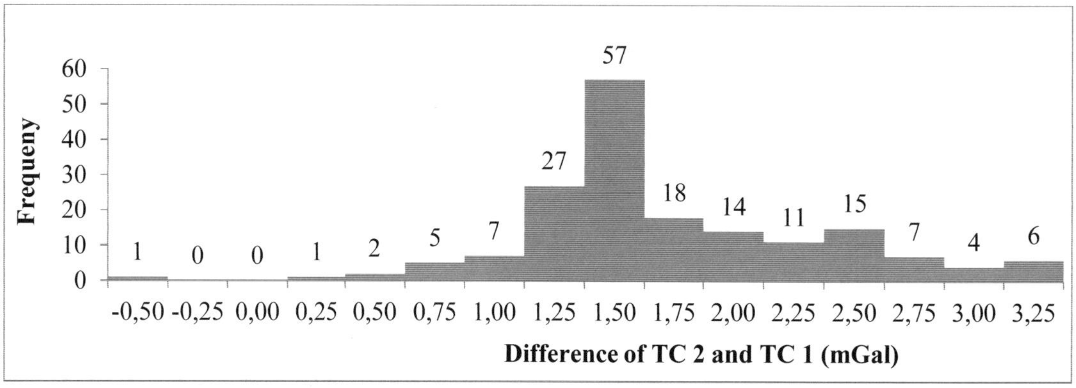

Figure 9. Histogram of TC 2 – TC 1 difference.

The results of terrain correction from the two methods are then processed to get the CBA map. The CBA maps are shown in figure 10. From the comparison, we can see there is continuity with the same orientation in two maps, with direction northeast-southwest and north–south. Whereas there is also a difference between the two maps, the value of CBA 2 is relatively greater than CBA 1, clearly visible in the northeast area. However, in modeling a structure, the value is likely not to have a significant effect on the result of the subsurface model.

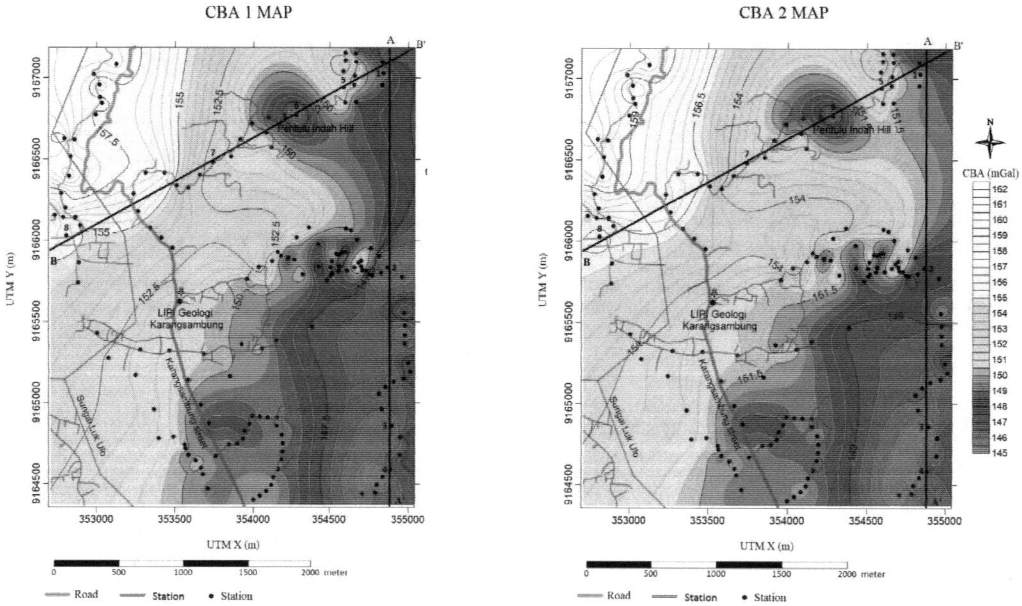

Figure 10. Comparison of calculated CBA maps between two method

Meanwhile, when viewed in cross-sections A-A' and B-B', it appears that the difference in CBA 1 and CBA 2 is relatively higher in high elevation areas, especially in the slope area. Cross-section A-A' and B-B' can be seen in figure 11. On the A-A' section, the projections of stations 1 and 2 on the slope area have a CBA difference of around 2-3 mGal. Meanwhile, station 3 and 4 projections, which are in relatively sloping areas, have a CBA difference of only about 1 mGal. The same thing is also found in the cross-section B-B'. In station projection 5 and 6, the difference in CBA is around 1.5 mGal, while in the projection of stations 7 and 8, the difference in CBA is only about 0.5 mGal. It happens because the conventional method relies on visual observations; the estimation of visible topographic height differences around the slope area will be far more complicated than the estimation on relatively flat areas.

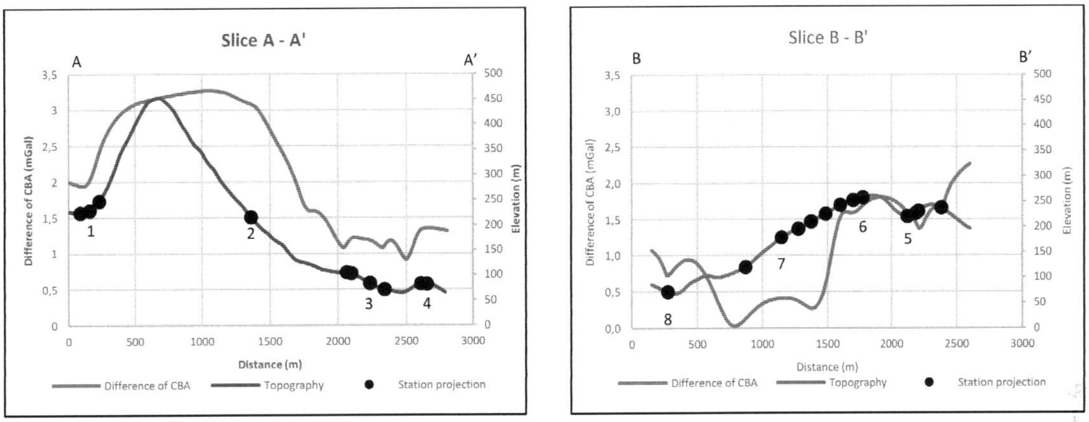

Figure 11. Comparison of CBA value from A – A' and B – B' slices

4. Conclusion

Based on the results and discussion, the following conclusions can be drawn:

1. South Serayu Mountains with a height of 1000 m at a distance of 20-30 km gives an effect of 0.05 mGal on terrain correction, while the Quaternary Volcano with an altitude of 3000 m at a distance of 30-40 km gives an impact of 0.1 mGal,
2. Calculations using the square pattern and sloped triangle methods can get better terrain correction values than simplified Hammer chart methods up to about 3 mGal,
3. CBA maps using the square pattern method and the Hammer chart method have a similarity in the anomaly pattern with the orientation of northeast-southwest and north-south,
4. The difference in the calculation of the two methods is higher in the slope because there are more significant topographic variations, as well as estimation of topographic height differences that are increasingly difficult to do.

Acknowledgements

The author would like to thank the Department of Geophysical Engineering ITB for supporting the data collection at the Karangsambung 2018 field study, and for allowing this study to be used.

References

[1] Nowell D 1999 Gravity terrain corrections – an overview *J. Appl. Geophysics.* **42** 117-34
[2] Kane M F 1962 A comprehensive system of terrain correction using a digital computer *Geophysics.* **27** 455-62
[3] Nagy D 1966 The prism method for terrain correction using digital computers *Pur. App. Geophysics.* **63** 31-9
[4] Salam R A, Wahyudi E J, Setianingsih 2020 Study of calculation of terrain correction using square pattern and sloped triangle method in Karangsambung area *J.Geofisika.* **17** 9-15
[5] Plouff D 1976 Gravity and magnetic fields of polygonal prisms and application to magnetic terrain correction *Geophysics.* **41** 727-41
[6] Hammer S 1939 Terrain corrections for gravimeter stations *Geophysics.* **4** 184-94

IOP Conf. Series: Earth and Environmental Science 1031 (2022) 012021 doi:10.1088/1755-1315/1031/1/012021

Resource Assessment of Ungaran Geothermal Field Using Numerical model and Monte Carlo Simulation

Bambang Wahyu Jatmiko[1,2], Muhamad Hasbi Assiddiqy[1], Prajamukti Ediatmaja[1,2], Ricky Prabowo[1,2], Sutopo[1], Heru Berian Pratama[1], Muhamad Ridwan Hamdani[1]
[1]Geothermal Engineering Master's Program, Faculty of Mining and Petroleum Engineering, Institut Teknologi Bandung, Jl. Ganesha 10 Bandung 40132, Indonesia
[2]PT. PLN (Persero), Jl. Trunojoyo Blok M-1 No. 135, Jakarta, Indonesia

E-mail: heru.berian@geothermal.itb.ac.id

Abstract. A careful decision needs a comprehensive and effective way to estimate the subsurface condition and the possible resource contained in the geothermal system. Numerical model and simulation sum up the result of geoscience study and its correlation to the behavior of the subsurface fluid flows. A well-developed numerical simulation could give a thorough analysis of a geothermal system and be utilized as input parameters to apply the resource assessment method. This study applied the numerical simulation to the Ungaran geothermal field using TOUGH2 reservoir simulation software. Several parameters were referring to these simulation results and used as the input parameters to the resource assessment method using Monte Carlo simulation. The result shows that the resource value with the highest confidence is lower than the earlier resource estimation of the Ungaran Geothermal System, which is only based on geoscience study. This study gives a careful analysis to support decision-making, especially in a geothermal green field.

Keywords: Resource assessment, Monte Carlo simulation, numerical model, Ungaran geothermal field.

1. Introduction

In general, a decision to estimate the subsurface condition in an exploration stage of a geothermal prospect is challenging to decide, especially in a greenfield area, because its potential resource is still not detailed yet. A decision to develop a geothermal prospect might be strengthened by presenting an estimated resource of a geothermal system. An assessment of geothermal resources could be done during the reconnaissance and exploratory stage prior to well drilling regarding a conceptual model, which is an integrated representation of geology, geochemistry, and geophysics analysis [1,2]. Furthermore, a reservoir simulation, which partially represents the conceptual model dimensionally and hydrothermal fluid flow within an enclosed geothermal system, would be better used as a resource estimation input parameter. However, Ungaran geothermal field (UGF) is still in the exploration stage; a numerical simulation and resource assessment would be necessary to be done.

At the exploration stage, with limited data available, reservoir numerical simulation is unlikely to provide a more realistic long-term. However, it has a value at that stage, but this method may be best for checking for consistency or updating the conceptual model. This method has applied in Arjuno-Welirang [3], Atadei [4–7], Patuha [8–11], Cisolok-Cisukarame [12], Kerinci [13], Tompaso [14,15],

Ulumbu [16–18], Mataloko [19,20], Karaha-Talaga Bodas [21–23], Songa-Wayaua [24], Danau Ranau [25,26], Lumut Balai [27].

In this study, a numerical model of the Ungaran geothermal system was built and used as a resource estimation input parameter. This method became the best practice in resource assessment. Several researchers [28–30] had applied a probabilistic approach using Monte Carlo simulation and geothermal reservoir numerical simulation to estimate the geothermal resources. Each block of this numerical model becomes the input to the reservoir simulation performed using the TOUGH2 program [31]. The resource estimation of the Ungaran geothermal field was conducted using a heat stored principal. It involves a probabilistic approach in the resource assessment.

2. Geosciences Review

A thorough geological, geochemical, and geophysical survey was recently executed by PLN in 2017 [3]. The appearance of geothermal manifestations simply indicates the occurrence of geothermal resources within the UGF (e.g., fumarole, hot springs, warm springs, and alteration zone). The possible main host for a geothermal system in UGF is Ungaran Volcano. Since it is situated at the subduction zone where the oceanic plate (Hindia Plate) subducting against the continental plate (Eurasia Plate), it gives a clue that Ungaran Volcano is most probably supplied by basaltic-andesitic magma. It has 1200°C - 750°C temperature as the possible heat source for this geothermal system. By considering the distribution of geothermal manifestations and each physical characteristic such as temperature, pH, debit, color of water, and the deposit around the manifestation, the geothermal system is categorized as a high terrain, high enthalpy geothermal system. The distribution of alteration minerals (advance argillic) and the occurrence of sulfur deposits on the Ungaran volcanic cone delineate the upflow zone and the possible location of clay cap of this geothermal system. The geological structure distribution gives a clue of the permeable pathway for the geothermal fluid to ascend, known as a discharge zone. Water and gas geochemistry also reveal that this geothermal system might be a two-phase reservoir with a 280°C maximum reservoir temperature. The geophysical survey results give the subsurface geometry model of the Ungaran geothermal system.

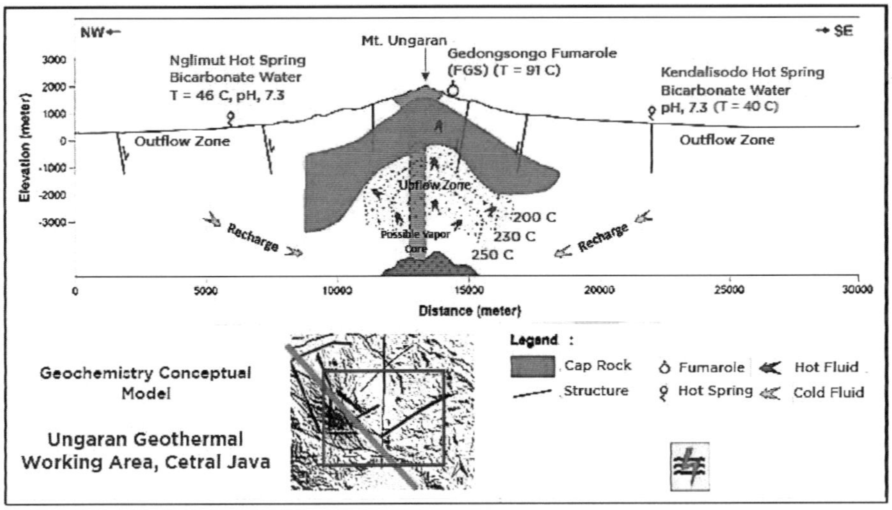

Figure 1. The conceptual model of Ungaran Geothermal System [3]

The results were compiled and represented by a simple conceptual model which explains that the geothermal system in UGF is a vapor core system. The term vapor core is explained as a chimney-like acidic vapor dominated zone, linking deep magmatic-high temperature zones to the floor of a crater [32]. Therefore, such a system should at least have a magmatic heat source that commonly occurs in a mountainous environment. More than a visualization of the subsurface dimension, it also describes the fluid flow and the temperature distribution within the geothermal system. Ungaran geothermal system is a high terrain, a volcanic hydrothermal system with a high enthalpy reservoir and a vapor core [33] linking the magmatic heat source to the surface. Based on the conceptual model (Figure 1), the upflow zone is located around the appearance of the Gedonsongo fumarole, which covers the peak of the Ungaran volcano and it's surrounding up to 500m radius.

Based on the conceptual model, the outflow zone covers most of the lower flank of the Ungaran volcanic cone, except the west and the northeast flank. The system has a 1000 – 2000 m thick advance argillic clay cap, from elevation 1918 masl to -1000 masl. The reservoir is under the clay cap, occurring from elevation 0 masl to -2200 masl, with various thicknesses from 1000 to 1500 m thick. The heat source lays beneath the upflow zone, at the deeper elevation than -3000 masl, which is presumed as a young magmatic body. The meteoric water infiltrates the reservoir through the shallow and deep mechanisms as the recharge mechanism. A vapor core is proposed to exist as a permeable pathway for reservoir fluid to ascend. This piece by piece information will be essentially regarded during the model building process.

3. Computer Model and Numerical Simulation

In the model building step, the conceptual model of the Ungaran geothermal system is being represented numerically. The term numerical could be elaborated as the value of several physical properties of the subsurface material. In the model, these values of physical properties are represented by blocks. Each block may have different physical property values to represent the heterogeneous subsurface. Generally, the physical property values configuration will be adjusted to resemble the material making up the components of a geothermal system, namely cap rock, reservoir, recharge, discharge area, and a heat source.

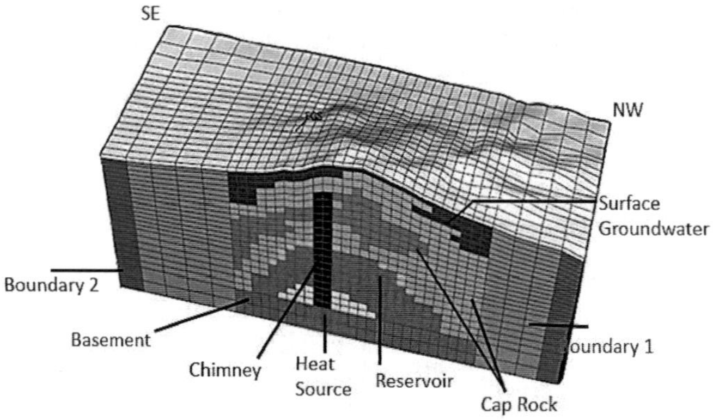

Figure 2. Numerical model of the Ungaran prospective production zone

The model should be constructed to simulate the cycle within a geothermal system as it could be unaffected by the condition outside the system. Therefore, a boundary condition was determined and applied in the model. Several boundary conditions were adjusted as a barrier of the system to the outside condition, involving impermeable material on every side and bottom of the model, and the atmosphere-like condition as the top boundary condition. Specifically, the side boundary may also allow contact to the lateral aquifer, and the bottom of the model is adjustable to represent the convective process inside the model [34]. A configuration of physical properties known as material (Table 1) is adjusted to be unique. As known in general application, the lateral permeability value of the material (Kxy) is set to be the same. In contrast, the vertical permeability value (Kz) is less than (generally half of) the lateral permeability value. The system has a 100 m thick surface with groundwater, followed by a 1364 - 2000 m thick cap rock and a 1200 m thick reservoir. Those layers are wrapped by layers defined as the boundary layer. On the top of the model, the atmosphere layer is defined as a layer with 1 bar pressure and 30°C temperature as the initial boundary condition, the basement layer at the bottom of the model, and boundaries 1 and 2 surrounding the model are set to be impermeable (Figure 2). As the initial boundary condition, the basement layer has 150–190 bar pressure and 255–270°C.

The assignment of these materials referred to the resistivity structure images from the magnetotelluric survey. The model dimension (8×10 km^2 and 5 km thick) was specifically created to cover only the prospective production zone on the lateral and vertical extent. The model is divided into five main layers: atmosphere, groundwater, caprock, reservoir, and basement. For better resolution, those five main layers are furtherly divided into sub-layers. The model is also divided laterally (mesh) with the smallest mesh size, 200×200 m^2.

As illustrated in the conceptual model, assigning subsurface materials with their physical properties is expected to accommodate how the hydrothermal fluid flows within the system. The permeability and porosity are happened to be the primary control of the fluid flow within the model. The system is modeled as it has a 100 m thick surface containing groundwater-like material, followed by a caprock with 1364 - 2000 m thick in the greater depth and a 1200 m thick reservoir beneath the caprock. The boundary layer wraps the model. The bottom of the model is set to be a high temperature, high pressure, and impermeable layer (150-190 bar, 255-270°C) to represent the basement layer. The top of the model is set to be adaptable with the atmosphere layer (1 bar, 30°C). The surrounding boundary is defined as boundaries 1 and 2, which are set to be impermeable layers.

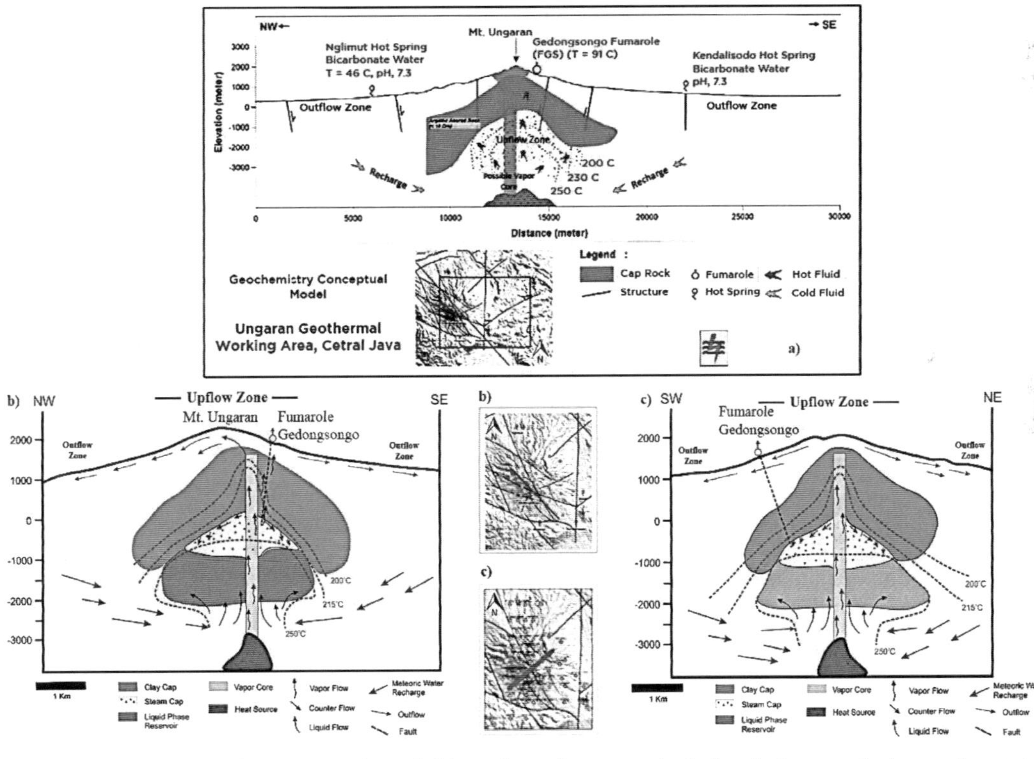

Figure 3. An updated conceptual model based on the numerical simulation result (natural state)

The numerical simulation result has been confirmed to be reaching its natural state by comparing the subsurface temperature from the simulation result to subsurface temperature estimation using an approach from Saputra [35] at the corresponding depth. The calculation result shows the estimated temperature beneath the Gedongsongo fumarole is 176°C at elevation 115 mbsl and compared with the temperature at the same position and depth in the model, which temperature is 168°C has an 8°C deviation. This similarity confirming the natural state of the model has been reached. The simulation

IOP Conf. Series: Earth and Environmental Science 1031 (2022) 012021 doi:10.1088/1755-1315/1031/1/012021

result also corresponds to the geothermometer result, which shows that the maximum reservoir temperature is 280°C. The simulation result also updates the conceptual model, specifically in the prospective area (Figure 3).

4. Resource Estimation

The resource estimation of the Ungaran geothermal field was conducted using the Monte Carlo simulation method principle. Monte Carlo simulation calculates the probabilistic model that simulates physical properties value distribution concerning the heat stored principal iteratively. In this simulation, the minimum, maximum, and most likely value of each parameter of geothermal resource was estimated. Reservoir temperature is determined based on the model's temperature at the top and bottom of the reservoir area. Each has a temperature 225°C at an elevation of approximately -250 msal and 280°C at an elevation of -2000 msal. The thickness of the reservoir is 1200 m approximately, with rock porosity around 0.05. All material properties values assigned to this model were calibrated and selected to control the mass and heat flow to match the natural state conditions and established conceptual model. The recovery factor is obtained based on the correlation value between recovery factor and porosity [36,37] around 0.375. The conversion factor is assumed to be 10%, and the project time is assumed to be 30 years. The input parameter and assumptions are shown in Table 1.

The result of the Monte Carlo simulation, with 60,000 Random Numbers, is presented as geothermal resource value, following its degree of confidence rank (P10, P50, and P90). The sensitivity analysis result for the calculation of Ungaran resource assessment is shown in Figure 4. The results show that reservoir area, initial reservoir temperature, and recovery factor are the most sensitive parameters. The recovery factor is affected by the permeable reservoir as well as the heat sweep efficiency of these permeable channels. This parameter is affected by the reservoir's thermodynamic and hydraulic properties [38]. As a result, these parameters must be determined carefully to avoid incorrect estimation. The Monte Carlo simulation results show that Ungaran geothermal field resource values are P10: 74,2 MW, P50: 116,4 MW, and P90: 173 MW. Based on this simulation, the Ungaran geothermal resource at the highest degree confidence is lower than the possible resource (100 MW) which was predicted by ESDM [39].

Table 1. Input parameters for Monte Carlo simulation in Ungaran resources assessment

Variable	Unit	Min.	Max.	Most Likely	Reference
Area	km2	8	21	15	Model at natural state condition
Thickness	m	1080	1320	1200	Model at natural state condition
Rock Density	kg/m3	2300	2500	2400	Model at natural state condition
Rock Heat Capacity	kJ/kg.°C	0.9	1.1	1	Model at natural state condition
Porosity		0.05	0.10	0.05	Model at natural state condition
Recovery Factor		0.25	0.50	0.38	Muffler and Cataldi, 1978
Electric Efficiency		0.9	1.1	1	SNI-13-6482-2000
Initial Water Saturation		0.6	0.8		Model at natural state condition
Final Water Saturation		0.3	0.5		Prediction of field condition after 30 years
Initial Temperature	°C	225	280	250	Prediction of field condition after 30 years
Final Temperature	°C	180	200	190	Prediction of field condition after 30 years
Life Time	year	30			SNI-13-6482-2000

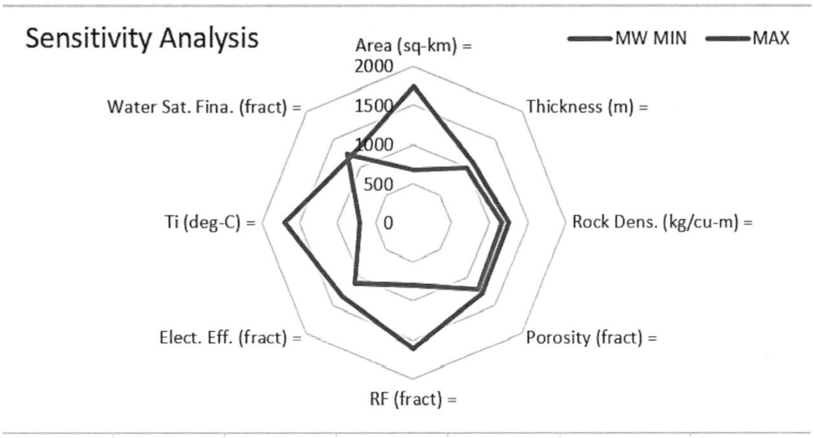

Figure 4. Sensitivity analysis for Ungaran resources assessment

5. Conclusion

The numerical simulation of the Ungaran geothermal system based on geoscience review results has been successfully reaching its natural state. The numerical simulation result combined with the values of blocks within the numerical model effectively represented the required input parameters for the resource estimation method. The estimated resource value (74.2 MW) gives a lower value than the initial estimation (100 MW). This approach provides a better perspective to determine a geothermal resource value and support the geothermal exploration stage decision-making.

6. Disclaimer

This study is based on the PLN Pre-Feasibility Study of Ungaran Geothermal Working Area data in 2017. Shall the updated data be accessible, the model might be reinterpreted.

References

[1] Sarmiento Z F and Steingrímsson B 2007 Computer Programme for Resource Assessment and risk evaluation using Monte Carlo simulation *Short course Geotherm. Dev. Cent. Am. Resour. Assess. Environ. Manag.*

[2] Sarmiento Z F and Steingrímsson B 2013 Volumetric resource assessment *Short course V on Conceptual Modelling of Geothermal Systems UNU-GTP* pp 1–15

[3] Putra R P M, Sutopo and Pratama H B 2019 Improved natural state simulation of Arjuno-Welirang Geothermal field, East Java, Indonesia *IOP Conf. Ser. Earth Environ. Sci.* **254**

[4] Pratama H B, Supijo M C and Sutopo 2020 Experimental design and response surface method in geothermal energy: A comprehensive study in probabilistic resource assessment *Geothermics* **87** 101869

[5] Supijo M C and Pratama H B 2021 Comparison Between Three-Level Box-Behnken Design and Three-Level Full Factorial Design for Probabilistic Resource Assessment : A Case Study in Atadei Geothermal Field , Indonesia *Proceedings World Geothermal Congress 2020+1* (Reykjavik, Iceland) pp 1–13

[6] Supijo M C, Pratama H B and Sutopo 2020 Response Surface Method Using Box-Behnken Design for Probabilistic Resource Assessment: A Case Study in Atadei Geothermal Field, Indonesia *IOP Conference Series: Earth and Environmental Science* vol 417

[7] Supijo M C, Pratama H B and Sutopo 2019 the Application of Experimental Design Using Three-Level Full Factorial Design for Probabilistic Resource Assessment in Atadei Geothermal Field, Indonesia 4 Publications 4 Citations See Profile *The 7th Indonesia International Geothermal Convention & Exhibition (IIGCE) 2019 Proceedings*

[8] Ashat A and Pratama H B 2018 Application of experimental design in geothermal resources

assessment of Ciwidey-Patuha, West Java, Indonesia *IOP Conference Series: Earth and Environmental Science* vol 103
[9] Ashat A, Ridwan R H, Judawisastra L H, Situmorang J, Elfajrie I, Atmaja R W, Iskandar C and Ibrahim R F I 2019 Conceptual Model and Numerical Simulation Update of Patuha Geothermal Field , West Java , Indonesia Conceptual Model and Numerical Simulation Update of Patuha *Proceedings 41st New Zealand Geothermal Workshop, Auckland, New Zealand*
[10] Pratama H B, Sutopo, Widiatmo J S and Ashat A 2021 Numerical Investigative Modeling of Changes Within the Patuha Geothermal Reservoir and Its Production Sustainability Under Two Different Conversion Technologies *Nat. Resour. Res.* **30** 2969–87
[11] Ashat A, Pratama H B and Itoi R 2019 Updating conceptual model of Ciwidey-Patuha geothermal using dynamic numerical model *IOP Conf. Ser. Earth Environ. Sci.* **254**
[12] Sumartha A G A, Kurniawan I, Wiradinata R, Nandaliarasyad N, Sutopo, Pratama H B and Prabata T W 2020 Updating the Conceptual Model of Cisolok-Cisukarame Geothermal field, West Java, Indonesia *IOP Conference Series: Earth and Environmental Science* vol 417
[13] Hidayat I, Sutopo and Pratama H B 2018 Probabilistic approach of resource assessment in Kerinci geothermal field using numerical simulation coupling with monte carlo simulation *IOP Conference Series: Earth and Environmental Science* vol 103
[14] Lesmana A, Pratama H B, Ashat A, Saptadji N M and Gunawan F 2019 An Updated Conceptual Model of the Tompaso Geothermal Field Using Numerical Simulation *Proceedings 41st New Zealand Geothermal Workshop*
[15] Lesmana A, Pratama H B, Ashat A and Saptadji N M 2021 Sustainability of geothermal development strategy using a numerical reservoir modeling: A case study of Tompaso geothermal field *Geothermics* **96** 102170
[16] Kurniawan I, Sutopo S, Pratama H B and Adiprana R 2018 An Improved Natural State Model of Ulumbu Geothermal Field, East Nusa Tenggara, Indonesia *Grand Renewable Energy 2018 Proceedings*
[17] Kurniawan I, Sutopo S, Pratama H B and Adiprana R 2019 A natural state model and resource assessment of Ulumbu Geothermal field *IOP Conf. Ser. Earth Environ. Sci.* **254**
[18] Kurniawan I, Sutopo and Pratama H B 2017 A Natural State Modelling of Ulumbu Geothermal Field, East Nusa Tenggara, Indonesia *Proceedings 39th New Zealand Geothermal Workshop* (Rotorua, New Zealand)
[19] Pradhipta Y D, Sutopo S, Pratama H B and Adiprana R 2019 Natural state modeling of Mataloko Geothermal field, Flores Island, East Nusa Tenggara, Indonesia using TOUGH2 simulator *IOP Conference Series: Earth and Environmental Science* vol 254
[20] Jatmiko B W, Pratama H B and Sutopo 2021 Updated Numerical Model of Mataloko Geothermal Field *IOP Conference Series: Earth and Environmental Science* vol 732
[21] Prabata W, Sutopo S and Pratama H B 2019 Experimental design and response surface method application in resources assessment: Case study Karaha-Talaga bodas, West Java, Indonesia *IOP Conf. Ser. Earth Environ. Sci.* **254**
[22] Prabata W, Sutopo and Pratama H B 2017 3D Natural State Model of Karaha-Talaga Bodas Geothermal Field, West Java, Indonesia *Proceedings 39th New Zealand Geothermal Workshop*
[23] Sutopo, Prabata W and Pratama H B 2019 The development study of Karaha–Talaga Bodas geothermal field using numerical simulation *Geotherm. Energy* **7**
[24] Hasbi H G F, Hamdani M R, Prasetyo A B T, Agung Reynaldi Khrisna W, Lampuasa M J, Sutopo, Pratama H B and Prabata T W 2020 Application of Numerical Simulation to Update Conceptual Model and Resource Assessment of Songa-Wayaua Geothermal Field *IOP Conf. Ser. Earth Environ. Sci.* **417**
[25] Afiat, Idianto O, Rera G F, Wardoyo G K, Sutopo, Pratama H B and Hamdani M R 2021 Updated Conceptual Model and Resource Assessment using Numerical Reservoir Simulation of Danau Ranau Geothermal Field Indonesia *IOP Conference Series: Earth and Environmental Science* vol 732

[26] Rera G F, Pratama H B and Saptadji N M 2021 Approach for Pre-Feasibility Study of Medium Enthalpy Geothermal Field, Case Study: Danau Ranau, Indonesia *IOP Conf. Ser. Earth Environ. Sci.* **732**

[27] Hamdani M R, Pratama H B and Sutopo 2020 Updating the Conceptual Model of Lumut Balai Geothermal Field, South Sumatera, Indonesia Using Numerical Simulation *IOP Conf. Ser. Earth Environ. Sci.* **417**

[28] Pratama H B and Saptadji N M 2021 Study of Production-Injection Strategies for Sustainable Production in Geothermal Reservoir Two-Phase by Numerical Simulation *Indones. J. Geosci.* **18** 25–38

[29] Pratama H B and Saptadji N M 2018 Study of sustainable production in two-phase liquid dominated with steam cap underlying brine reservoir by numerical simulation *IOP Conference Series: Earth and Environmental Science* vol 103

[30] Pratama H B and Saptadji N M 2016 Numerical Simulation for Natural State of Two-Phase Liquid Dominated Geothermal Reservoir with Steam Cap Underlying Brine Reservoir *IOP Conference Series: Earth and Environmental Science* vol 42

[31] Pruess K, Oldenburg C and Moridis G 1999 *TOUGH2 User's Guide*

[32] Reyes A G, Giggenbach W F, Saleras J R M, Salonga N D and Vergara M C 1993 Petrology and geochemistry of alto peak, a vapor-cored hydrothermal system, Leyte province, Philippines *Geothermics* **22** 479–519

[33] Bogie I, Lovelock B and Ussher G 2015 The Neutral High-Cl Thermal Springs of Java, Indonesia *World Geothermal Congress 2015* p 7

[34] O'Sullivan M J, Pruess K and Lippmann M J 2001 State of the art geothermal reservoir simulation *Geothermics* **30** 395–429

[35] Saputra M P, Suryantini, Catigtig D, Regandara R, Asnin S N and Pratama A B 2016 Geological, isothermal, and isobaric 3-D model construction in early stage of geothermal exploration *IOP Conference Series: Earth and Environmental Science* vol 42

[36] Muffler P and Cataldi R 1978 Methods for Regional Assessment Resources of Geothermal *Geothermics* **7** 53–89

[37] Muffler L J P 1978 USGS Geothermal Resource Assessment *1978 USGS Geotherm. Resour. Assess.* 3–8

[38] Ashat A, Pratama H B and Itoi R 2019 Comparison of resource assessment methods with numerical reservoir model between heat stored and experimental design: Case study Ciwidey-Patuha geothermal field *IOP Conference Series: Earth and Environmental Science* vol 254

[39] EBTKE 2017 *Potensi Panas Bumi Indonesia Jilid 2* (Ditjen EBTKE dan Badan Geologi)

Application of Vertical Gravity Gradient to Delineate Groundwater Damage at Bandung Basin

Dadi Abdurrahman[1*],Setianingsih[1], Djoko Santoso[1], Wawan Gunawan A. Kadir[1], Darharta Dahrin[1], Noor Fauzi Isniarno[2], Rian Amukti[2]

[1]Institut Teknologi Bandung
[2]Universitas Islam Bandung
[3]Indonesian Institute of Sciences

*corresponding author's email : abdurrahman@mail.gf.ac.id

Abstract. Bandung basin is the center of the population at West Java, and several Industrial areasis. In supporting its sustainability, water needs is one of the factors that must be fulfilled. The water needs are obtained by taking water directly from groundwater. Excessive extraction of groundwater can cause damage, the impact of which is the more difficult it is to obtain groundwater and cause environmental disaster in the form of land subsidence. Using the vertical gravity gradient technique in the Bandung basin area is expected to map the critical groundwater damage zone. The results from the vertical gravity gradient area of the Bandung basin map, the region of groundwater with critical status are found in almost all cities, except in the Baleendah area to Ciparay and Soreang to the south of Cimahi.

1. Introduction

Bandung basin has an altitude between 650 MSL and 2300 MSL with a low elevation in the center and has a high elevation around it. Bandung Basin is Tangkubanprahu Mt., Bukittunggul Mt., Manglayang Mt., Patuha Mt., Malabar Mt located around.

In Bandung basin, it lies in several large cities such as Bandung city, Bandung regencies, the Cimahi city, and West Bandung regencies. Bandung basin has a population around 7.6 million people, with around 40% of the people in the Bandung regency area. There are about 30 thousand industries in the Bandung basin.

IOP Conf. Series: Earth and Environmental Science 1031 (2022) 012022 doi:10.1088/1755-1315/1031/1/012022

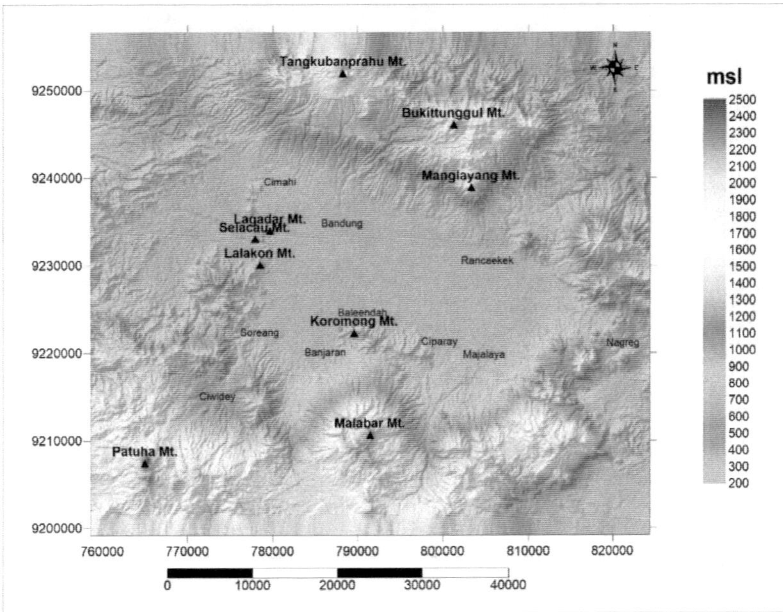

Figure 1. Geomorphology of Bandung Basin.

2. Method

2.1. Geological Map

There are several vital formations in the Bandung Basin, such as Kosambi Fm., Cibeureum Fm., Cikapundung Fm. And Beser Fm. In the research area, Cibeureum Fm. is the primary aquifer, while Kosambi Fm. is aquitard, and bedrock is Cikapundung Fm.

Cibeureum Fm. Consisting of a low level of consolidation of breccias and tuffs and several basal lava inserts, it has age around Late Pleistocene – Holocene. Breccia, in this formation, is a volcanic breccia compiled by scoring fragments of basalt andesite igneous rocks and pumice. Kosambi Fm. Consists of claystone, siltstone, and sandstones that have not been compact with the age of the Holocene. This formation has interfinger relation with top Cibeureum Fm.. Cikapundung Fm. Consists of conglomerates and breccia compact, tuff, and andesite lava. The age of this formation belong to Early Pleistocene. The compactness of the lithology constituent of this formation can be used as a differentiator to Cibeureum Fm., and the basis for determining the role of this formation as a hydrogeological bedrock in the study area. Besar Fm. In this area is the oldest formation and occupies a basement.

IOP Conf. Series: Earth and Environmental Science 1031 (2022) 012022 doi:10.1088/1755-1315/1031/1/012022

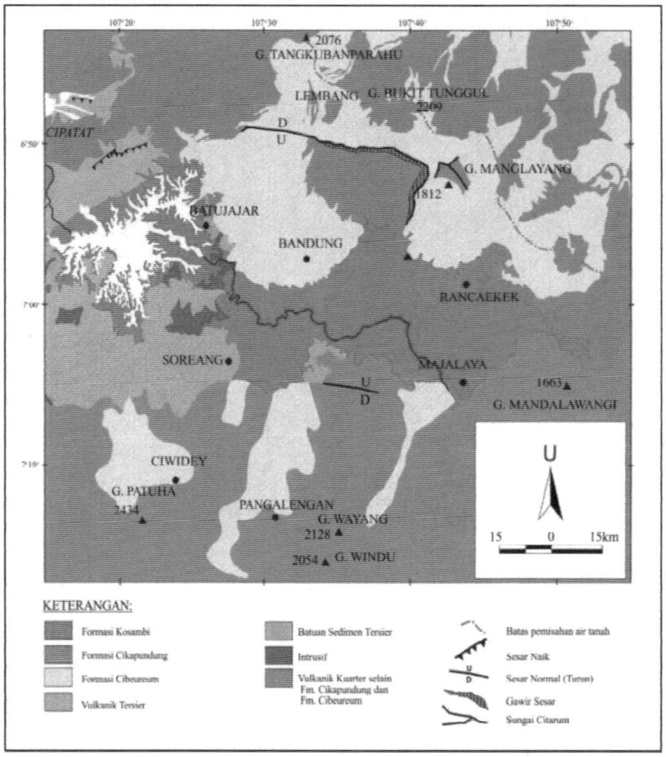

Figure 2. The geological map of the study area, modification results from [2].

2.2. Groundwater Conservation Map (2000)

The following is a groundwater conservation map in the Bandung basin area published by the Directorate of Geology and Environmental Management (Figure 3). It shows that the groundwater level in the Bandung basin area has an altitude between 600 above MSL and 1200 above MSL. Polygon area shows status zones, red for critical areas, yellow for prone zones, blue, orange, and green zones for safe spaces.

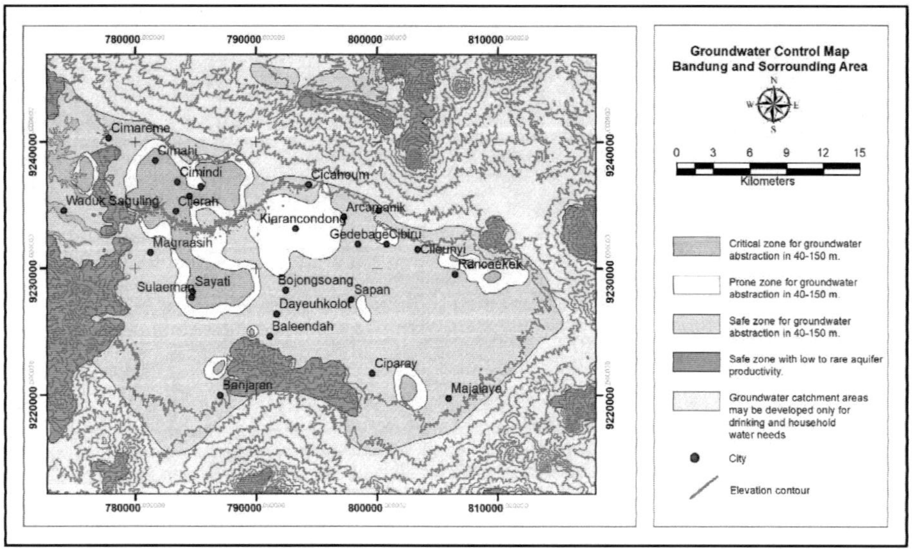

Figure 3. Groundwater conservation map in Bandung basin modified from the Directorate of Geology and Environmental Management in 2000 [3].

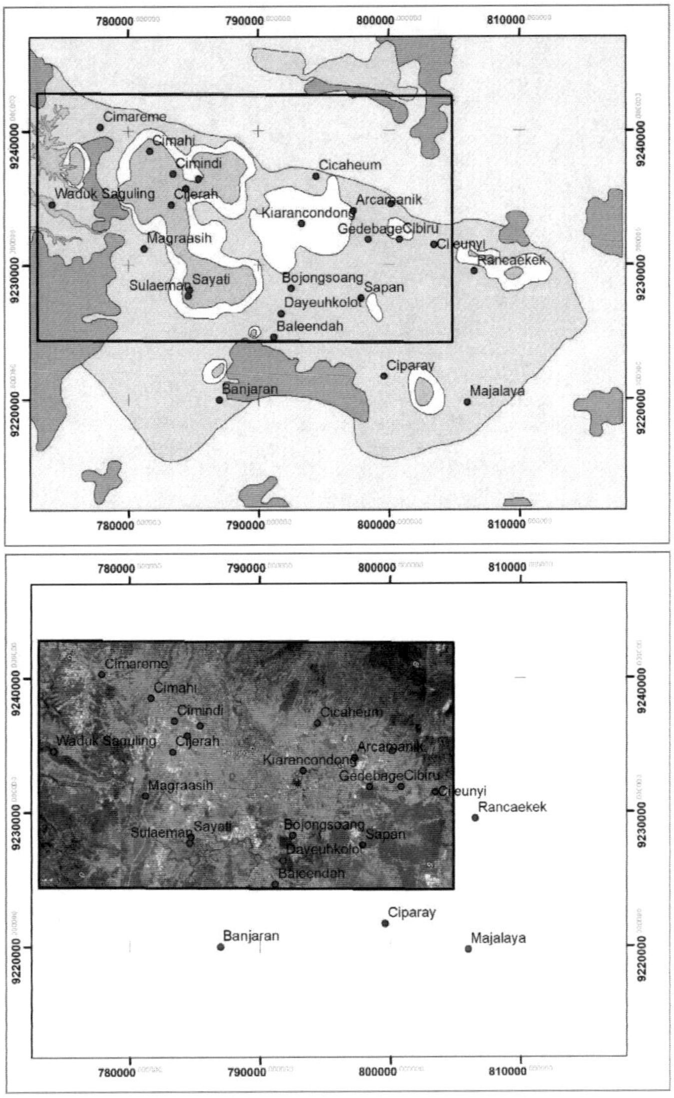

Figure 4. (above) Groundwater conservation maps and (below) sightings of measurement points from google earth.

From the map of groundwater conservation, we can see that general condition of groundwater in the Bandung basin area is good enough, but in some areas, it has critical situations, such as Cimahi, Margahayu-Kopo, Dayeuhkolot, Majalaya, and Rancaekek. If we look at Google Earth (Figure 4), it turns out that the critical groundwater zone in the Bandung basin area is very carefully related to the existence of the industry. In order to this fact, known that the industry exploits groundwater for their needsand it makes groundwater become the critical zone.

2.3. Subsidence rate map (14 January 2007–12 March 2011) generated with ENVISAT ASAR data
Using the remote sensing method come from Envisat Ansar data to determine subsidence rate per year [1]. It can be seen in Figure 5 that almost all areas of the Bandung basin have subsidence, especially in the middle area. The subsidence can be 250 mm/year. The most significant land subsidence is in Cimahi, Dayeuhkolot, Majalaya, and Rancaekek areas.

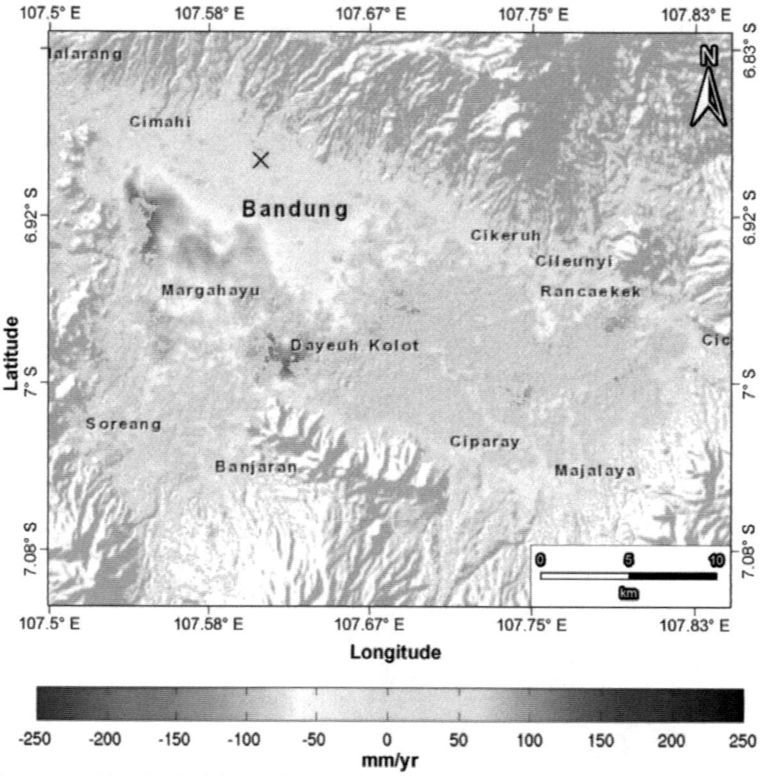

Figure 5. The rate of land subsidence in Bandung basin come from ENVISAT ASAR data [1].

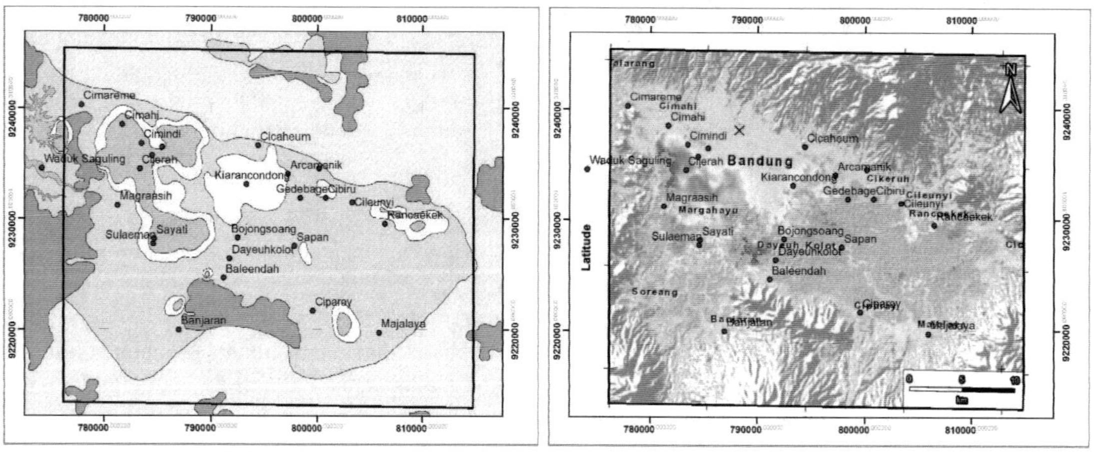

Figure 6. (left) map of extending groundwater, and (right) map of land subsidence.

Figure 6 shows the comparison of groundwater (left), and land subsidence (right). It shows that there is a positive correlation between the critical zone of the groundwater (red in the left image) and the magnitude of subsidence (red in the right). So we can conclude that a decrease in groundwater at Bandung basin makes decrease pores pressure at aquifer, it makes rock has imbalances stress in it. The reduction in pore pressure makes compression (decrease in volume). Compression of the rock continues from the aquifer zone to the surface which results in subsidence.

IOP Conf. Series: Earth and Environmental Science 1031 (2022) 012022 doi:10.1088/1755-1315/1031/1/012022

2.4. Vertical Gravity Gradient (VGG)

Gravity acquisition techniques by measuring at the same point with varying heights (at least two different levels) are called a vertical gravity gradient technique. The purpose of this measurement is to find out the gravitational gradient value vertically at the measurement point. The formula for calculating vertical gravity gradient written in equation (1). g_1, and g_2 for gravity value at the lower elevation and the high elevation, and z_1, and z_2 for the low elevation and the high elevation.

$$VGG = \frac{g_2 - g_1}{z_2 - z_1}$$

Measuring the gravity value is using a gravimeter Scintrex Cg-5, and measuring the elevation value is using a geodetic tripod shown in Figure 7. In general, the vertical gravity gradient value in Indonesia is -0.3085 mGal/m, so anomaly gravity vertical gradient is defined as a different vertical gravity value with this value.

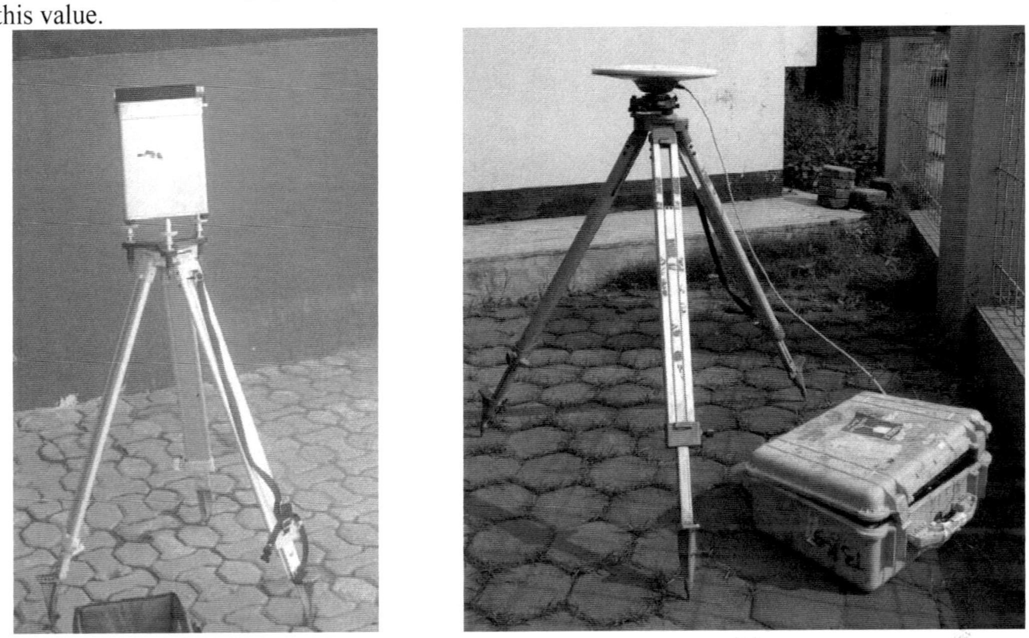

(a) (b)

Figure 7. Vertical gravity gradient measurements using a gravimeter Scintrex CG-5 (a), and a geodetic tripod (b).

To find out the phenomenon of groundwater damage, the author tries to measure the vertical gravity gradient with 63 scattered stations, shown in Figure 8. The distribution of vertical gravity gradient stations concentrated in the central-western area of the Bandung basin.

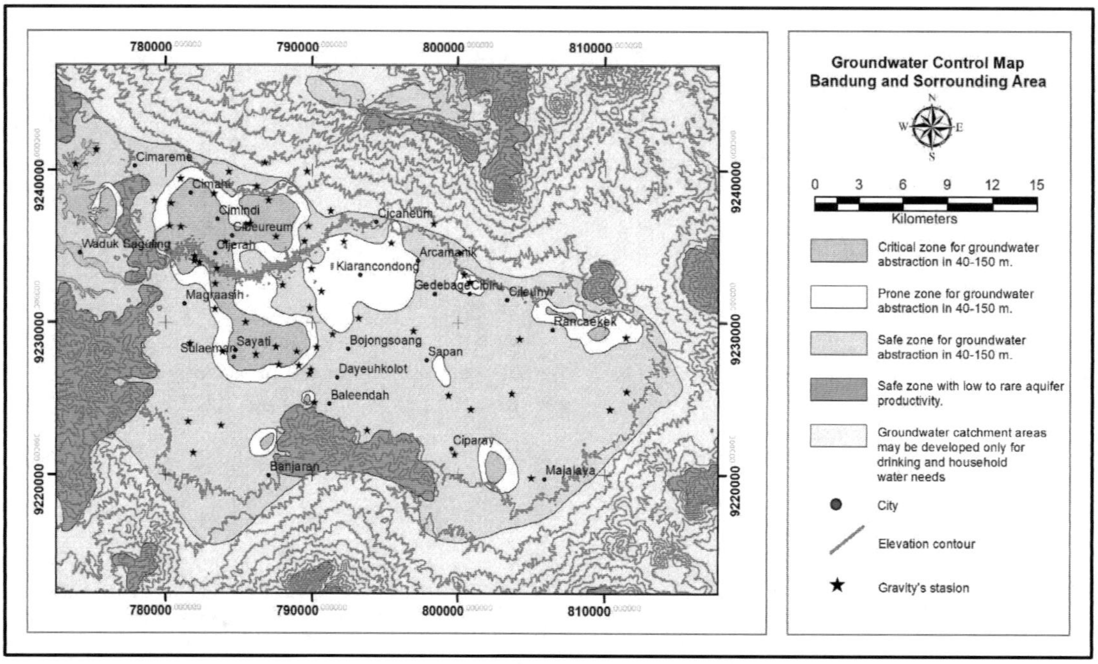

Figure 8. A symbol plus indicates the distribution of the gravity station.

Figure 9 is results vertical gravity gradient measurement in the Bandung basin. In Figure 9, the red color shows high values of vertical gravity gradient (> -0.3085 mGal/m), and blue indicates a low vertical gravity gradient (<-0.3085 mGal/m). In Figure 9, it can be seen that in the Cimahi area to Bandung, it has a high vertical gravity value. The area of Bandung city to the city of Cimahi, as explained above, shows a critical zone of groundwater. From this, we can conclude that a high vertical gravity gradient shows the critical zone of groundwater.

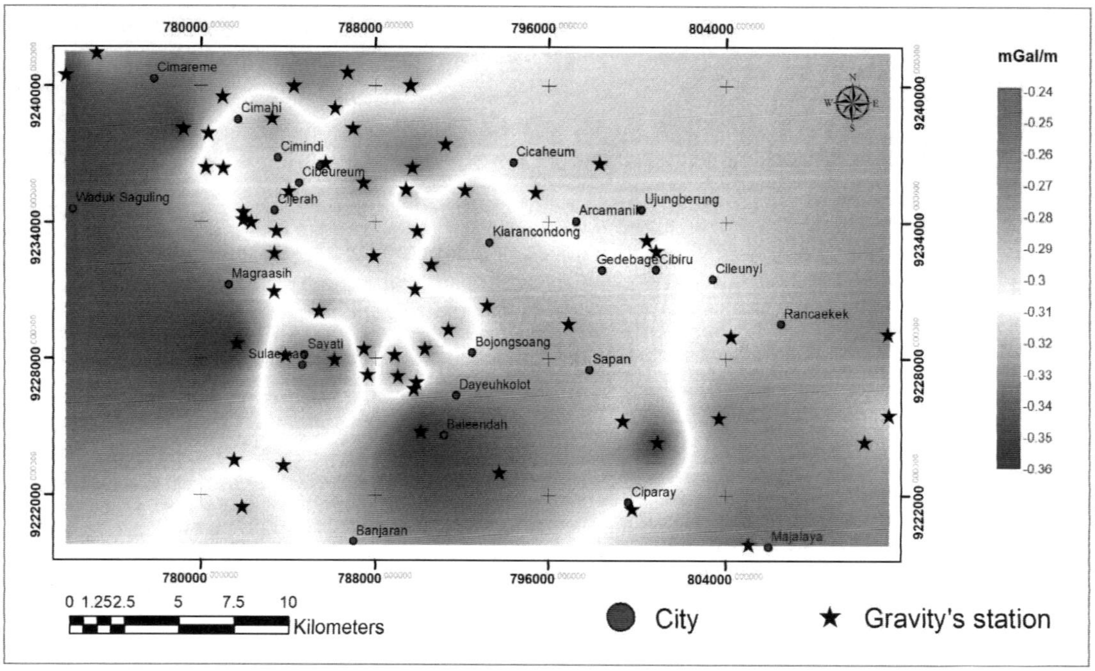

Figure 9. Vertical gravity gradient at Bandung Basin.

Based on the correlation between critical groundwater zones and large values of vertical gravity gradients, we can apply to other areas in the Bandung basin. Based on vertical gravity gradient maps in general Bandung basin suffered groundwater damage at a critical stage except in the Baleendah area to Ciparay and Soreang to the south of Cimahi.

3. Conclusions

By obtaining a positive correlation between the zone of groundwater damage and the value of the vertical gravity gradient, this method can delineate the critical area of groundwater in the Bandung basin. The value of the critical area on the map of the vertical gravity gradient is higher than <| -0.3086 | mGal/m. Based on vertical gravity gradient maps in general Bandung basin suffered groundwater damage at a critical stage except in the Baleendah area to Ciparay and Soreang to the south of Cimahi.

References

[1] Ge L, Ng A H, Li X, Abidin H Z and Gumilar I 2014 Land subsidence characteristics of Bandung Basin as revealed by ENVISAT ASAR and ALOS PALSAR interferometry *Remote Sensing of Environment* **154** 46–60

[2] Hutasoit L M 2009 Kondisi Muka Airtanah Dengan dan Tanpa Peresapan Buatan Daerah Bandung: Hasil Simulasi Numerik *Jurnal Geologi Indonesia* **4** 177-188

[3] Taufiq A and Iskandar N 2000 Peta Pengendalian Pengambilan Airtanah *Directorate of Geology and Environmental Management*

Petrophysics Analysis for Determination of Density Porosity and Neutron-Density Porosity on Carbonate Rock in East Java Basin

D Julianti[1*], Fatkhan[2], E Dinanto[3], A S Murtani[4]

[123]Geophysical Engineering, Faculty of Mining and Petroleum Engineering, Institut Teknologi Bandung, Jl. Ganesha 10, Bandung, 40132 Indonesia
[4]SAKA Indonesia Pangkah Limited

*Correspondence email: dwijulianti30@gmail.com

Abstract. Indonesia has much potential in energy resources. The exploration and exploitation activities in Indonesia help to fulfill the energy needs in Indonesia. The exploration activities in Indonesia help to find the potential energy resources such oil and gas. Reservoir characterization with petrophysics can be applied to define the lithology, porosity, water saturation, and permeability the rock under the ground. One of the important aspect to define the reservoir is to understand the porosity of the targe reservoir. In this paper, the author will calculate the porosity in the reservoir using petrophysics analysis with density and neutron-density. The result is the neutron-density porosity giving better result than the density porosity.
Keywords: density; neutron; petrophysics; porosity

1. Introduction

Indonesia has much potential in energy resources. The exploration and exploitation activities in Indonesia help to fulfill the energy needs in Indonesia. The exploration activities in Indonesia help to find the potential energy resources such oil and gas. Reservoir characterization with petrophysics can be applied to define the lithology, porosity, water saturation, and permeability of the rock under the ground. One of the important aspect to define the reservoir is to understand the porosity of the target reservoir.

Porosity can be determined from density log, neutron log, neutron and density log, and many other methods. The neutron porosity log on formation evaluation used to measure hydrogen index that contain in the formation. Hydrogen index of a material is defined as the partial concentration of hydrogen per unit volume relative to water. A lower number of neutron porosity is caused by a lower number of hydrogen atoms. Density porosity measures the total number of formation electrons, where a lower number of electrons is equivalent to a lower formation density, or a higher formation porosity. In this paper, author will emerge porosity from density log and porosity from neutron-density log and do the comparison for both results to define which one better for defining the porosity of target reservoir.

2. Theory

2.1. Porosity

Porosity is the ratio of void volume to total volume. There are three porosity: porosity sonic, density porosity, and neutron porosity [1]. Below the illustration sketched by author for matrix, wet clay, and pore.

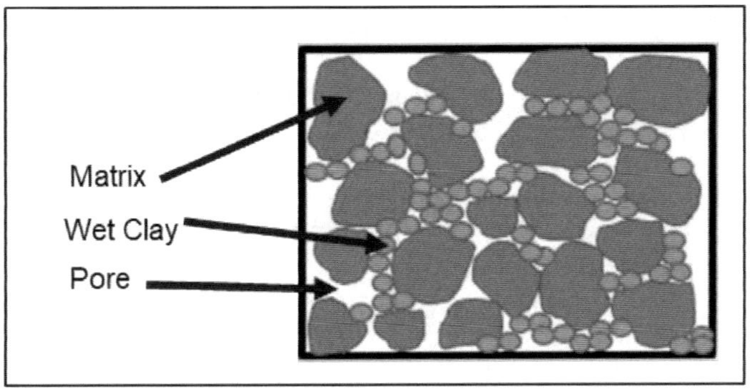

Figure 1. The Illustration of the assumption of Matrix, Wet Clay, and Pore in the Rock.

Density porosity will measure bulk porosity from the formation. Density porosity(ϕD) can be calculated with equation (1) using parameters: matrix density(ρma), input bulk density log(ρb), wet clay volume(Vcl), wet clay density(ρcl), filtrate density(ρfl), apparent hydrocarbon density(ρHyAp) from hydrocarbon density calculation(ρHy), flushed zone water saturation(Sxo) from water saturation(Sw) and invasion factor calculation, while neutron porosity calculated (ϕN) with equation (4) using parameters: input neutron log(ϕ neu), neutron wet clay value(NeuCl), neutron hydrocarbondydrogen index(NeuHyHI), neutron matrix correction(NeuMatrix), neutron formation salinity correction(NeuSal), neutron excavation factor(Exfact).

$$\phi D = \frac{\rho ma - \rho b - Vcl(\rho ma - \rho cl)}{\rho ma - (\rho fl * Sxo) - (\rho HyAp * (1 - Sxo))} \quad (1)$$

Where,

$$Sxo = \frac{Sw + Invasion\ Factor}{1 + Invasion\ Factor} \quad (2)$$

$$\rho HyAp = \frac{5.5 * \rho Hy * (4 - \rho Hy) - 3}{16 - 2.5 * \rho Hy} \quad (3)$$

While ρHy is the entered true downhole hydrocarbon density, ρHyAp is the density tool apparent hydrocarbon density in gm/cc that can calibrate the density tool from electron density to apparent density and will be applied to the outer zone of flushed zone water saturation. Neutron porosity model are shown on the equation below.

$$\phi N = \frac{\phi_{neu} - Vcl * NeuCl + NeuMatrix + Exfact + NeuSal}{Sxo + (1 - Sxo) * NeuHyHI} \quad (4)$$

Where,

$$NeuHyHl = 9 * \rho_{hden} \frac{(4 - 2.5 * \rho_{Hy})}{(16 - 2.5 * \rho_{Hy})} \quad (5)$$

There are two correction: Neutron formation salinity correction(NeuSal) and neutron matrix correction(NeuMatrix). The salinity correction calculated from the Neutron Tool look-up table that is defined for a particular neutron logging tool, selected with the Neu Tool Type parameter. The entry into the Look-up table will be formation porosity, flushed zone salinity, and matrix density and the result will be an interpolation between the values in the Look-up table. The matrix type in the Look-up table will depend on the input matrix density and the Mineral Model parameter. For example, if the mineral model chosen was Sand/Dolomite and the input matrix density was 2.75 g/cc, the correction would be half way in-between Sand and Dolomite correction. The salinity correction will assume the same salinity correction for the special mineral as for the main mineral in special mineral models.

The Neutron matrix correction(NeuMatrix) on the Matrix parameters tab will be calculated from the Neutron tool look-up tables for the defined tool. The result will depend on the actual formation porosity, the input matrix density, and the mineral model. To define the overall neutron mineral response the same mixture will be used if the matrix density indicates a mixture of minerals.

Neutron-density porosity is a combination between neutron log(φN) and density log(φD). neutron-density porosity(ϕND) can be calculated with equation below.

$$\phi ND = \frac{\sqrt{\varphi D^2 + \varphi N^2}}{2} \tag{6}$$

2.2. Volume of Clay

Volume of clay can be determined from single or more indicator. Gamma ray clay indicator calculate using equation (7) using gamma ray(Gr).

$$VclGr = \frac{Gr - Grclean}{GrClay - GrClean} \tag{7}$$

2.3. Water Saturation

Water saturation is the fraction of formation water in the undisturbed zone. Water Saturation can be calculated using Archie Equation below using water saturation, input resistivity(Rt), Formation water resistivity(Rw), cementation factor(m), and saturation exponent(n).

$$\frac{1}{Rt} = \frac{\phi^m \, x \, Sw^n}{a \, x \, Rw} \tag{8}$$

3. Data and Methodology

Data on this research are log data on well that located in East Java Basin consist of Gamma Ray, Density, Neutron Porosity, Core Data, Resistivity (Rt), and Formation Water Resistivity (Rw).

The workflow of this research shown on the figure below. From the well data with following log used for determining the volume of clay. From the volume of clay, will be defined where the clean zone to be the area of matrix density estimation using log data. After the matrix density estimated, the result then validated with matrix density from core data. Wet clay density and wet clay neutron will determined in shale zone using log data. Water saturation will be defined using Archie Equation. When all paremeter for calculation defined, density porosity and neutron-density porosity could be calculated.

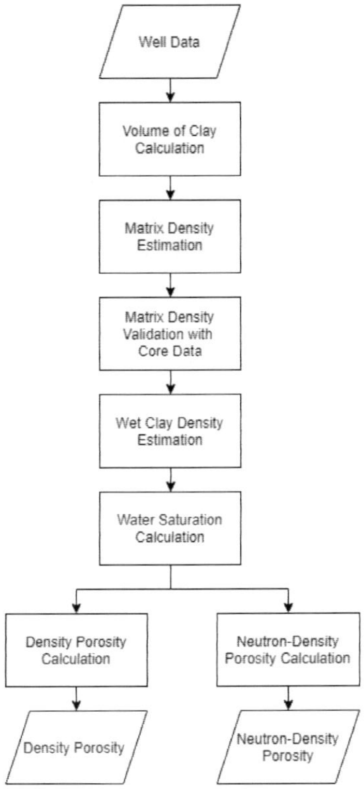

Figure 2. Workflow of the research.

4. Result and Analysis

The formation of target on this research is "Juli" Formation where located offshore in the southwest part of East Java Basin. The "Juli" field is in Pangkah Block. The located of East Java basin is along the southeastern edge of Sundaland (Figure 3).

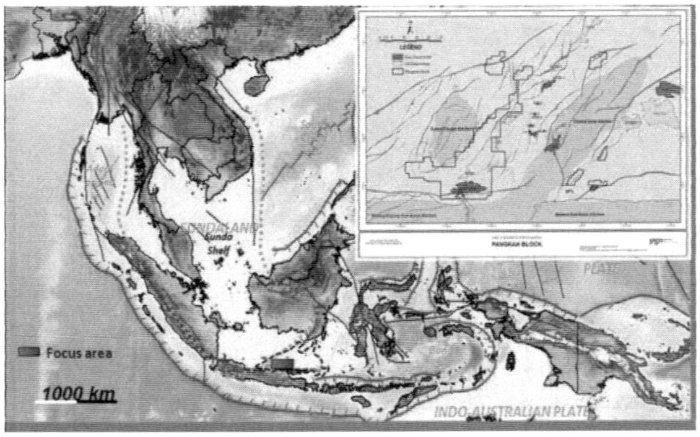

Figure 3. Location of Pangkah PSC, Offshore East Java, North East Java basin [2].

The regional stratigraphy in the area of study is divided into Paleogene and Neogene sediments as the figure below. Paleogene sediment deposition is related to the rifting or tensional processes during Paleocene to Eocene time where the accommodation space was influenced by the basin configuration at that time. Where the Neogene sediment deposition related to compressional processes where tectonically influenced by the activity of subduction zone to the south of Java Island [2].

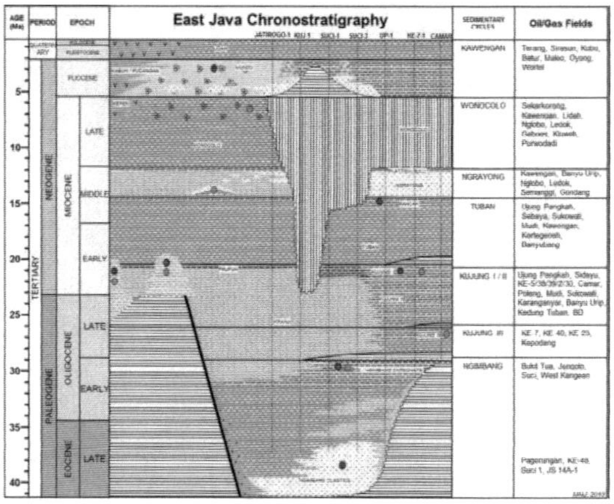

Figure 4. East Java regional stratigraphy [2].

The "Juli" Formation devided into 3 zones. Juli Zone 1 filled with shale, Juli Zone 2 filled with high density Carbonate, and Juli Zone 3 filled with Low density carbonate. In order to determine density porosity and neutron-density porosity we need to determine the parameter such as volume of clay, density matrix, and density wet clay.

4.1. Volume of Clay Calculation

In this study there are shale and carbonate area. The two lithologies can be devided using volume of clay. Volume of Clay is a fraction of the volume of clay in the formation to the total volume of the formation. Volume of clay can be determined from single or more indicator. In this study, will use the single clay indicator gamma ray to calculate the volume of clay.

Gr Clean and Gr Clay shown in Table 1 and the result of volume of clay for the well is on figure 5 below.

Table 1. Gr Clean and Gr Clay of each zone.

Zone	Gr Clean	Gr Clay
Juli Zone 1	40	118
Juli Zone 2	33	100
Juli Zone 3	33	100

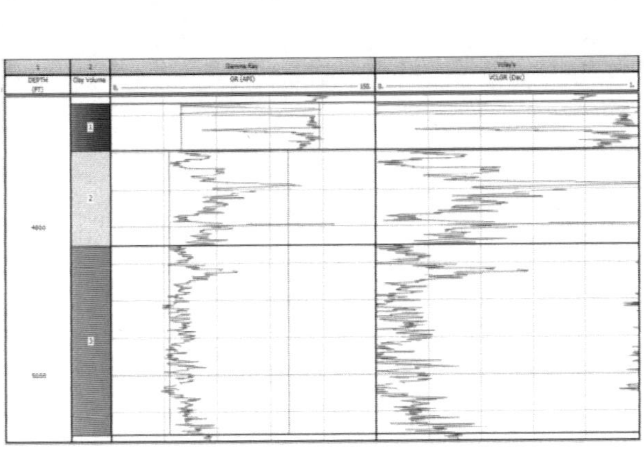

Figure 5. Volume of Clay.

4.2. Matrix Density Estimation and validation with data core

Matrix density can be estimated using crossplot of RHOB and PHI at the clean zone. The regression line for the point will be the estimate value of matrix density. In figure 6, the estimation of matrix density is 2.71gr/cc in the clean zone at depth 4670-4671ft. The crossplot on two other zone that are 4708-4709ft and 5016-5017ft and the matrix density estimation is 2.71 gr/cc too. The Matrix density data resulted from the estimation showed the similarity with the the grain density from core analysis result. So, the matrix density that will be used to the porosity calculation is 2.71 gr/cc.

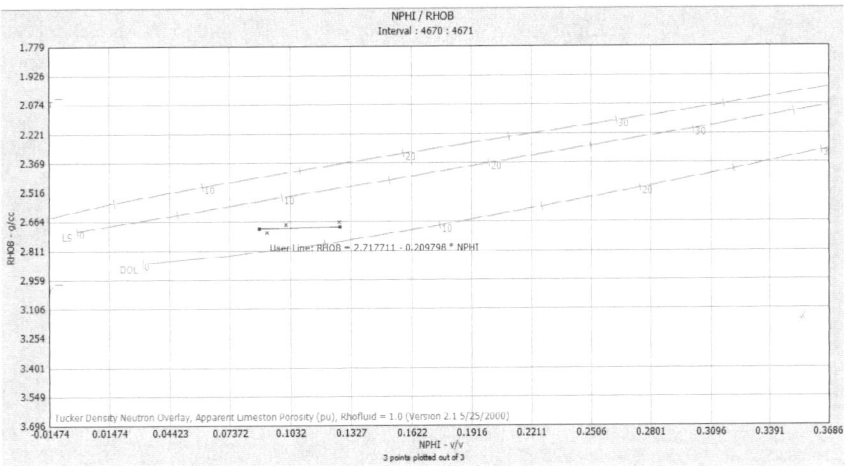

Figure 6. Crossplot to estimate density matrix at depth 4670-4671ft.

4.3. Wet Clay Density and Wet Clay Neutron Estimation

Wet clay density and wet clay neutron can be estimate using crossplot of RHOB and PHI at the shale zone. Wet clay density and wet clay neutron will be estimate from value of the intersection between the line from 100% porosity to the outer point of clustered data with the line from matrix density (2.71gr/cc) to the outer point of clustered data. In figure 7, the estimation of matrix density is 2.49gr/cc in the shale zone at depth 4633.77-4697.22ft. Shale zone found in Juli 1 Zone 1 and two other zone are carbonate area, so wet clay density will be estimated only using this area. The wet clay density that will be used to the porosity calculation is 2.49gr/cc and wet clay neutron value is 0.373.

IOP Conf. Series: Earth and Environmental Science 1031 (2022) 012023 doi:10.1088/1755-1315/1031/1/012023

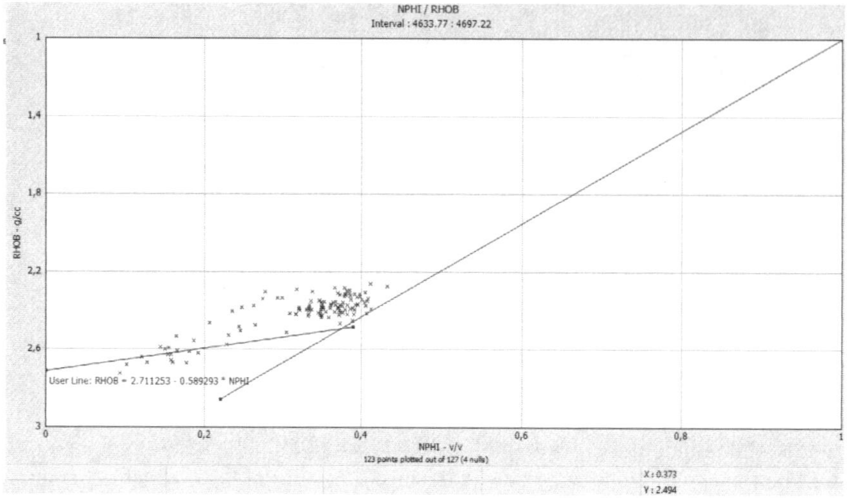

Figure 7. Crossplot to estimate wet clay density and wet clay neutron at depth 4633.77-4697.22ft.

4.4. Water Saturation

Water Saturation can be calculated using Archie Equation with parameters Rw=1.2 at temperature 180 F, the temperature gradient data applied from other well near the well used in this paper, also parameter m=1.87, n=1.93, and n=1 from company data. Parameters ρfl=1.03 and parameters ρHy=0.793 from company data. Some of parameter are given by company data, some are being assumption, and some are estimated.

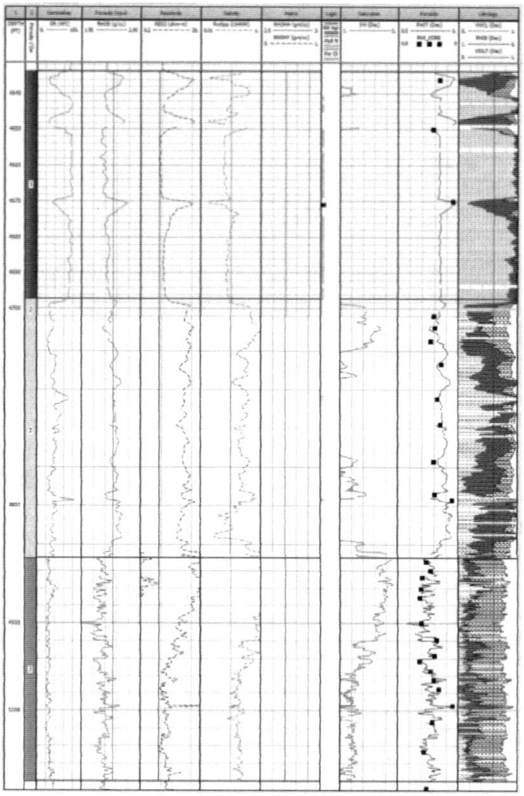

Figure 8. Density porosity Log.

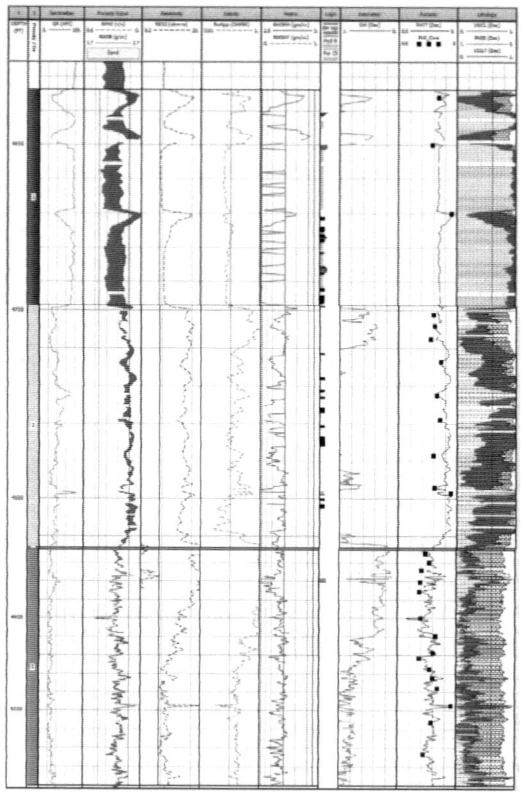

Figure 9. Neutron-Density Porosity Log.

After defining all the parameters for the calculation, both density porosity and neutron-density porosity could be calculated. The result of the density porosity shown in the figure 8 and neutron-density porosity shown in the figure 9. We could see that both results of density porosity and neutron-density porosity are having good correlation with the core data. Core data shown in black square, while density porosity and neutron-density porosity in blue line.

Density porosity and neutron-density porosity have been determined from the research. Almost all the core data fit with the porosity. But we could see at some depth the density porosity and neutron-density porosity have different value with core data. We know that density porosity measures porosity from the number of formation electrons, when it log into gas-filled formation, the result will be a porosity estimate that higher than the true porosity. Meanwhile the neutron porosity mainly calculated the number of hydrogen atoms in formation, that when a gas-filled formation logged, which has a lower number of hydrogen atoms than a water filled formation of the same porosity, the porosity estimate will be lower that the true porosity. These information bring to the conclusion that it is better for using both density and neutron to define the porosity of formation. By properly combining the two reasons above, it is possible to arrive more accurate porosity than would be possible by interpreting each of them separately. We could also see from the figure 8 and figure 9 that the neutron-density porosity bigger than density porosity at the Juli Zone 1 which is in carbonate rock. It may because of the bulk density in the area close to matrix density which interpreted as zone filled with matrix with low porosity. However, this proven to be difficult to have 100% porosity model that fit to data core.

From the result of neutron-density porosity model, shown that some of result have different value with the core data. Below the table showing the error from the porosity log and porosity core data. The average error from 26 core data is 0.03883 and showing the result of neutron-density porosity model having good correlation with core data.

Table 2. The Error Calculation Between Neutron-Density Porosity Log and Core Data.

No	Depth	Porosity from Core Data	Porosity from Log Data	Error
1	4636	0.157	0.12	0.037
2	4650	0.213	0.225	0.012
3	4670	0.047	0.0747	0.0277
4	4702	0.205	0.159	0.046
5	4708	0.195	0.202	0.007
6	4715	0.23	0.183	0.047
7	4727	0.142	0.146	0.004
8	4745	0.177	0.163	0.014
9	4758	0.148	0.157	0.009
10	4777	0.204	0.137	0.067
11	4794	0.197	0.185	0.012
12	4797	0.054	0.163	0.109
13	4830	0.266	0.228	0.038
14	4840	0.233	0.232	0.001
15	4848	0.298	0.243	0.055
16	4861	0.307	0.289	0.018
17	4871	0.317	0.283	0.034
18	4900	0.305	0.426	0.121
19	4919	0.179	0.284	0.105
20	4937	0.2	0.217	0.017
21	4943	0.321	0.308	0.013
22	4965	0.204	0.126	0.078
23	4976	0.161	0.258	0.097
24	4995	0.047	0.038	0.009
25	5013	0.217	0.227	0.01
26	5047	0.282	0.304	0.022
Average				0.03883

IOP Conf. Series: Earth and Environmental Science 1031 (2022) 012023 doi:10.1088/1755-1315/1031/1/012023

5. Conclusion

The results from this study can be summarized as follow,

1. Porosity can be determined from density porosity and neutron-density porosity model using log data which in this paper both density porosity and neutron-density porosity show good correlation with porosity from core data. However, we could still see at some depth, some density porosity and neutron-density porosity show different value with core data.
2. Neutron-density porosity have better result for defining porosity rather than density porosity. It may drive from the neutron log that help to do the correction to the density porosity. In this paper the neutron-density porosity has been validated with core data and has average error of 0.03883.

References

[1] Asquith G B and Gibson CR 1982 *Basic Well Log Analysis for Geologists* (Tulsa: The American Association of Petroleum Geologists (AAPG))

[2] D Setiawan M N Juliansyah and I W A Darma 2014 *Ann. Conv. Proc. on IPA AAPG 2014 – 38th*

Gravity - Depth Regression Method for 3D Modelling of Basement Geometry

Edy Wijanarko and Hendra Grandis*

Geophysical Engineering Program, Faculty of Mining and Petroleum Engineering, Institut Teknologi Bandung (ITB), Bandung, Indonesia.

*e-mail: grandis@geoph.itb.ac.id

Abstract. Gravity method is often used to estimate the sedimentary basin configuration at the preliminary stage of oil and gas exploration. This paper aims at the determination of 3D basement topography based on gravity data and known depth at several points. The gravity-depth relationship is used to determine the initial model and to adjust the model iteratively from the residuals. The algorithm is relatively simple and fast, since it does not involve matrix inversion nor multiplication. The proposed method also allows density contrast estimation. Tests using synthetic data showed satisfactory results in both synthetic model recovery and low RMS misfit after a few iterations.

1. Introduction

The basement geometry or depth estimation is an important phase in hydrocarbon prospecting. Gravity method is of particular interest for mapping basement topography. One of conventional approaches to estimate the basin configuration from gravity data is the well-known Bott's method [1]. The algorithm employs the Bouguer plate formula to estimate the initial depth model and its iterative modifications from the residuals or misfits. Bott's method can be applied for both 2D and 3D basement modelling by using relevant gravity model response calculation or forward modelling [2,3].

In Bott's method, an estimation of the constant density contrast between the basement and the overlying sedimentary rocks has to be assumed. The choice of the density contrast is a critical issue. Erroneous density contrast will give a significantly different model. We followed Florio [4,5] who proposed a similar iterative algorithm incorporating the density contrast estimation. The regression between gravity data and known basement depth at several points is used to convert the gravity anomalies to depths as the initial model. We used the 3D vertical prism model as building blocks of a 3D basin model. The density contrast is estimated from the regression of the observed gravity with the theoretical response of the initial model with a unit density. The model is updated iteratively based on residuals by using the initial gravity-depth relationship. We apply the algorithm to synthetic gravity data associated with a 3D basement model having a constant density contrast.

2. Data and Method

Data

To illustrate and to test the method we used a 3D synthetic basin model that covers an area of 132 x 104 km^2 discretized into 33 x 26 blocks of vertical prism with 4 x 4 km^2 in horizontal size. The maximum sediment thickness is around 11 km with a constant density contrast of -0.5 gr/cm^3. The theoretical gravity effect at the centre of each grid was calculated by using a 3D forward modelling code modified from Blakely [2]. We applied the iterative modelling method to synthetic data without noise and also with Gaussian noise, i.e. zero mean and 0.5 mGal standard deviation.

Method

The iterative procedure to obtain the final model is relatively simple as there is no matrix inversion nor multiplication involved. By assuming a constant density contrast, a linear relationship between gravity data and known basement depth at several points (from well, seismic or the other geophysical data) can be used to convert gravity data to basement depth. Since we deal with synthetic data, the regression was done by using basement depths from synthetic model and their associated gravity response. The result of the regression from synthetic data without noise is presented in Figure 1. This linear relationship is used to obtain the initial model by converting the gravity data to basement depth.

The gravity response of the initial basement model is calculated by using a unit density, i.e. 1 g/cm^3.or 1000 kg/m^3. The linear relationship between observed and calculated gravity data determines the density contrast (see Figure 2). The results for synthetic data with 0.5 mGal Gaussian noise (not shown) are nearly similar to those from noise-free data, since the noise only introduced small alterations in the synthetic data. From both noise-free and noisy data, the density contrasts were estimated to be -0.4844 and -0.4845 g/cm^3 respectively, which are very close to the true value of -0.5 g/cm^3.

The initial gravity-depth relationship is applied to the residual, i.e. difference between observed and calculated gravity at each grid, to update iteratively the model. The basin depth at k-th iteration is formulated by,

$$b^k(x_i) = b^{k-1}(x_i) + \left(g_{obs}(x_i) - g_{calc}(x_i, \Delta\rho_{est}, b^{k-1}) \right) m \tag{1}$$

where $g_{obs}(x_i)$ is the observed gravity anomaly field at observation points x_i, $g_{cal}(x_i, \Delta\rho_{est}, b_{k-1})$ is the computed gravity anomaly as function of basement model from previous iteration b_{k-1} and the estimated density contrast $\Delta\rho_{est}$ and m is the slope of the line that best approximates the gravity-depth relationship. The iteration is stopped when the misfit is less than a threshold value or after a maximum number of iterations. In this case, we performed systematically up to 8-th iteration for all tests without other criteria for convergence.

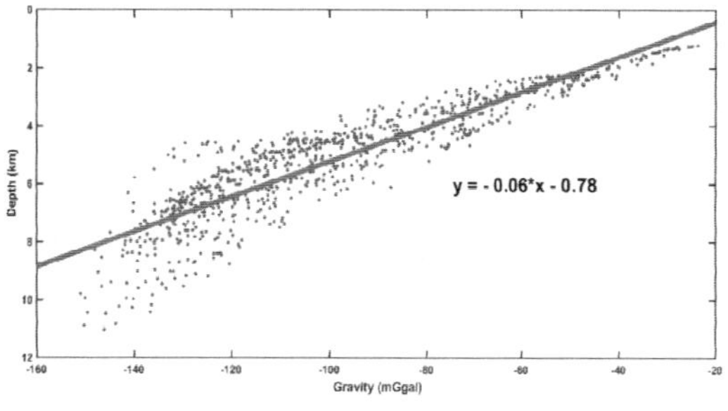

Figure 1. Depth-gravity relationship from the regression between basement depths of the synthetic model and their associated gravity response.

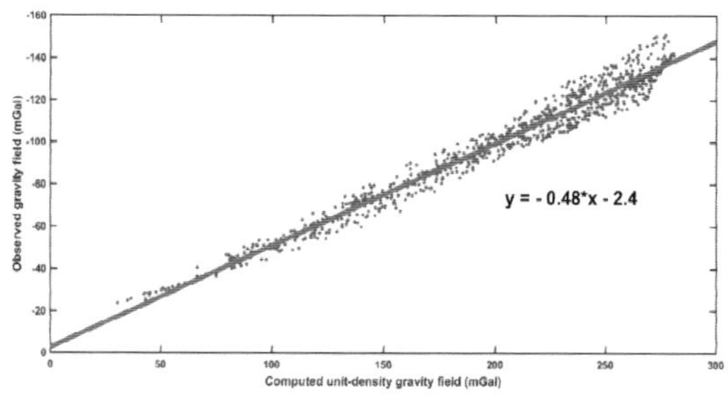

Figure 2. The regression between observed and calculated gravity response of the initial model with unit density to obtain estimated density contrast.

3. Results and Discussion

The synthetic model and its associated synthetic gravity data are shown in Figure 3a. There are five sub-basins: three of them are in the central part with a NW-SE trend, while two other sub-basins with similar trend are in SW and NE parts of the area. The gravity-depth regression technique was applied to the synthetic data without noise. Figure 3b presents the recovered basin model with calculated gravity anomaly map. A similar set of figures is presented in Figure 4 as profiles crossing diagonally the area in SW to NE direction. The initial model from gravity-depth conversion and the model from final iteration are compared with the synthetic model. The recovered and synthetic models are quite similar in both map and mostly in profile representations. The RMS error is 0.396 mGal after 8 iterations.

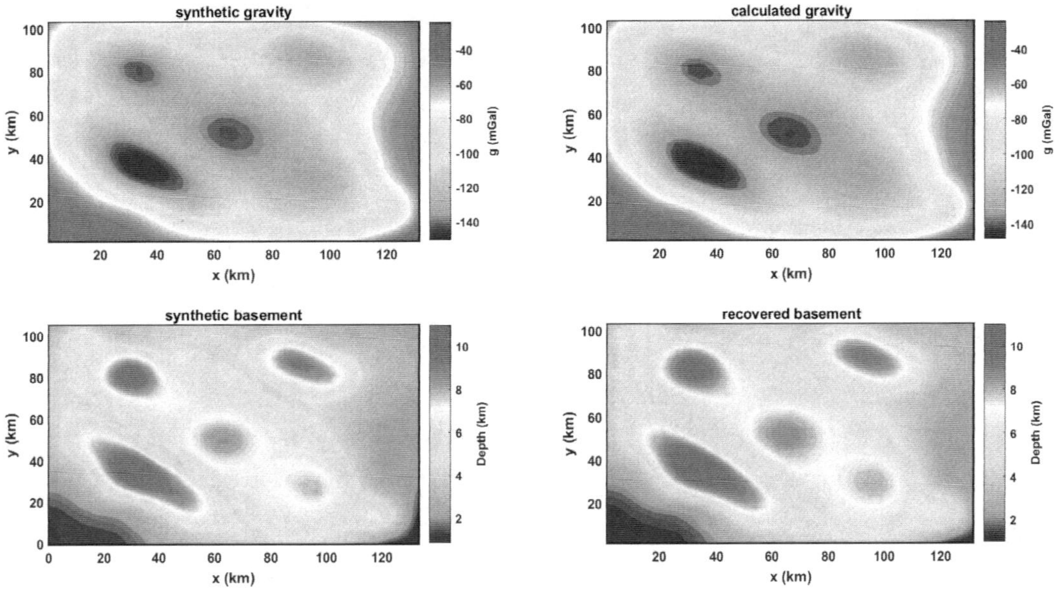

(a) (b)

Figure 3. (a) Synthetic model and data without noise, (b) recovered model and calculated gravity.

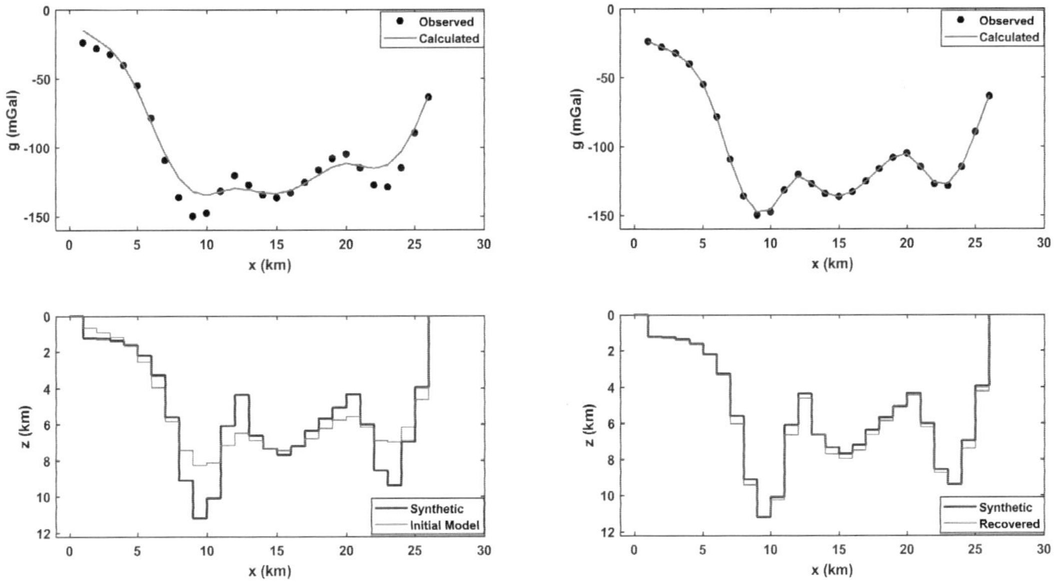

(a) (b)

Figure 4. Comparisons of synthetic data and calculated gravity anomaly for profiles crossing diagonally SW- NE the central part of the area shown in Figure 3, (a) for initial model and (b) for final model, both shown at the bottom of the figure compared with the synthetic model.

IOP Conf. Series: Earth and Environmental Science 1031 (2022) 012024 doi:10.1088/1755-1315/1031/1/012024

The second case deals with the same synthetic model where Gaussian noise of 0.5 mGal was added to the synthetic data. The results are presented, with similar set of figures as previously done with noise-free synthetic data, in Figure 5 and Figure 6.

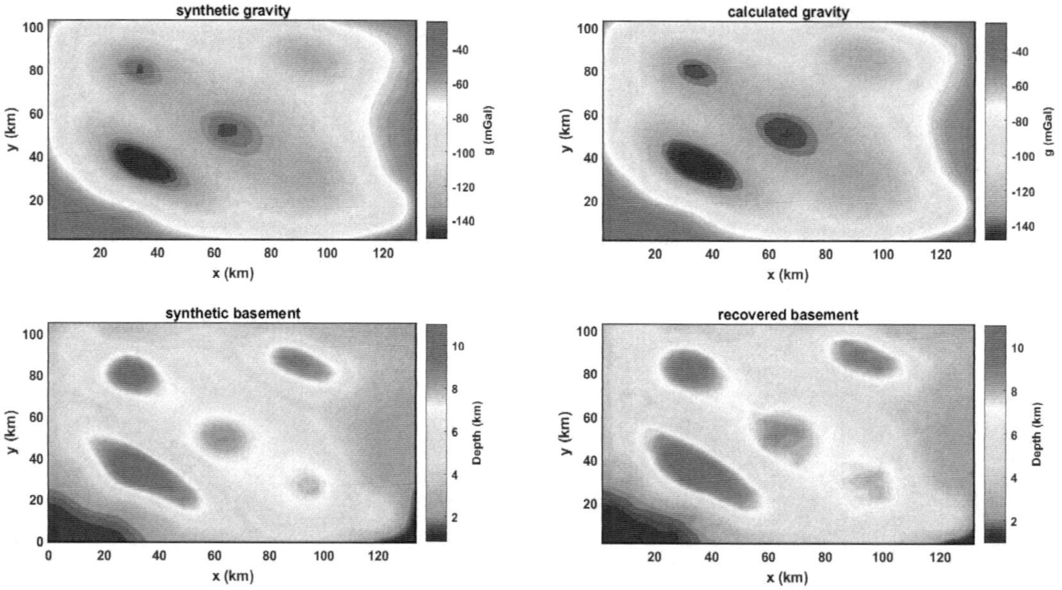

(a) (b)

Figure 5. (a) Synthetic model and data with 0.5 mGal Gaussian noise, (b) recovered model and calculated gravity.

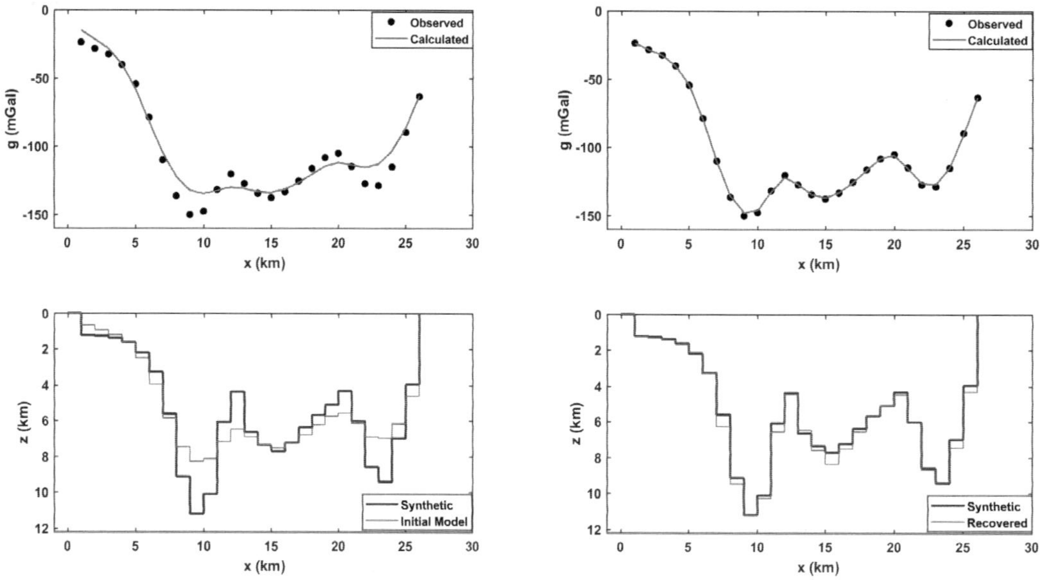

(a) (b)

Figure 6. Comparisons of synthetic data and calculated gravity anomaly for profiles crossing diagonally the central part of the area shown in Figure 5, (a) for initial model and (b) for final model, both shown at the bottom of the figure compared with the synthetic model.

IOP Conf. Series: Earth and Environmental Science 1031 (2022) 012024 doi:10.1088/1755-1315/1031/1/012024

The recovered model exhibits only minor differences, i.e. only in the central part of the basin model. The RMS error for the final model is 0.5 mGal after 8 iterations. The addition of 0.5 mGal Gaussian noise does not alter much the synthetic data as shown in profiles (Figure 6). In general, tests using both noise-free and noise-added synthetic data resulted in satisfactorily similar recovered models. The misfit of the final model is also in the same level of the noise that was introduced to the synthetic data. However, a higher level of noise may result in more oscillations of the recovered model.

4. Conclusion

The iterative 3D basement modelling based on gravity-depth regression was presented with satisfactory results. The problem of determining the density contrast is also resolved by this method. The algorithm was applied for synthetic data associated with synthetic model having only a constant density contrast for simplicity. In fact, the original algorithm proposed by Florio [4,5] can also be applied for the density contrast varying with depth. In such case, the gravity-depth relationship becomes non-linear.

References

[1] Reynolds JM 2011 An introduction to applied and environmental geophysics, Wiley.
[2] Cordell L, Henderson RG 1968 *Geophysics* **33** 596-601.
[3] Blakely RJ 1996 Potential theory in gravity and magnetic applications, Cambridge.
[4] Florio G 2018a SEG Technical Program Expanded Abstracts 2018 1415-1419.
[5] Florio G 2018b *Journal of Geophysical Research: Solid Earth* **123** 9101-9120.

Comparative Study of Gradient Gravity Method Between Signal Processing and Direct Measurement

Purwaditya Nugraha[1*], Wawan Gunawan A. Kadir[2], Setianingsih[2]

[1]Geophysical Engineering, Institut Teknologi Sumatera
[2]Faculty of Mining and Petroleum Engineering, Institut Teknologi Bandung, Basic Science Center B 2nd Floor, Jalan Ganesha 10, Bandung 40132, Indonesia.

*corresponding author's email: purwaditya.nugraha@tg.itera.ac.id

Abstract. The vertical gradient gravity method is one of the developments from the conventional gravity method. The advantage of this method is that it is able to detect the presence of shallow anomalies below the soil surface. In addition, this method is also able to minimize errors resulting from processing and correcting data. This research was conducted by testing the vertical gradient gravity method which is carried out with signal processing or with a vertical gradient gravity filter and the vertical gradient gravity method which is carried out by direct measurement. It was found that the vertical gradient gravity method which was carried out by direct measurements resulted in a clearer pattern in determining the presence of shallow anomaly sources.
Keywords: gravity, vertical gradient, shallow anomaly

1. Introduction

The gravity method is one of the important geophysical methods for solving many problems that involve mapping the subsurface of the earth, besides this method is also the main method in certain geological studies[1], [2]. Currently, several techniques have been developed in the utilization of gravity to determine subsurface density variations with better results, one of these techniques is the vertical gradient gravity technique[3]–[8]. The vertical gradient gravity technique can be used to determine and strengthen the source of gravity anomaly at shallow depths where it is difficult to obtain using the Bouguer gravity method only.

2. Gravity Method

The Bouguer gravity method is a geophysical method that can identify subsurface conditions based on variations in density which is then called the Bouguer anomaly. Bouguer anomaly is defined as the difference between the observed gravity from the measurement results and the theoretical gravity based on the specified earth model. The value of the observed gravity measured at the earth's surface is the total value of the gravity suffered at one measurement point due to various sources. So that to be able to use the gravity data from the measurement results, the data needs to be corrected and reduced to eliminate unwanted values [1], [2]. Mathematically, to get a complete Bouguer anomaly you can use the following equation:

$$CBA = g_{obs} - g_N + FAC - BC + TC$$

Where *CBA*, g_{obs},g_N, *FAC*, *BC*, and *TC* is Complete Bouguer Anomaly, Observed Gravity, Theoritical Gravity, Free Air Correction, Bouguer Correction, and Terrain Correction respectively.

The vertical gradient gravity method is a development of the Bouguer gravity method where this method can detect local heterogeneity in the distribution of shallow depths[5], [8]. One application of the vertical gradient gravity method is to detect the presence of geological formations at shallow depths. In the response to the gravity anomaly of the vertical gradient, an anomaly source in the form of an object with a density contrast at shallow depth is stronger, because this method is very sensitive to anomaly sources with shallow depths. The main factor that influences the measurement of the vertical gradient gravity method is the free air factor because in the measurement process this method requires two measurements at one measurement point location. The first measurement is carried out above the ground or the height of the gravimeter is close to 0 cm, then at the same point the second measurement is carried out by lifting the gravimeter with a tripod at a certain height[9]. The scheme for measuring the gravity of the vertical gradient can be seen in Figure 1.

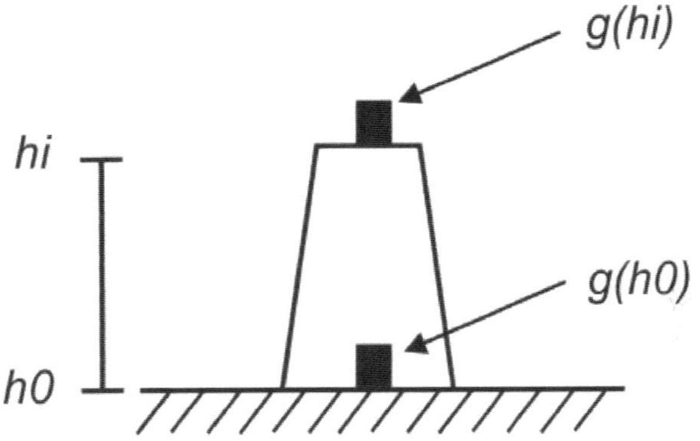

Figure 1 Schematic measurement of the vertical gradient gravity technique, where h0 is the height of the gravitimeter at ground level and hi is the measurement by lifting the gravitimeter to the height level i, then g (h0) and g (hi) are the measurement of gravity on the ground and measurement raised at height i.

Where h0 is the height of the gravimeter at ground level and hi is the height of the gravimeter that is lifted at a certain height of i. Then g (h0) and g (hi) are the measurement of gravity above the ground and the measurement raised at a certain height, respectively. Mathematically, the calculation of the vertical gradient gravity method can be written as follows:

$$\frac{\partial g}{\partial z} = \lim_{h \to 0} \frac{\Delta g}{\Delta h}$$

$$g_{zz} = \frac{g(h0) - g(hi)}{h0 - hi}$$

So that based on the schematic and mathematical equations it can be concluded that the vertical gradient gravity technique is the difference in the value of gravity at the same point at two different heights where based on the free air correction the value is 0.308 mGal / m [8], [10].

3. Simulation

The vertical gradient gravity method is tested using synthetic data. The test aims to determine the pattern and characteristics of the vertical gradient gravity method. This test is carried out on the vertical gradient gravity method which is carried out by direct measurement and the gravity method which is carried out using the vertical gradient gravity filter.

IOP Conf. Series: Earth and Environmental Science 1031 (2022) 012025 doi:10.1088/1755-1315/1031/1/012025

The test scenario is carried out by making a simple model with several sources of anomaly at a shallow depth, then calculating the gravity of the vertical gradient using a vertical gradient gravity filter and direct calculations. The use of a vertical gradient gravity filter was carried out using Potensoft [11] based on the equation obtained by the potential gravity [12]. anomaly response at different elevations, so that the vertical gradient gravity anomaly can be calculated.

The first model is carried out on a model with one anomaly source as shown in Figure 2. The model was made with one source of anomaly with a density contrast of 0.2 g / cc at a depth of 1 meter below the ground surface. Based on this model, the gravity anomaly response is calculated so that it is obtained as shown in Figure 3. The next calculation is carried out by the same process by changing the elevation by one meter so that it is obtained as shown in Figure 4.

Based on the gravity anomaly data at an elevation of 0 meters and an elevation of 1 meter, the vertical gradient gravity anomaly is calculated by using the equation to get the gzz, then a direct measurement vertical gradient gravity anomaly map is obtained as shown in Figure 5

The next processing is processing the vertical gradient gravity anomaly using a vertical gradient filter. This process was carried out using Potensoft software[11]. The vertical gradient gravity anomaly is obtained as shown in Figure 6.

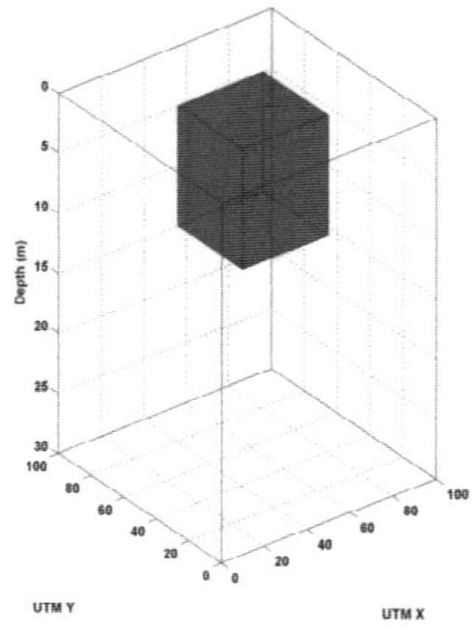

Figure 2. model with 1 source of anomaly

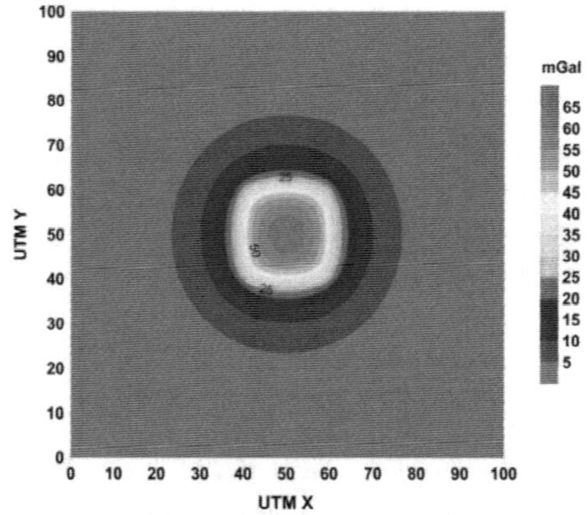

Figure 3. Gravity Anomaly Response due to 1 source of anomaly at 0 m Elevation

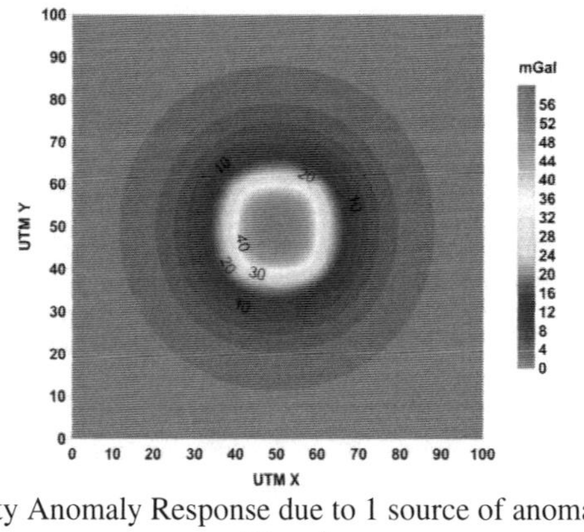

Figure 4. Gravity Anomaly Response due to 1 source of anomaly Source at 1 m Elevation

4. Result and Discussion

Based on the test results of the vertical gradient gravity method using the direct measurement vertical gradient gravity method and using a vertical gradient gravity filter, it was found that the results of both methods were able to show the presence of anomalous source at shallow depths quite well as in Figure 5 and Figure 6.Only the results based on the vertical gradient gravity filter, a gross pattern or error around the boundary of the anomaly source is greater than using the direct measurement of the vertical gradient gravity method, so to obtain subsurface anomalies based on the gravity method, we can use the vertical gradient gravity method, but for getting better results is recommended to use the gravity method by taking direct measurements.

5. Conclusion

The vertical gradient gravity method is able to determine the presence of shallow anomalies. The vertical gradient gravity anomaly from the vertical gradient filter results in a larger error value than the vertical gradient gravity anomaly from the direct measurement. The recommended application of the vertical gradient gravity method to get good results can use the vertical gradient gravity method by taking direct measurements.

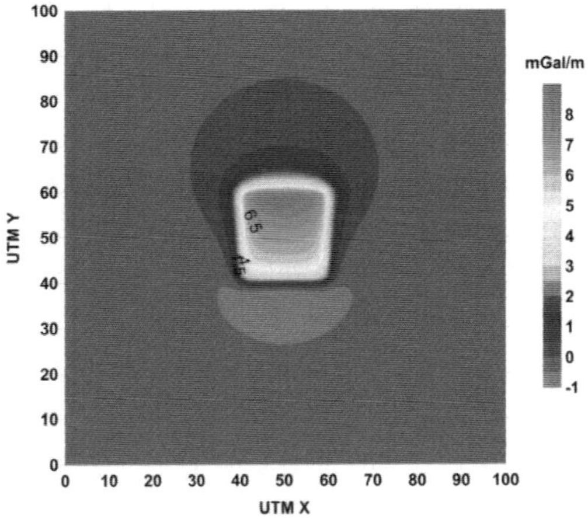

Figure 5 Vertical Gradient Gravity Anomaly Response of Direct Measurement Results

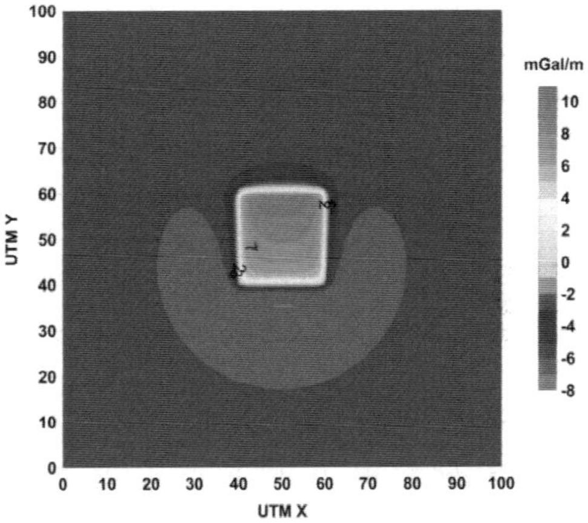

Figure 6 Vertical Gradient Gravity Anomaly Response Result of Signal Processing

References

[1] Hinze W J, von Frese R R B and Saad A 2013 *Gravity and magnetic exploration: principles, practices and exploration.*

[2] Telford W M, Geldart L P and Sheriff R E 2004 Applied Geophysics 2nd Edition_ Telford Geldart Sheriff.pdf.

[3] Ager C A and Liard J O 2002 Vertical gravity gradient surveys: Field results and interpretations in British Columbia, Canada *Geophysics* **47** no 6 pp 919–925 doi: 10.1190/1.1441358

[4] Butler D K 2002 Microgravimetric and gravity gradient techniques for detection of subsurface cavities *Geophysics* **49** no 7 pp 1084–1096 doi: 10.1190/1.1441723

[5] Fajklewicz Z, Glinski A and Sliz J 2002 Some applications of the underground tower gravity vertical gradient *Geophysics* **47** no 12 pp 1688–1692 doi: 10.1190/1.1441318

[6] Hunt T, Sugihara M, Sato T, and Takemura T 2002 Measurement and use of the vertical gravity gradient in correcting repeat microgravity measurements for the effects of ground subsidence in geothermal systems *Geothermics* **31** no 5 pp 525–543 doi: 10.1016/S0375-6505(02)00010-X

[7] KLINGELÉ E E, MARSON I and KAHLE H-G 1991 Automatic Interpretation of Gravity Gradiometric Data in Two Dimensions: Vertical Gradient *Geophys. Prospect.* **39** no 3 pp 407–434 doi: 10.1111/j.1365-2478.1991.tb00319.x

[8] Valley T 1976 GRAVITY VERTICAL GRADIENT MEASUREMENTS FOR THE

DETECTION **41** no 5 pp 1016–1030

[9] Wibowo S N E, Mamuaya G E and Djamaluddin R 2018 Land Subsidence Analysis of Reclaimed Land using Time-Lapse Microgravity Anomaly in Manado, Indonesia *Forum Geogr.* **32** no 1 doi: 10.23917/forgeo.v32i1.5882

[10] Dykowski P 2012 Vertical gravity gradient determination for the needs of contemporary absolute gravity measurements–first results *Reports Geod.* **92** no 1 pp 23–36 [Online] Available: http://yadda.icm.edu.pl/baztech/element/bwmeta1.element.baztech-50441301-2fcb-4e73-8da2-6853629d193f.

[11] Özgü Arisoy M and Dikmen Ü 2011 Potensoft: MATLAB-based software for potential field data processing, modeling and mapping *Comput. Geosci.* **37** no 7 pp 935–942 doi: 10.1016/j.cageo.2011.02.008.

[12] R. J. Blakely 1995 *Potential Theory in Gravity and Magnetic*

IOP Conf. Series: Earth and Environmental Science 1031 (2022) 012026 doi:10.1088/1755-1315/1031/1/012026

GMODL: An Indonesian MATLAB-based ground-penetrating radar data modeling and processing software

Widodo Widodo[1], Kurnia Anwar Ra'if[2], Muhammad Aldi Firdaus[1], Ibnu Thoriq Hidayatullah[2]

[1]Department of Geophysical Engineering, Faculty of Mining and Petroleum Engineering, Bandung Institute of Technology, Bandung, West Java, Indonesia.

[2]Department of Physics, Faculty of Mathematics and Natural Sciences, Bandung Institute of Technology, Bandung, West Java, Indonesia.

Email: widodo@gf.itb.ac.id

Abstract. Ground-penetrating radar (GPR) data modeling and processing is crucial to near-surface geophysics. Its use in geohazard mitigation, construction, shallow hydrocarbon contamination, and other shallow subsurface detection is undeniable due to its high-resolution imaging. GPR software (MATGPR) that we used currently requires access to MATLAB which not everybody can use because of its licensing prices. Thus, we developed this program (GMODL) that uses finite-difference time-domain (FDTD) and split-step Fourier algorithms. To test the software, we created synthetic models. The synthetic model used for the testing is a river model for flood mitigation that consists of a layer of freshwater with ρ = 20 Ωm, k = 81 and μ_r = 1 of depth 5 m, two layers of sandstone with ρ = 850 Ωm, k = 2.5 and μ_r = 1 of total depth 4 m, and a layer of claystone with ρ = 120 Ωm, k = 11 and μ_r = 1 of depth 1 m. The GPR antenna frequency is set to 250 MHz. The testing algorithm of GMODL shows that it can create a synthetic radargram for the river model. The boundary layers are obvious to identify. Also, the freshwater thickness can be determined from the radargram.

1. Introduction

GPR is a non-intrusive, near-surface geophysics method that uses electromagnetic (EM) waves to image the subsurface. GPR used two antennas, a transmitter that transmits EM waves and is reflected by a reflective surface into the receiver. The two-way travel time (TWTT) is then plotted against the scan-axis, creating a radargram. The radargram's TWTT is then converted to depth estimates using 1D EM wave propagation velocity structure [1].

To produce a radargram from synthetic modeling or field data acquisition, there needs to be software that encompasses the algorithms necessary for GPR modeling. GMODL provides the tools necessary to model radargrams such as FDTD and split-step. The program is executable and doesn't need to be accessed from MATLAB. For testing purposes, we have decided to make a synthetic radargram of the river model to determine freshwater thickness because river monitoring using GPR is used for flood mitigation.

2. Methods

2.1. Finite-Difference Time-Domain (FDTD)
The FDTD modeling in GMODL uses a concept from Irving and Knight[4]. It is used by applying the TM (Transverse Magnetic) concept to an antenna perpendicular to the x-z plane. To prevent reflection from the edges of the model, we used the coordinates for the absorption boundary at a perfectly matched layer (PML) [2].

For TM cases on the x-z plane, the magnetic waves are in the x-direction (H_x) and the z-direction (H_z). Then the electric waves in the y-direction (E_y) are as follows:

$$\mu \frac{\partial H_x}{\partial t} = -\frac{1}{k_z}\frac{\partial E_y}{\partial z} - \zeta_z(t) * \frac{\partial E_y}{\partial z} \quad (1)$$

$$\mu \frac{\partial H_z}{\partial t} = -\frac{1}{k_x}\frac{\partial E_y}{\partial x} - \zeta_x(t) * \frac{\partial E_y}{\partial x} \quad (2)$$

$$\sigma E_y + \epsilon \frac{\partial E_y}{\partial t} = \frac{1}{k_x}\frac{\partial H_z}{\partial x} - \frac{1}{k_z}\frac{\partial H_x}{\partial z} + \zeta_x(t) * \frac{\partial H_z}{\partial x} - \zeta_z(t) * \frac{\partial H_x}{\partial z} \quad (3)$$

σ, σ, ϵ, and μ are each physical parameters in the form of electrical conductivity, dielectric constant or permittivity, and magnetic permeability. *k* is stretching the coordinates in the x,y, and z-direction. The equation above is modeled with finite-difference using leapfrog. With this approach, the electric field and the magnetic field become balanced as in the partial derivative approach on the same grid using a fourth-order finite-difference for space and second-order for time.

FDTD uses an algorithm proposed by Kane Yee in 1966. This algorithm uses second-order on central difference [3]. The algorithm is as follows:

1. Ampere's Law and Faraday's Law in Maxwell's equation are modified, the differential form is converted into the finite difference form. Discretization of space and time so that the electric field and the magnetic field are both in space and time.
2. Then the problem of the modified equation is solved to obtain the "update equation" as the shape of the plane of the object that is not known from the shape of the plane of the object that is known.
3. Evaluate the magnetic field at one time on an unknown object so that we find that unknown object (effectively this is obtained from a known object).
4. Do the same step as step 3 but this time we evaluate the electric field.
5. Then, repeat the previous two steps until the field obtains the desired results.

Within GMODL, there exists 2 different data set to store data. The first one is called Old Data (OD). OD dataset is used for storing imported data and processed data that will be processed further. The second one is called New Data (ND). This ND data set is used to store data resulted from the process within GMODL. In GMODL field data processing, imported data from field acquisition gets saved to the OD. When field data is processed, the results will be stored temporarily in the ND dataset. The user then can review the data within ND and choose whether they want to keep the data within ND or to discard it. If the user keeps the data in ND, then the data within OD will be replaced by the new data from ND. Then the ND data set will be emptied and ready to store another data resulted from the process within GMODL. The same process happened for modeling. The results from modeling will be stored in the ND data set and can be discarded or be stored in OD for further processing.

2.2. Split-Step Fourier Method
In GMODL, there are two main concepts used for Split Step Fourier Method, constant reference slowness and perturbation term [4]. Fourier transformation methods are used to solve the acoustic wave equation with constant lateral velocity. Within the split-step Fourier method, the reciprocal of

interval velocity is divided into two parts: a constant reference slowness and a perturbation term. Reference slowness is used to migrate across intervals for the wave-frequency domain as a form of the constant-phase value for phase-displacement.

Consider a wave equation with the propagation of compressional waves within an acoustic constant density medium:

$$\nabla^2 p - \frac{u^2 \partial^2}{\partial t^2} p = 0 \tag{4}$$

With p = p(x,y,z,t) is the pressure dan u = u(x,y,z) is the medium slowness, which is defined as the inverse of half the propagation velocity u(x,y,z) = 2/v(x,y,z) with *exploding reflector model*. After equation (4) has been transformed, the domain frequency can be obtained as:

$$\nabla^2 P + \omega^2 u^2 P = 0 \tag{5}$$

with,

$$\boldsymbol{P}(r, z, \omega) = \int_{-\infty}^{\infty} p(\boldsymbol{r}, z, t) e^{-i\omega t} \, dt \tag{6}$$

Where r is the horizontal position vector which is defined as the following:

$$r = |\boldsymbol{r}| = \sqrt{x^2 + y^2}$$

With *slowness field value* u(**r**,z) into 2 components:

$$u(\boldsymbol{r}, z) = u_0(z) + \Delta u(\boldsymbol{r}, z) \tag{7}$$

By defining $t_o(z)$ as reference slowness value, equation (7) is then substituted to equation (5):

$$\nabla^2 P + \omega^2 u^2 P = -\omega^2 (2u_0 \Delta u + \Delta u^2) P$$

With,

$$\boldsymbol{S}\,(\boldsymbol{r}, z, \omega) = \omega^2 \big(2u_0 \Delta u(\boldsymbol{r}, z) + \Delta u^2(\boldsymbol{r}, z)\big) P(\boldsymbol{r}, z, \omega)$$

We then obtain,

$$\nabla^2 P + \omega^2 u^2 P = -\boldsymbol{S}\,(\boldsymbol{r}, z, \omega) \tag{8}$$

Therefore, homogeneous acoustic wave equation (5) is transformed into the inhomogeneous wave equation (8) because there is a source S(**r**,z,ω) during the varied slowness process.

The solution of equation (8) is then used for the split-step Fourier method by ignoring Δu^2 as a contributing factor. We can conclude that in the previous Fourier transformation, the wave is reflected as a field wave at depth z_n, $\boldsymbol{P}_{-}(\boldsymbol{r}, z_n, \omega)$, from **r** to **k**. Then,

$$\boldsymbol{P}_{-}(\boldsymbol{r}, z_n, \omega) = \int_{-\infty}^{\infty} \boldsymbol{P}_{-}(\boldsymbol{r}, z_n, \omega) \, e^{i k_r . r} \, dr, \tag{9}$$

Where $\boldsymbol{k}_r$ is the horizontal wave vector with $\boldsymbol{k}_r = k_x \boldsymbol{i}_{k_x} + k_y \boldsymbol{i}_{k_y}$.

$$k_r = |\boldsymbol{k}_r| = \sqrt{k_x^2 + k_y^2}$$

Apply a phase shift based on the vertical wavenumber and compute using reference slowness for all frequency and wavenumber,

$$\boldsymbol{P}_1(\boldsymbol{k}_r, z_n, \Delta z, \omega) = \boldsymbol{P}_{-}(\boldsymbol{k}_r, z_n, \omega) \, e^{i k_{z_0} \Delta z}$$

with,

$$k_{z_0} = \sqrt{\omega^2 u_0^2 - k_r^2} = \omega u_0 \sqrt{\left(1 - \frac{k_r}{\omega u_0}\right)^2} \tag{10}$$

And u_0 is the average slowness value for interval Δz with the assumption of very small Δz value, inverse Fourier transform as data shift $P_1((\mathbf{k_r}, z_n, \Delta z, \omega))$ from $\boldsymbol{k}_r$ to r :

$$\boldsymbol{P_1}(\boldsymbol{r}, \mathrm{z_n}, \Delta z, \omega) = \frac{1}{4\pi^2} \int_{-\infty}^{\infty} P_1(\mathbf{k_r}, z_n, \Delta z, \omega) e^{-i k_r . r} \, d\boldsymbol{k}_r \tag{11}$$

Then, we use the second phase shift for perturbation inside slowness, Δu (**r**,z) = u(**r**,z) – $u_0(z)$, within interval Δz :

$$\boldsymbol{P}(\boldsymbol{r}, z_{n+1}, \omega) = e^{i\omega \Delta u(r,z) \Delta z} \, \boldsymbol{P_1}(\boldsymbol{r}, \mathrm{z_n}, \Delta z, \omega) \tag{12}$$

$P_{-}(\boldsymbol{r}, z_{n+1}, \omega)$ is then integrated within the frequency, from ω_1 to ω_2 to obtain the displacement data for every depth z_{n+1} :

$$\boldsymbol{P}(\boldsymbol{r}, z_{n+1}, 0) = \frac{1}{4\pi^2} \int_{\omega_1}^{\omega_2} P_{-}(\boldsymbol{r}, z_{n+1}, \omega) \, d\omega \tag{13}$$

IOP Conf. Series: Earth and Environmental Science 1031 (2022) 012026 doi:10.1088/1755-1315/1031/1/012026

We do the first identical phase shift for the constant-velocity phase shift migration value. The secondary phase shift functions as a correction factor provided that it acts as a time shift based on the difference between the actual value and the reference slowness of each spatial position.

3. Results

GMODL can make synthetic models for forward modeling and process field data. The synthetic model is set with an antenna frequency of 250 MHz. First, a homogeneous model is made to check the fitness of the grid and radargram size. The model is based on the standard river stratigraphy. The model's parameter values are reference values[5,6]. The model consists of a layer of freshwater with $\rho = 20\ \Omega$m, $k = 81$ and $\mu_r = 1$ of depth 5 m, two layers of sandstone with $\rho = 850\ \Omega$m, $k = 2.5$ and $\mu_r = 1$ of total depth 4 m, and a layer of claystone with $\rho = 120\ \Omega$m, $k = 11$ and $\mu_r = 1$ of depth 1 m. The model's radargram is made using FDTD and split-step. The models and their radargrams can be seen in figures 2 and 3.

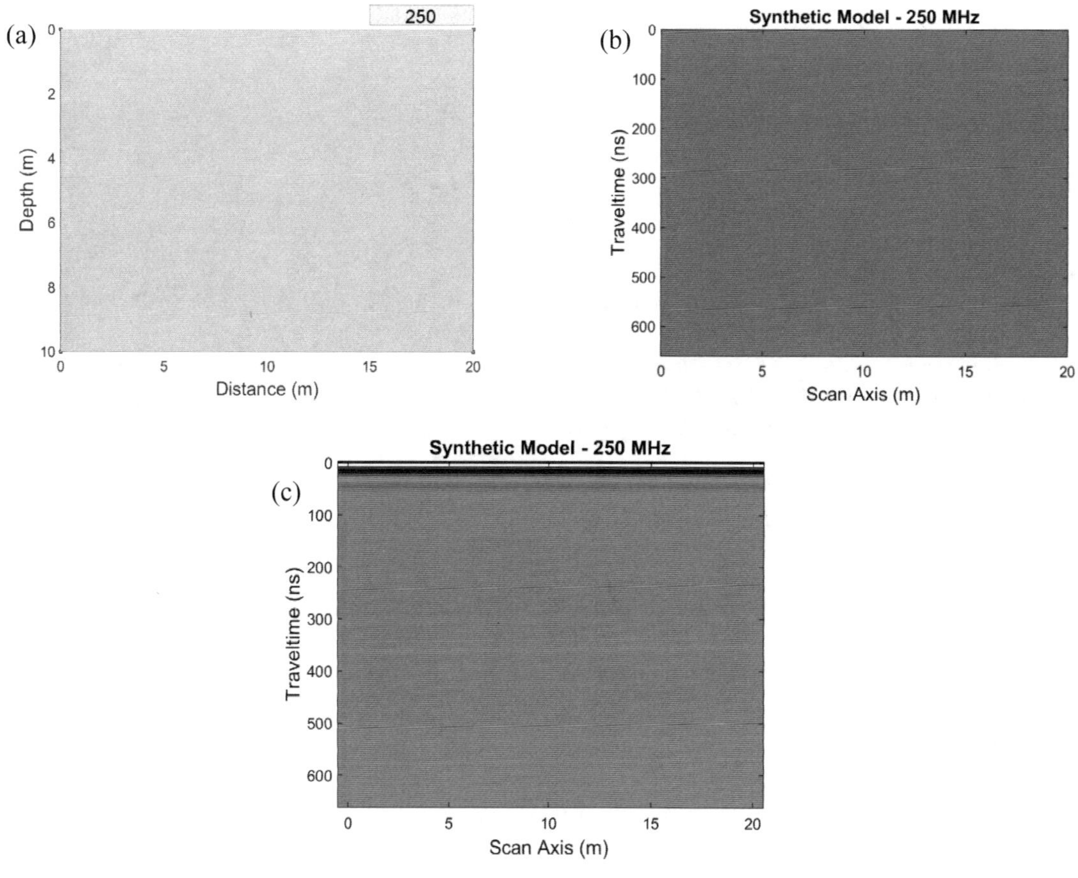

Figure 1. (From left to right) a.) Geological homogeneous model. b.) Split-step radargram of the model. c.) FDTD radargram of the model

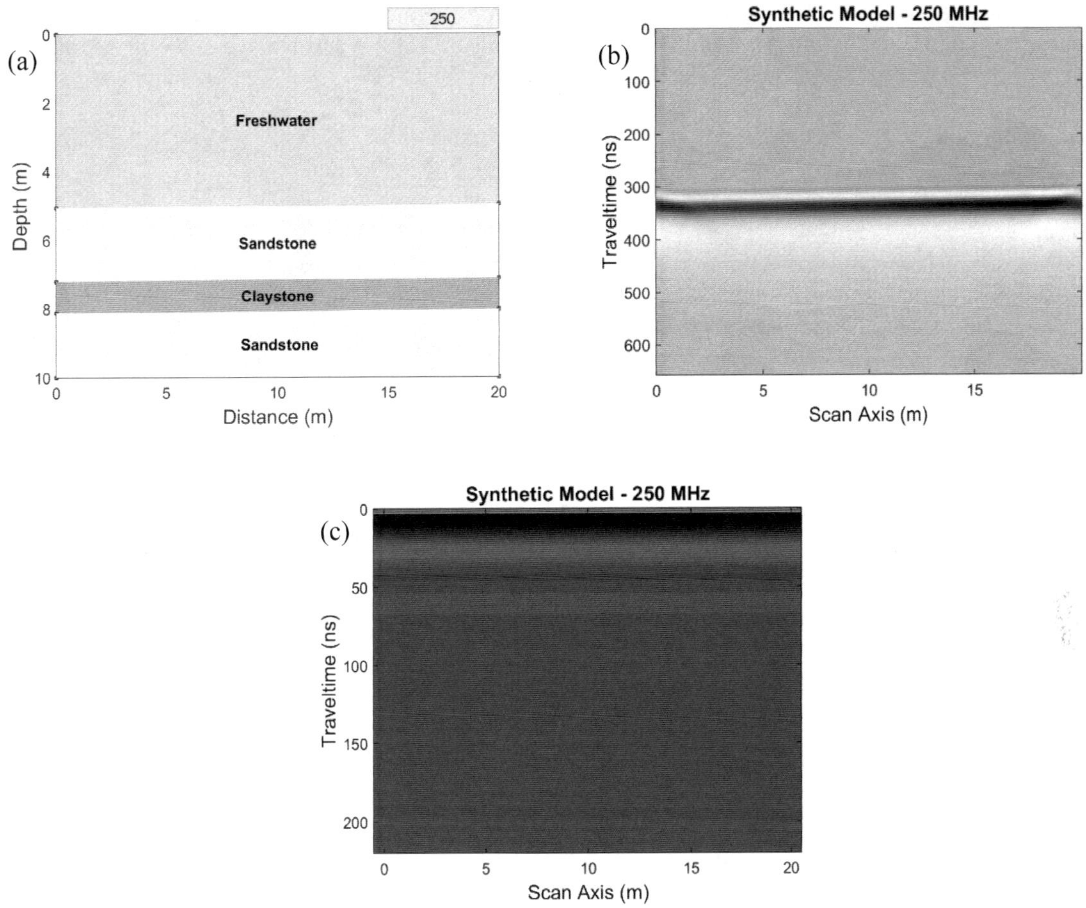

Figure 2. (From left to right) a.) Geological model of the standard river. b.) Split-step *radargram* of the model. c.) FDTD radargram of the model

The split-step Fourier method's radargram is closer to an actual radargram than its FDTD counterpart. The boundary layers of the split-step radargram are distinguishable (figure 1b and 2b). And the layering of the homogenous and river model (figure 1b and 2b) is perfectly aligned with the geological model (figure 1a and 2a) Based on the TWTT (top TWTT = 0 ns, bottom TWTT = 300 ns) and velocity structure [1] of the freshwater layer (calculated to be 0.033302 m/ns) and using the layer thickness equation [7] the freshwater thickness of the split-step layer is calculated to be 4.9953 m. The FDTD radargram's TWTT range of 0 – 15 ns (figure 1c and 2c) produces a layer of fresh water with a calculated thickness of 0.249765 m. The distinct difference of the FDTD radargram's TWTT is likely caused by the velocity pull-up caused by the contrast claystone layer. The split-step radargram is closer to the geological model than the FDTD radargram. Despite the differences, the FDTD radargram of the homogeneous and river model (figure 1c and 2c) shows correctly the layering of the geological model (figure 1a and 2a)

4. Conclusion

The program that used in GMODL is executable and doesn't need to be accessed from MATLAB. GMODL provides the tools necessary to model radargrams such as FDTD and split-step. The actual radargram result is closer to an actual radargram result when using split-step Fourier method's than using FDTD. For the boundary layers of the split-step radargram are distinguishable and obvious to see clearly. So, the geological model is clearly can be interpreted more by using split-step radargram than the FDTD radargram. GMODL is successful in creating a synthetic radargram of an underwater geological model. GMODL can be used for a variety of GPR modeling and we hope that Indonesian geophysicist can utilized GMODL for other GPR-related investigations.

Acknowledgment
This research received financial support from LPPM ITB and DIKTI.

References
[1] Bano M 1996 Constant dielectric losses in ground-penetrating radar waves *Geophys.J.Int.* **124** 279-288
[2] Ramadan O 2007 Second-order split-step envelope PML algorithm for 2D FDTD simulations *Electron. Lett* **43** 792-793
[3] Schneider J B 2020 Understanding the finite-difference time-domain method www.eecs.wsu.edu/~schneidj/ufdtd
[4] Stoffa P L, Fokkema J T, de Luna Freire R M and Kessinger W P 1990 Split-step Fourier migration *Geophysics* **55** 410-421
[5] Davis J and Annan A 1989 Ground-penetrating radar for high-resolution mapping of soil and rock stratigraphy *Geophys. Prospect.* **37** 535
[6] Palacky G 1988 Resistivity characteristics of geologic targets *Electromagnetic Methods in Applied Geophysics* vol 1 ed M Nabighian (Tulsa: Society of Exploration Geophysicists) chapter 3 pp 53-129
[7] Lachhab A, Booterbaugh A and Beren M 2015 Bathymetry and sediment accumulation of Walter Lake, PA using two GPR antennas in a new integrated method *JEEG* **20** 245-255

IOP Conf. Series: Earth and Environmental Science 1031 (2022) 012027 doi:10.1088/1755-1315/1031/1/012027

Analysis of Ground Shear Strain (GSS) In Mapping, Liquefaction Vulnerability Potential Using HVSR (Horizontal To Vertical Spectral Ratio) Method. Case Study: Sepaku Subdistrict, North Penajam Paser Regency, East Kalimantan

Wira Hadi Kusuma[1], **T.A. Sanny**[2]

[1] Graduate Bachelor Geophysical Engineering, Faculty of Mining and Petroleum Engineering, Institut Teknologi Bandung

[2] Exploration and Engineering Seismology Expertise Group, Department Geophysical Engineering, Faculty of Mining and Petroleum Engineering, Institut Teknoogi Bandung

Abstract. East Kalimantan has been designated as the new Capital Province of Indonesia. The criteria for determining the relocation of the nation's capital have passed various studies in the placement of a new state capital, one of which is by reviewing disaster aspects such as the risk of earthquakes, volcanoes and tsunamis. Based on historical records of earthquakes that have occurred, East Kalimantan is not a completely safe area from potential earthquakes or tsunamis. By using the HVSR method, the authors tried to determine the value of Ground Shear Strain (GSS) in the research area, namely Sepaku District, Penajam Paser Utara Regency. The result show that there were three zones namely Zona A, Zone B and Zone C had GSS values in Sepaku Subdistrict with landslides, soil compaction and liquefaction potential, namely 0.4 - 0.54. The result of HVSR interpretation curve are based on the phenomenon of the resulting ground shear strain value in accordance with the geological conditions in the research area of the Sepaku Subdistrict, Penajam Paser Utara Regency, East Kalimantan in characterizing the soil profile in the study area.
Keyword: Amplification Factor, Dominant Frequency, GSS, HVSR method, Liquefaction.

1. Introduction

Administratively, the research area is located in the province of East Kalimantan in North Penajam Paser Regency, Sepaku Subdistrict. The government has determined the relocation of the candidate for the nation's capital through various studies, such as assessing aspects of disaster, the risk of earthquakes, volcanoes, and tsunamis [1]. The area is located in the Kutai Basin. The Kutai Basin was formed due to basin subsidence from the Early Oligocene to the Early Miocene. The deposited material comes from the southern, western and northern parts of the Kutai Basin. The Adang Fault is located in the southern part of the study area which is thought to have formed during the Eocene period. Almost all of the rocks in this area are deformed, from Pre - Tertiary to Late Tertiary. As a result of this process an anticline and syncline were formed known as the Samarinda Anticlonorium and fault. During the Early Oligocene

IOP Conf. Series: Earth and Environmental Science 1031 (2022) 012027 doi:10.1088/1755-1315/1031/1/012027

to Early Miocene periods, there is a continuous decline. The deposited material comes from the southern, western and northern of the Kutai Basin areas (**Fig.1.**).

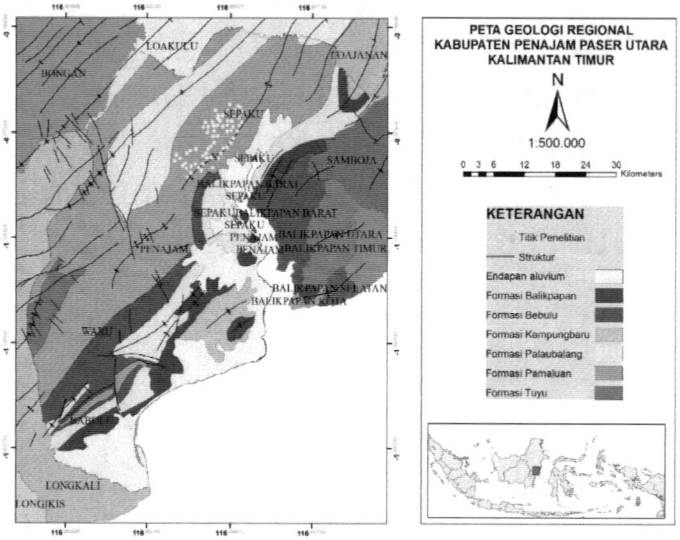

Figure 1. The regional geological map of the study area of Penajam Paser Utara Regency, East Kalimantan was modified from (Hidayat et al, 1994 and Supriatna et al, 1995).

2. Methodology

2.1 Amplification Factor
The value of the amplification factor of the soil is related to the contrast ratio of the impedance of the surface layer to the layer below it [2]. If the impedance contrast ratio of the two layers is high, the gain factor is high, and conversely. The energy from earthquake waves can be trapped in a soft layer near the surface.

2.2 Dominant Frequency
The dominant period (1 / dominant frequency) is the period of vibration that corresponds to the maximum value of the Fourier amplitude spectrum [3]. The dominant frequency value can be obtained through the HVSR curve where the amplitude (H / V) is the maximum value. The characteristic frequency of an area can be estimated through the dominant frequency obtained from the HVSR curve.

2.3 Ground Shear Strain
Ground shear strain is the ability of a soil layer material to stretch and shift during an earthquake. If the area has a high GSS value, potentially have a high risk of ground motion due to the earthquakes, such as soil compaction, ground vibrations, and soil stretching to liquefaction [2][4].

Table 1. The relationship between strain and soil dynamic properties (Nakamura, 1997).

Size of Strain γ	10^{-6} 10^{-5}	10^{-4} 10^{-3}	10^{-2} 10^{-1}
Phenomena	Wave, Vibration	Crack, Settlement	Landslide, Soil Compaction, Liquefaction
Dynamic Properties	Elasticity	Elasto-Plasticity	Collapse
		Repeat-Effect, Speed-Effect of Loading	

3. Result and Discussions

The distribution of the amplification factor value in the study area is 1,5 – 8 (**Fig.2**). The random distribution of values is proven when the dominant frequency value is low, the amplification value is high and conversely. The greater of the amplification factor value, so that the softer of the sediment in the area, conversely if the amplification factor value is low, the sediment layer will be harder.

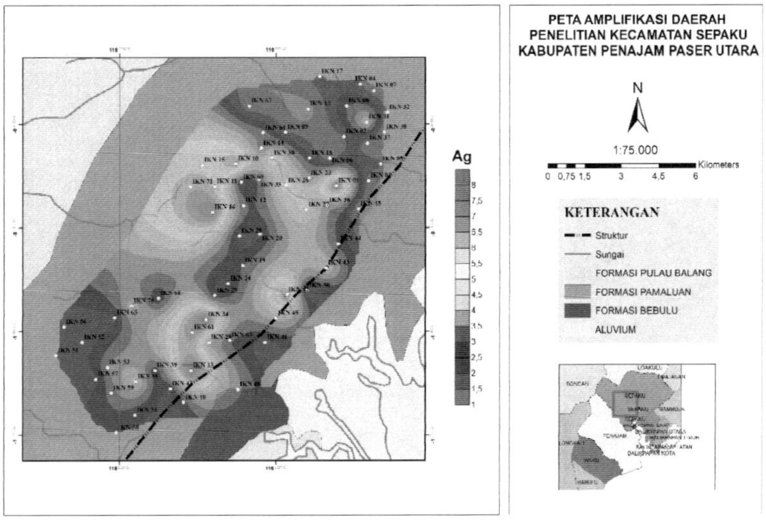

Figure 2. Amplification map of research areas in Sepaku District.

The dominant frequency value has a value in 0,5 – 14,5 Hz (**Fig.3.**). The surface sediment thickness of the study area was spread from low to high levels. Low dominant frequency is identified as having soft soil type. The areas that have a low dominant frequency value are very susceptible to multi-reflection or trapping of the earthquake waves in sediments. It causes of the large potential for damage to allow liquefaction [2].

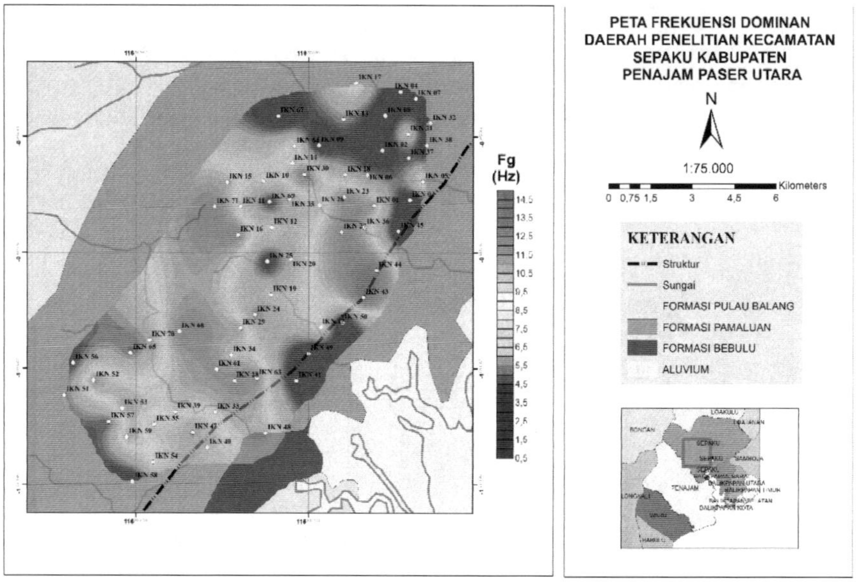

Figure 3. Dominant Frequency Map of research areas in Sepaku District.

The measured value of ground shear strain in the study area show the level of strain in the constituent materials during an earthquake. The amount of shear strain in a place by multiplying the seismic susceptibility index based on the microtremor with the amount of PGA in bedrock [2]. So that the values of the shear strain distribution in the study area is obtained, namely 0.02 - 0.54. The ground shear strain value classification can be determined in Table 1. If the strain 10^{-6}, the soil conditions can only occur vibrations, but in the strains 10^{-2} - 10^{-1}, the soil layers can be occur landslides to liquefaction. Based on the map that has been produced, there are three zones which can be seen in Figure IV divided into Zone A, Zone B and Zone C which are considered to have the potential for landslides, soil compaction to liquefaction based on the relatively high ground shear strain value, namely 0,4 - 0,54 (**Fig.4.**). The potential for this relatively high strain value is causing of soft soil types, shallow groundwater depths and weak plane zones. Based on the topography in the field, the geological cross-section analysis using HVSR parameters, Zone A and Zone B have the potential for soil compaction to liquefaction because it located in the lower Kutai Basin area which has a very thick surface sediment thickness. This zone is an indication of a weak plane zone which if the earthquake occurs, it will have high compression and extension forces. Whereas in Zone C, it can be potential for soil compaction to landslides. Because the resistance of slopes in high topography will decrease when the soil movement increases so that it will be accumulated by soil with high water content. This occurs is causing of high strain in the event of an earthquake or high pressure due to the compaction of sediment on the surface which can allow collapsible. The area is part of the mining and agricultural land that still operating with most of the densest forest located in the Tengin Baru area, Sepaku District.

IOP Conf. Series: Earth and Environmental Science 1031 (2022) 012027 doi:10.1088/1755-1315/1031/1/012027

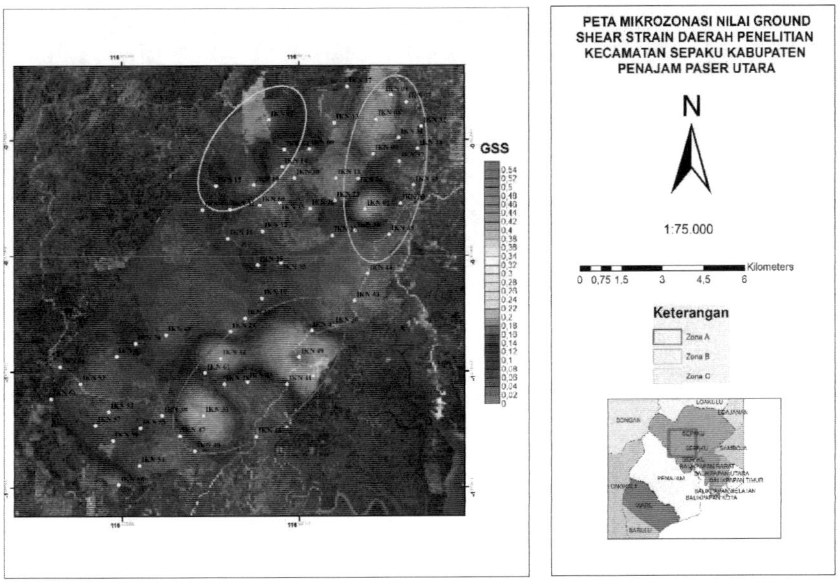

Figure 4. Microzonation map based on the ground shear strain value of the research area in Sepaku District.

With a small PGA value, the potential for liquefaction in the study area can be caused by other factors. One of factor is loading (gravity) on the sediment below the surface. It will be occur if the activity of the sedimentation process from the delta of the study area is very high [5]. The soft soil produced from the delta sedimentation process has a high water content. So that it can flow if there is an earthquake event loading or magnitude in the Sepaku Subdistrict area. [5]. The phenomenon due to the gravity in the study area can allow mud diapir due to the compressional system of the surrounding sedimentary rock that is moving through the sedimentation process from the delta [5]. In addition, various passive structures and faults can gradually cause landslides to liquefaction phenomena. So that it needs further study of the aspects of engineering geology and drill data in more detail detail.

4. Conclusions

Based on the analysis that has been carried out on the research results of this final project, it can be concluded:

- The values of the ground shear strain in the research area ranges from 0.02 to 0.54. The values of the ground shear strain is caused by the seismic susceptibility index and the PGA value of the research area in Sepaku Subdistrict, Penajam Paser Utara Regency.
- Based on the resulting ground shear strain value, the research area of Sepaku Subdistrict, Penajam Paser Utara Regency has the potential to occur the phenomena of landslides, soil compaction to liquefaction. However, the criteria for ascertaining the potential liquefaction still require more detailed data such as drill data in characterizing subsoil facies. However, by having the characterization of thick and soft sediments, the potential for mudflows from the sedimentation activity of the delta can occur.
- There are three areas which are divided into Zone A, Zone B and Zone C are considered has the potential for landslides, soil compaction to liquefaction based on the relatively high ground shear strain values, namely 0.4 - 0.54. It caused of the soil type, groundwater depth and weak field zone factors. Zone A and Zone B has the potential for soil compaction to liquefactions.

While Zone C have the potential for soil compaction to landslide. The area is part of the mining and agricultural land that still operating with most of the densest forest located in the Tengin Baru area, Sepaku District.

Acknowledgments
I thank the Geological Survey Center for regulary gave the microtremor data. Mr. Yayan taught me how to operate the raw data with a several software and also the second authors Mr. T.A Sanny as my supervisor already guide me to establish this research study. Institut Teknologi Bandung especially Geophysical Engineering program study as my major always gave me a favorable conditions and last but not least my thesis fighters fadil, hendro, devan and memeh who always support and help when I got stuck to finished it. I couldn't stand like this without your help. For all of that, thank you very much.

References
[1] Kementerian PPN/ Bappenas 2019 Dampak Ekonomi dan Skema Pembiayaan Pemindahan Ibu Kota Negara *Materi Presentasi Menteri Perencanaan.*
[2] Nakamura Y 2000 *Real Time Information Systems for Seismic Hazards Mitigation UREDAS, HERAS and PIC* (Japan: Quarterly Report of RTRI, Vol. 37, No. 3, 112-127)
[3] Kramer S L 1996 *Geotechnical Earthquake Engineering* (United States: Prentice-Hall)
[4] Nakamura Y 2008 *On The H/V Spectrum* (Beijing: The 14th World Conference on Earthquake Engineering)
[5] Bachtiar A 2020 Aspek Geologi Untuk Pembangunan Ibukota Negara (IKN) *Webinar IAGI*

Geothermal Potential in a Small Volcanic Island

Rahim Achmad[1,a)] and Ramdani Salam [2,b)]

[1] Physics Dept of Khairun University, Indonesia
[2] Geography Dept. of Khairun Universit , Indonesia
Correspondence: [a] diditernate71@gmail.com; [b] danipalopo72@gmail.com

Abstract. In this paper we report the continuation of previous observation of hot spring phenomenon in northern Ternate island. The hot spring distribution spread unevenly with temperature ranging from 29 °C to 35.5 °C along shore line. Geomagnetic measurements, chemical water and physical water properties have been carried out to find the caused of unevenly hot spring distribution . This research intends to reveal geothermal potential in a small volcanic island. The result of geomagnetic measurements, showed low magnetic anomalies as demagnetization rock by heat. Magnetic anomaly map shown any fault structure alignment with the northeast – southwest. The fault structures create path of heat. The result of chemical and physical water properties also agree with geomagnetic measurement that show indication of geothermal system. The heat source comes from magma chamber and passes through rock fractures (fault) as known as up flow geothermal system.
Key words: Geothemal: Geomagnetic; Small volcanic island

1. Introduction

Geothermal energy, is known as renewable energy and environmentally friendly, will be targeted as an alternative national energy that supplied about 5% until 2025, or about 9500 MW [1]. This policy should be followed by data supporting and technical readiness. [1], records about 203 geothermal areas associated with quarternary volcanic pathways. Generally, geothermal area is dominated by large islands in Indonesia, such as Sumatra, Jawa, Sulawesi and Papua. A small volcanic islands (in the island arcs) also has the potential of geothermal or hydrothermal.

Ternate island as a small volcanic island, become the study area is a part of unseparated tectonic system of eastern Indonesia. Ternate island is situated above the oblique subduction path to the east with a small angle. Gamalama as a strato volcano with a height about 1715 m above mean sea level and 11 km in diameter [1],[3]. As physiography, cone-shaped volcanoes can be divided into: Central Zone, Proximal Zone, Medial Zone and Distal Zone. [6]
The existence of Gamalama volcano indirectly provided geothermal potential in Ternate Island. The manifestations evidence of geothermal is the presence of hot springs. Geothermal potential mapping is done by geomagnetic measurements to determine variations in magnetic susceptibility of rock

below the surface. Physical and chemical properties of groundwater are also measured to support geothermal properties.

2. Material and methods

This research was conducted around hot springs site, northern of Ternate island (see Fig.1). The average temperature of groundwater at ranging from 26 to 28 °C. At the site of hot spring temperature ranges from 30 to 35 °C. Geologically, the rocks in the northern of Ternate island are young Gamalama (Gm) [6]. Morphologically, in the northern, is generally hilly with a little plain. The dug wells is only found around the shoreline.

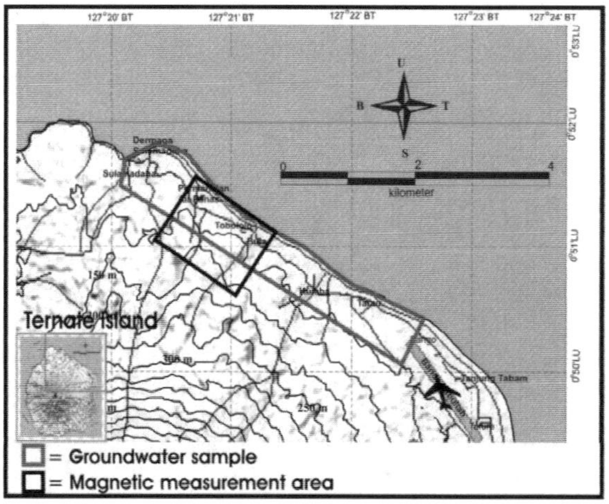

Figure 1. Research area

Magnetic measurements in this research using a handheld fluxgate magnetometer (It measures the three components of a dc magnetic field vector with a resolution of 1 nT over its range of ± 100,000 nT (± 1 Gauss by MEDA Inc.), which is focused at Tobololo hot springs. Magnetic measurement was carried out in the grid with a distance 150 to 200 m each point. Before interpretation of measurement field results, it should be corrected by daily value of magnetic variation and IGRF. Total magnetic of anomaly map is presented then due to analyze goethermal situation.

Measurements of physical and chemical groundwaters are also carried out along coastline . Data physics and chemistry of groundwater is measured using a water quality checker from Horiba instruments. The depth of groundwater level was measured manually to describe groundwater flow patterns.

3. Results and discussion

3.1 Groundwater Properties

Figure 2 shows the results of the physical measurements of groundwater such as temperature of water distribution and groundwater flow patterns. A small volcanic island has a radial pattern of groundwater flow which means, groundwater flows to the lowest area or coastal areas [7],[9],[12]. Groundwater flow (see Figure 2) was modeled based on phreatic data of dug wells and drilled wells. In general, dug wells are located in the lowest region or near shore line. Groundwater table elevation follows the topography of the mountain. Groundwater table is getting deeper according to the distance of shor line. To discribe groundwater position and movement (groundwater flow pattern), the groundwater data then ploted as contour of elevation . According to [5], to described the groundwater flow direction can be used the contour lines and make drawing a line perpendicular to contour lines.

Groundwater flows through the induced rock by geothermal then will become hot water and flowed out to the recharged area. The fault structure with the direction of the northeast - southwest controled the flow of hot water to the village of Tobololo [3]. Groundwater temperature distribution looks normal except in the village of Tobololo and in the village of Sulamadaha. In these two villages, groundwater temperatures range between 33-35 ° C. The pH value of the hot water in the village of Tobolo ranging between 6.9 to 7.6, while the pH value of hot water in the village Sulamadaha ranging between 7.6 - 8.2 with sodium chloride water type. The electrical conductance value (ranging from 3.3 to 4.4 mS / cm) and Salinity (ranging from 1.7 - 2.3 ppt) for hot water near the shoreline, can be under influenced of saline water. Mg^{2+} content of water samples from Sulamadaha higher than the other samples. This indicates of mixing process of meteoric water to geothermal fluid . The dilution volcanic rock by groundwater containing high elements Mg^{2+} [11],[10].

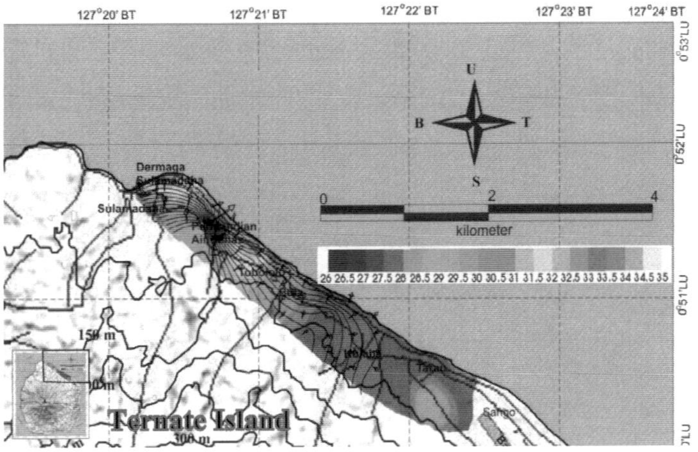

Figure 2. Groundwater thermal distribution and flow model

3.2 Geomagnetic Measurement

The results of total magnetic measurements are shown in Figure 3. Residual magnetic anomaly value shows the ranging between 7000 nT to 33000 nT. Magnetic anomaly contour map shows the value of anomaly magnetic rocks can be classified into three groups, namely: low anomaly (less than 12000 nT); moderate anomaly (between 13000 nT to 23 000 nT); and high anomaly (more than 23000 nT). High – Low patterns of magnetic anomaly in the coast, is interpreted as a rock with high suseptibility (andesitic-basaltic lava rocks) and low suseptibility as a rock which has experienced with strong weathering. Some of rocks is seen as outcrop.

Moderate to low magnetic anomaly, which is seen to spread and dominate the research areas, is interpreted as the low suseptibility rocks to non-magnetic rock. These rocks have experienced weathering and demagnetization due to the heat influence and close related to manifestation of hot water presence.

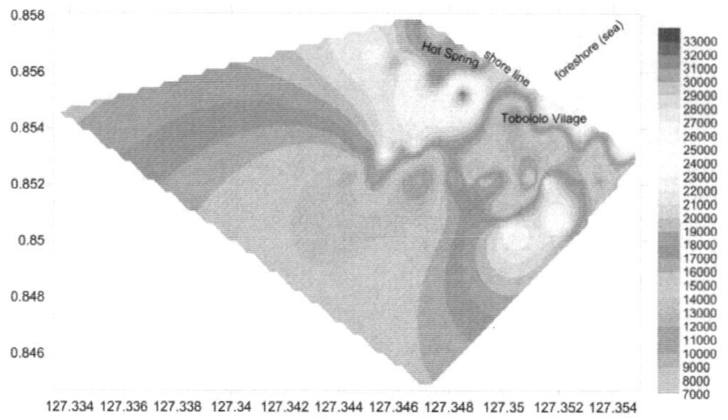

Figure 3. Total anomaly map of geomagnetic measurment

The results of geomagnetic measurement, indicates that the flow of hot water is driven by fault toward to the shore in Tobololo vilage [4]. Direction of fault is northeast – southwest. Visualization of the geothermal potential in Ternate Island can be illustrated by tentative model as shown in the picture below.

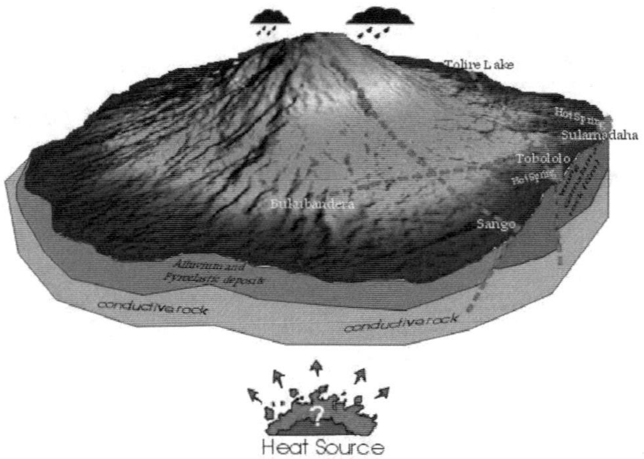

Figure 4. Tentative model of Ternate island geothrmal

4. Conclusion

In summary, we have studied about hot spring presence and geothemal potensial in a small island. Based on the results of chemical analysis and physical groundwater and the results of magnetic measurements can be conclude that Ternate Island also has a moderate geothermal potential. Although energy needs are still far from ideal conditions, further research is needed considering the geological conditions of the island. At least the geothermal potential can still be used for drying by farmers / fishermen and tourism.

Acknowledgments

This work was supported by research grants and fellowships from DIKTI. The author is grateful to M. Rusli, and Lab. Of Fishery Faculty of Khairun University, for loaning the relevant device (fluxgate magnetometer and Horiba Water Quality). For assistance in fieldwork from Geography Student are also acknowledged.

References

[1] Kasbani Suhanto and Dahlan, 2005 *Readiness of Indonesia's Geothermal Potential Data In Supporting the Preparation of Work Areas* . Geothermal Research Program Group. Geological Resource Center p 66

[2] Hamilton W 1979 *Tectonics of the Indonesian region.* US Geological Survey *Professional Paper,* 1078
[3] Geology Agency 2011 *Data Dasar Gunung Api Indonesia*, Second Edition, Ministry of Energy and Mineral Resources p 426
[4] Geology Agency 2009 *Geophysical Investigation of Gamalama Vulcano in Ternate* , Ministry of Energy and Mineral Resources p 22
[5] Fetter CW 2004 *Applied Hydrogeology*, 4th edition Mac millan Publishing New York,
[6] Bronto S 2006 *Volcanic facies and its applications* Indonesian Journal of Geology Vol.1 No.2: p 59-71
[7] Achmad R 2016 Ternate Island Hydrogeological Study , Dissertation. Faculty of Geography, UGM, p 127
[8] Fournier RO and Potter RW 1979 Magnesium correction to the Na-KCa chemical geothermometer, *Geochim. Cosmochim. Acta* 43 p 1543-1550
[9] Falkland AC 1993 Hydrology and Water management on Small Tropical Island, *Proceeding* of *Yokohama Symposium*, IAHS p 263-304
[10] Giggenbach WF 1988 Geothermal Solute Equilibria Deviation of Na-KMg-Ca Geo-Indicators. *Geochemica Acta* 52 : 2749–2765
[11] Giggenbach WF and Goguel RL 1989 *Collection and analysis of geothermal and volcanic water and gas discharges*. Report No. CD 2401. Department of Scientific and Industrial Research. Chemistry Division. Petone, New Zealand
[12] Hadi MP Salam R and Achmad R 2019 *Groundwater resources mapping for small island using Geoelectrical Technique* Indonesian Journal of Geography Vol. 51 No. 1 p 49-53

IOP Conf. Series: Earth and Environmental Science 1031 (2022) 012029 doi:10.1088/1755-1315/1031/1/012029

Determination of Q-factor of Transmission Ground-Penetrating Radar Waves in Attenuating Media

Wahyudi W. Parnadi[1]

[1]Department of Geophysical Engineering, Institut Teknologi Bandung, Jl. Ganesha 10 Bandung 40132, Indonesia

[*]Corresponding author: wahyudi@gf.itb.ac.id

Abstract. Electromagnetic wave attenuation in a medium is the primary parameter for interpreting Ground-Penetrating Radar (GPR) data. This attenuation manifests in wave energy absorption and velocity dispersion, which have different behavior depending on the rock type. Knowledge of the attenuation behavior of GPR waves in distinct media is, therefore, a success key to distinguishing rock types. The Q that is inversely proportional to the absorption coefficient represents its attenuation behavior. To observe the attenuation behavior of rocks, we conducted two different stages. The first stage is simulated the attenuation behavior by creating synthetic traces as a convolution between the wavelet and the corresponding impulse response. A Q-constant model and Futterman's velocity dispersion were used for the impulse response calculation. In the second stage, Q was determined using three different Q determination methods: the amplitude decay method, the spectral ratio method, and the Equivalent Bandwidth (EBW) method. Our research shows that due to the absorption, the wave amplitude becomes lower with increasing distance at the same Q, while at the same distance, its wavelength is longer, and its amplitude becomes low. The longer the distance at the same Q value, the narrower is the bandwidth of the transmission signals. We got a similar result with a small Q at the same distance length. At the same time, its peak frequency shifted to lower frequencies. Moreover, the applicability of the three methods for the determination of Q was investigated and tested on field transmission data acquired at the test field in the Reiche-Zeche Mine Shaft near Freiberg in Germany. Our investigation result shows that the three Q-methods were successfully applied to field transmission data resulting in comparable Q values ranging from 31 to 36 for gneiss rocks.

1. Introduction

Electromagnetic (EM) waves will undergo an attenuation process during their propagation in geological media which causes the wave energy to decrease [1]. In that process, the waves will experience energy absorption and velocity dispersion. The absorption of EM waves is reflected by lower wave amplitudes with longer distance, while its dispersion effect causes the wavelength to increase.

How the EM waves experience attenuation during their propagation in a medium is illustrated in Figure 1. The output signal is a product of the input signal convoluted with the impulse response of the medium. The phase velocity v and the absorption coefficient α are two quantities describing the propagation of electromagnetic waves in a medium, which are frequency-dependent. For Ground-Penetrating Radar (GPR) waves, which are in the frequency range 1 MHz – 1000 MHz, the absorption coefficient α is determined by the ratio between the real component (ε_r') and imaginary component (ε_r'') of the complex relative permittivity ε_r.

The impulse response g(t) is affected by petrophysical parameters like porosity Ø, density γ, grain size G_S, permeability ς, degree of mineralization S_M, and water saturation Sw, which have significant meaning in geology and geotechnical fields [2] as well as in the exploration field [3]. Therefore, measuring the attenuation behavior of EM waves is appropriate for differentiating rocks. Thus the determination of both phase velocity v and absorption coefficient α can lead to structural as well as rock-type interpretations. GPR waves behave similarly to seismic waves so that their propagation rules follow the rule of seismic waves. In line with the linear dependence of the frequency of the seismic wave absorption coefficient, the absorption coefficient of the GPR wave is also frequency-dependent, such that the Q factor of the GPR wave is frequency-independent [1,4,5]. The Q factor has an inverse relationship with the absorption coefficient α and clearly explains the absorption power of each different rock to the EM waves propagating on it. Based on our previous studies, we adopted the concept of constant Q model and dispersion velocity according to Futtermann (1962) in this study [6]. Q values of rocks in GPR frequencies lie in the range 2 – 30 [7], and we used it from 2 – 50 in this study.

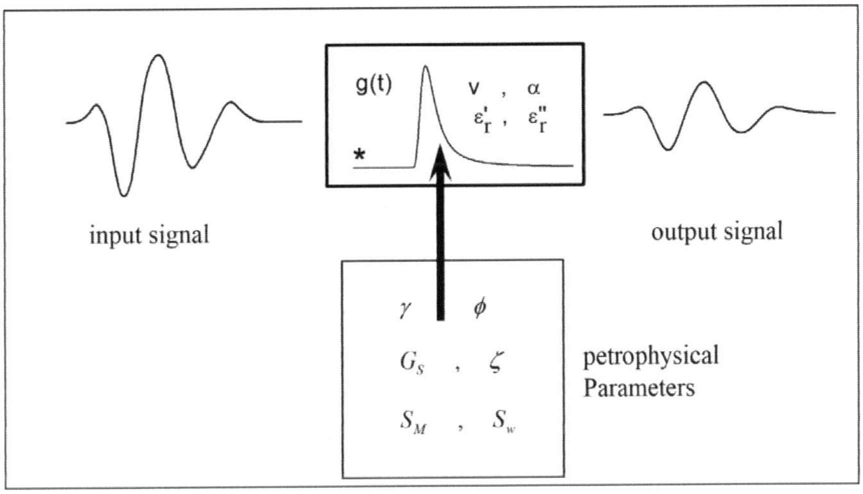

Figure 1. Sketch of a physical process in attenuating media and its effect in the waveform. The sign * refers to a convolution.

In this research, we investigated the behavior of the attenuation process in the form of absorption and dispersion in two stages. In the first stage, we conducted a simulation of synthetic traces and reviewed three different methods for Q determination, namely the amplitude decay method, the spectral ratio method, and Q determination through the absorption coefficient α. The latter method, we named the equivalent bandwidth method (EBM), is our technique developed at our laboratory. Q values contained in the resulted synthetic traces were then inversely computed using the three different methods. In the second stage, we tested the applicability of the three methods on field data from a test site.

2. Method

2.1. Impulse response

Electromagnetic waves propagation in attenuation geological media can be describe as a linear system, as shown in figure 2 below.

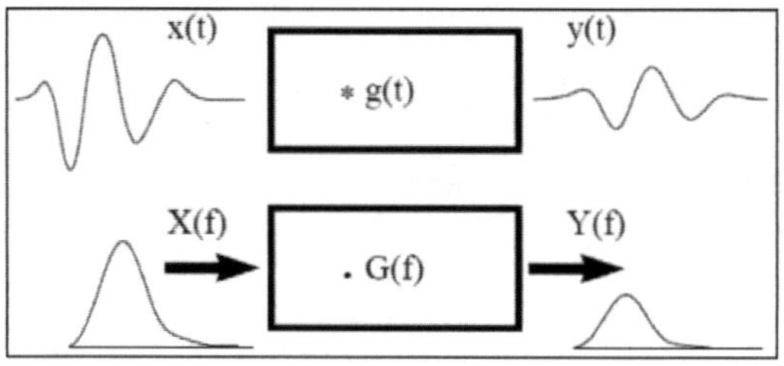

Figure 2. Physical processes of an EM wave in attenuating media. describes in time domain (top) and in frequency domain (bottom)

In the time domain, the output signal y(t) is the product of the input signal x(t) convoluted with the medium impulse response g(t). In the frequency domain, the output signal is the product of the spectrum of the input signal X(ω) multiplied by the transfer function of the medium G(ω):

$$y(t) = x(t) * r(t) \tag{1}$$

$$Y(\omega) = X(\omega). R(\omega) \tag{2}$$

where ω denotes the angular frequency and $\omega = 2\pi f$.

The transfer function R(ω) of attenuating media with causal impulse response g(t) has a form [1]

$$G(\omega) = e^{-\frac{\omega t_0}{2Q}}. e^{-j\omega t_0}. e^{j\frac{\omega t_0}{\pi Q} ln \frac{\omega}{\omega_{NY}}} \tag{3}$$

where t_0 and Q are travel time and Q factor, respectively. $\omega_{NY} = 2\pi f_{NY}$ and f_{NY} is Nyquist frequency.

Equation (3) describes 3 factors during the propagation of EW waves in media, i.e. the absorption factor in the first term, time factor in the second term, and dispersion factor in the third term. The dependency of the Q factor on the amplitude and the phase spectra is recognizable (see figure 3). In particular with decreasing Q, absorption and dispersion effect increase.

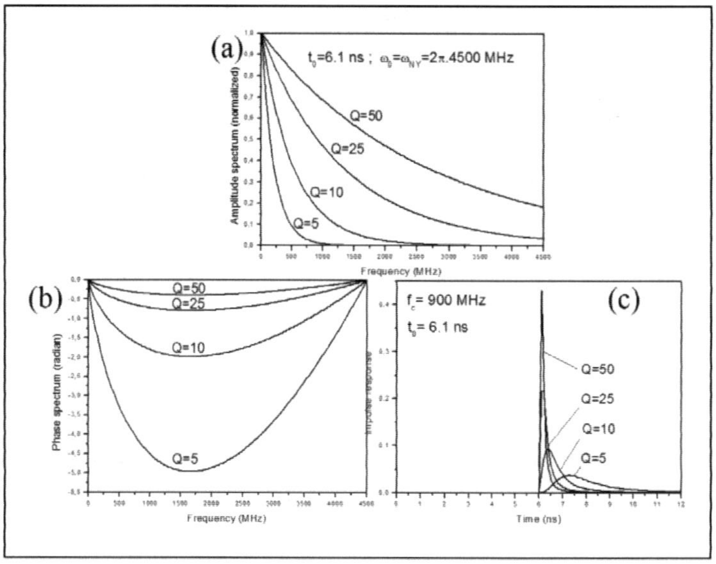

Figure 3. Amplitude spectrum (a) and phase spectrum (b) of transfer function G(ω) in equation (3) and their corresponding impulse response g(t) (c). Parameters used in the calculation: t_0 = 6.1 ns, working frequency f_c = 900 MHz.

The medium impulse response g(t) is calculated by the inverse of Fourier Transformation of G(ω):

$$g(t) \overset{IFT}{\longleftarrow} G(\omega) \tag{4}$$

2.2. *Forward modeling*

Synthetic traces forward modelings were conducted using equation 1 with an input signal w(t) and traveltime data listed in table 1 below.

Table 1. Input signal and travel time for forward modelling

	Input signal w(t)	Traveltime in ns
Case 1	Input signal A	3
Case 2	Input signal A	6
Case 3	Input signal A	9
Case 4	Input signal A	12
Case 5	Input signal B	6
Case 6	Input signal B	12

The input signal A is a wavelet obtained from measurements in the air with a nominal frequency of 900 MHz whilst input signal B is the one with a nominal frequency of 450 MHz [1].

2.3. *Q determination from GPR transmission signal*

There are some methods to calculate Q values. Tonn (1991) investigated 10 different methods for Q determination using VSP data. It is concluded that there is no single method that is superior [8]. In this research 3 different methods were reviewed and tested: the amplitude decay method, the spectral ratio method, and the Equivalent Bandwidth Method [1, 8].

2.3.1. *The Amplitude decay method*

This is the easiest way to Q determination and it is calculated in the time domain. The value of Q is directly proportional to the ratio of the amplitude of the wave over different distances x_1 and x_2 with travel time $t_1=x_1/v$ and $t_2=x_2/v$:

$$Q = \frac{\pi f \Delta x}{v} \ln\left[\frac{a(x_1)}{a(x_2)}\right] = \pi . f . t_0 . \ln\left[\frac{a(x_1)}{a(x_2)}\right] \tag{5}$$

where f refers to dominant frequency, $t_0=t_2-t_1$ is traveltime different and $\Delta x=x_2-x_1$. $a(x_1)$ and $a(x_2)$ are signal amplitudes at x_1 and x_2, respectively.

The limitation of this method is that it can only be applied when actual amplitudes are present. The use dominant frequency here is only an estimated value.

2.3.2. *The Spectral ratio*

The spectra ratio method is the popular technique in determining Q values, which is calculated in the frequency domain. First, we calculate the logarithmic ratio of amplitude spectra of two different signals y(t) and x(t). The ratio of the corresponding Futtermann's transfer function [1, 8, 9] appears to be

$$\ln \frac{|Y(f)|}{|W(f)|} = const. - \frac{\Delta t . \pi . f}{Q} \tag{6}$$

where Δt refers to the traveltime difference between y(t) and w(t).

The Q value is calculated from the linear line steepness specified in equation (6) as a function of frequency. A general problem in this method is the estimation of frequency intervals.

IOP Conf. Series: Earth and Environmental Science 1031 (2022) 012029 doi:10.1088/1755-1315/1031/1/012029

2.3.3. The Equivalent Bandwidth (EBW) method

This method estimate the Q value in time domain. Q is calculated directly using

$$Q = \frac{2\pi}{1-\frac{E_2}{E_1}} \cdot t_0 . f_R \tag{7}$$

where E_1 as well as E_2 describe wave energies, subscript numbers "1" and "2" denote input and output signal, respectively and f_R is reference frequency.

The reference frequency f_R (in MHz) was determined experimentally from the cross plot between reference frequency and the ratio between amplitude density value of equivalent rectangle spectrum of input signal and output signal $\tilde{A}_{21}$ [1]

$$f_{R,900} = 771.6(\tilde{A}_{21})^3 - 325.5(\tilde{A}_{21})^2 + 611.3(\tilde{A}_{21}) + 78.5 \tag{8}$$

$$f_{R,450} = 0.5\, f_{R,900} \tag{9}$$

where $f_{R,900}$ and $f_{R,450}$ are reference frequency for 900 MHz and 450 MHz, respectively.

3. Results and discussion

3.1. Forward modelling

Figures 4 and 5 show the behaviour of absorption and dispersion through the simulation of synthetic traces. Some impulse responses at the same travel time t_0 = 3 ns are shown in figure 4. For a smaller Q value – means higher attenuation - the impulse response becomes smaller and wider. Smaller Q denotes higher energy absorbed. A wider but smaller impulse response has the meaning that more time is needed to deliver wave energy. It is means also that the primary signal energy moves at a later time. Figure 5 shows the course of Impulse responses at the same travel time t_0, output signals, and their corresponding frequency spectrum at the same travel time t_0 = 3ns but with different Q values. These figures 4 and 5 show clearly that the peak frequencies are shifted to lower frequencies with decreasing Q values.

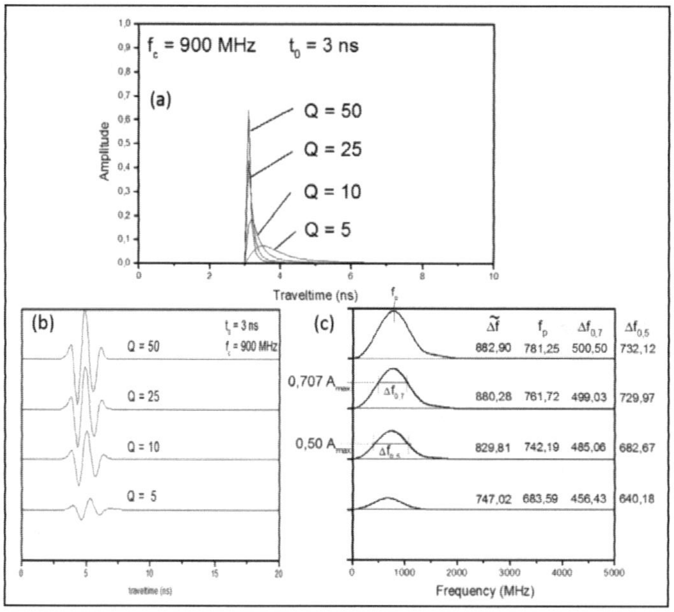

Figure 4. (a) Impulse responses at the same travel time t_0 (t_0 = 3 ns) with different Q value; (b) output signals using impulse response in (a); and (c) their corresponding frequency spectra. t_0 used is 3 ns.

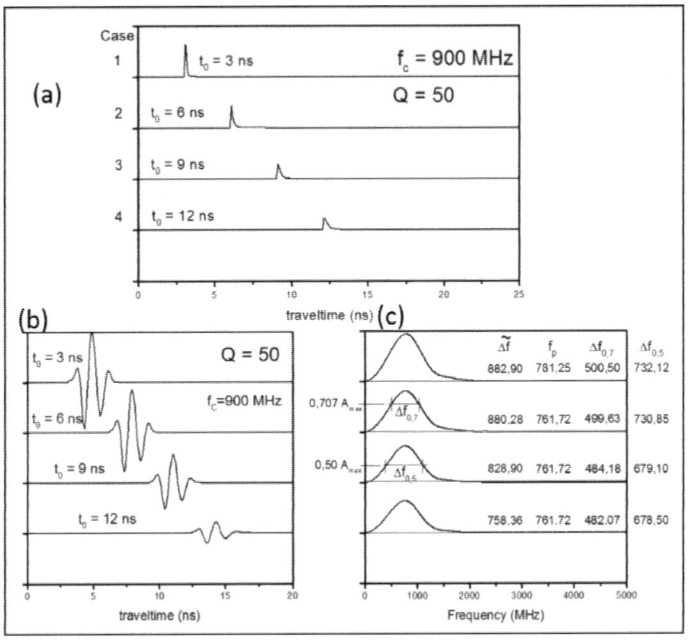

Figure 5. The course of impulse responses y(t) at the same Q value with a variety of travel time t0; (b) Output signals using impulse responses in (a) and (c) their corresponding frequency spectra. Q used is 50.

3.2. Investigation on field data

We investigated Q determination with GPR technique on gneiss, which is a crystalline rock, due to two ideal conditions: gneiss has low attenuation and thus great penetration depth for GPR waves [1]. The object as shown in figure 6 is located in the Reiche-Zeche shaft in the "Schwarzer Hirsch" corridor at a depth of around 180 m in Germany. The test object was selected in such a way that it represents a quasi-homogeneous medium.

Two different measurement configurations were realized: horizontal measurement and vertical measurement. In horizontal measurement, both transmitter antenna and receiver antenna were moved together with 0.5 m forward at a fixed height, from O to T for transmitter antenna and from O to R for receiver antenna. The measurements were conducted three times, at a fixed height of 1.20 m to 1.50 m. In vertical measurement, the transmitter antenna at T position and the receiver antenna at R position at the same height were moved together vertically from bottom to top parts of the wall with 20 cm step upwards, beginning at 0.8 m from tunnel floor (see Figure 6.b).

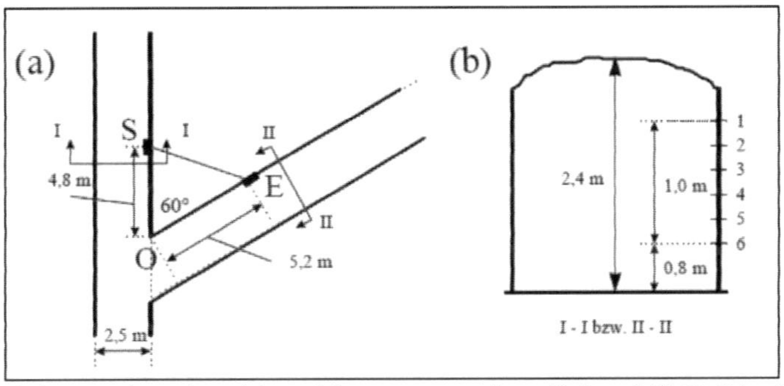

Figure 6. Sketch of the test location (a) and its cross-sections I–I and II–II (b). T and R denote the position of transmitter as well as receiver antennas, respectively.

Figure 7 shows field transmission signals acquired with 900 MHz centre antennas for the vertical measurement. To get velocity and absorption coefficient values for the calculation of Q using the Equivalent Bandwidth method, some traces from horizontal measurement were used.
Results of Q determination using three different methods are listed in table 2. Q values using three different methods showed comparable values with Q values determined by the EBW method lie between that of the amplitude decay method and that of the spectral ratio method. It is concluded that the results from three different methods agree well with Q values ranges from 31 to 36. Our results with these methods and compared with the values determined in the literature for gneiss [1] are listed in table 3.

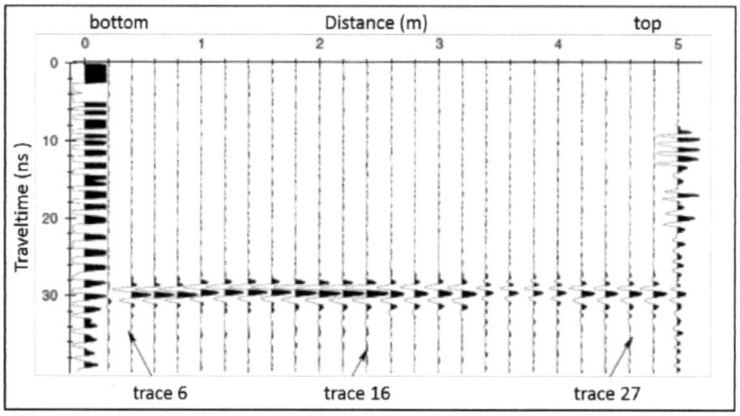

Figure 7. Field transmission signals acquired with 900 MHz antennas for vertical measurement

Table 2. Q determination from vertical measurement

Input signal	Output signal	Δt_0 (ns)	Calculated Q Amplitude decay	Spectral ratio	EBW
Trace 5	Trace 10	27.8	32.40	33.46	32.88
Trace 5	Trace 11	27.7	32.53	33.59	32.92
Trace 5	Trace 12	27.7	32.65	33.72	33.04
Trace 6	Trace 13	27.7	32.80	33.86	32.98
Trace 6	Trace 14	27.8	33.85	34.96	34.14
Trace 6	Trace 15	27.9	33.97	35.08	34.57
Trace 6	Trace 16	27.9	33.88	34.97	34.21
Trace 6	Trace 17	27.8	34.32	35.44	35.02
Trace 6	Trace 28	28.0	31.08	32.12	31.68

Table 3. Resulted v, α, and Q values of gneiss rocks

No.	v (m/ns)	α (dB/m)	Q	f_c (MHz)	Depth (m)	source
1	0.17 – 0.18	1.7 – 2.1	31 - 36	900	180	This study
2	0.10 – 0.12					Grasmueck [11]
3	1.12 – 0.12				up to 30	
4	0.17	2.6				
5	0.12	6.9				
6	0.17	1.5 – 4.5	30			
7	0.13	9.0 – 27.0	7			
8	0.08 – 0.11					

4. Conclusion

The attenuation of electromagnetic waves is the primary parameter for interpreting Ground-Penetrating Radar (GPR) data, which is manifested in the form of energy absorption and velocity dispersion. Measuring its attenuation behaviour is therefore appropriate for differentiating rocks type. To observe the attenuation behaviour of GPR waves, two different stages were undertaken in this study: simulating wave propagation and determining the attenuation level through the Q factor.

Based on the above basic understanding, our simulation results using synthetic traces reveal the behaviour of the attenuation process. It process results in lower wave amplitudes with longer distances at the same Q values and leads to lower wave amplitudes and longer wavelengths with decreasing Q values at the same travel time. The former is known as the absorption process, and the latter is the dispersion process. At the same Q values, the longer the distance is, the narrower the bandwidth of the transmitted signal. Similar behaviour is observable. The smaller the Q values at the same time distance, the narrower is its frequency bandwidth. Its peak frequency shifts to lower frequencies.

The applicability of the three Q methods was then investigated using field data acquired at the test field in the Reiche-Zeche Mine Shaft, near Freiberg in Germany. The results showed that among the above methods, including our proposed formula to determine Q values, all were applicable to field transmission GPR data, resulting in Q values ranging from 31 to 36 for gneiss rocks.

Acknowledgments
This research was partially supported financially by Institut Teknologi Bandung through the P3MI2020 Program.

References

[1] Parnadi W W 2001 Kennwert-Schätzung aus Georadar-Transmissionsdaten *PhD thesis*: TU Bergakademie Freiberg
[2] Parnadi W W, Warsa W, Laesanpura A and Santoso D 2021 Multi frequency Ground-Penetrating Radar method for mapping underground pipes and man-made objects: *IOP Conference Series: Earth and Environmental Science* **873**(1) 012090
[3] Antonio T, Parnadi W W and Heriawan M N 2021 Integrating Ground Penetrating Radar, Induced Polarization and Aerial Photograph to Analyze Land Subsidence in Borehole Mining Operation Area: A Case Study from South Bangka Island: *IOP Conference Series: Earth and Environmental Science* **873**(1) 012081
[4] Economou N and Kritikakis G 2016 Attenuation analysis of real GPR wavelets: The equivalent amplitude spectrum (EAS): *J. Applied Geophysics* **126** pp 13–26
[5] Bano M 2004 Modelling of GPR waves for lossymedia obeying a complex power law of frequency for dielectric permittivity: *Geophysical Prospecting* **52** pp 11–26
[6] Parnadi W W, Warsa W and Santoso D 2019 Water content delineation using Ground_penetrating Radar Q tomography: *Proc. Internat. Symp. Earth Science and Technology* Fukuoka
[7] Turner G and Siggins A F 1994 Constant Q attenuation of subsurface radar pulses*: Geophysics* **59**(8) pp 1192-1200
[8] Tonn R 2001 The determination of the seismic quality factor Q from VSP data: a comparison of different computational methods: *Geophysical Prospecting* **39** pp l-27
[9] Futtermann W I 1962 Dispersive body waves: *J. Geophysical Research* **67**(13) pp 5279-5291
[10] Widess M B 1982 Quantifying resolving power of seismic system: *Geophysics* **47**(8) pp 1160-1173
[11] Grasmueck M P 1995 Development of a georadar system for three-dimensional imaging of the subsurface and its application to studies of crystalline rock bodies *PhD thesis*: ETH Zürich

Fabrication of DSSC photoanode based on TiO_2 produced by caustic fusion of local ilmenite

S Wijayanti[1*], **N M Nursam**[2], **S Soepriyanto**[1], **D Suhendar**[1]

[1] Department of Metallurgical Engineering, Institute of Technology Bandung, Bandung 40132, Indonesia
[2] Research Center for Electronics and Telecommunications, Indonesian Institute of Science, Kampus LIPI Building 20 Level 4, Cisitu, Bandung – Indonesia 40135

*Corresponding author's email: suwastikaw@students.itb.ac.id

Abstract. Photoanode is an essential part in dye sensitized solar cells (DSSC). Semiconductor materials, such as TiO_2, ZnO, and SnO_2, are commonly used as photoanode material. In this study, DSSC photoanode was fabricated using TiO_2 obtained from Bangka ilmenite concentrate by caustic fusion method followed by water leaching, acid leaching, and sol-gel reactions. The synthesized TiO_2 powder has an anatase phase with a crystal size of 9.55 nm, band gap energy of 3.0 eV, spherical morphology with an average particle size of 0.53 μm, and a specific area of 138.39 m^2/g. The synthesized TiO_2 powder was found to have Fe impurities. The DSSC with TiO_2 photoanode synthesized in this work produced current-voltage characteristics with photoconversion efficiency of 0.75%.

Keywords: DSSC, ilmenite, TiO_2, caustic fusion

1. Introduction

Dye sensitized solar cell (DSSC) is a semiconductor solar cell that can convert solar energy into electric current with the principle of photoelectrochemistry. DSSC was first developed by Professor Michael Grätzel and Dr. Brian O'Regan in 1991. DSSC is different from other generations of solar cells, where light is absorbed by dye molecules attached to the semiconductor layer [1]. DSSC has a layered structure (sandwich) consisting of transparent conductive glass coated by a thin layer of semiconductor (photoanode), dye solution, electrolyte solution, sealing material, and counter electrode [2].

One of the most commonly used semiconductors in DSSC that has a wide band gap energy value is titanium dioxide (TiO_2). TiO_2 is frequently used because it is relatively inexpensive, easy to synthesize, abundant, harmless, and stable in acidic or basic environments [3]. TiO_2 mostly has three phases of crystal structure in nature, namely anatase, rutile, and brookite. Among the three phases, rutile is the most stable phase, while anatase is a metastable phase and tends to transform to rutile at a temperature range of 600-1000 °C depending on the size of the crystal and the level of purity. The band gap energy value for anatase is 3.2 eV, while for rutile is 3.0 eV. Rutile has a slightly lower band gap energy compared to anatase and can absorb several percent of the light in the area near infrared, but electron excitation causes holes to occur, causing long-term instability of solar cells [4,5]. One of the raw materials in nature that could be processed to obtain TiO_2 is ilmenite. This type of material is abundant in Indonesia, yet, ilmenite has not been utilized maximally to date. Mulyono and Soepriyanto investigated the TiO_2 synthesized from Bangka ilmenite, using caustic fusion process and applied in photocatalytic activity [6]. In this study, we aim to obtain TiO_2 powder from Bangka ilmenite concentrate by utilizing caustic fusion method combined with a water leaching, acid leaching, and sol-gel process. The synthesized TiO_2 powder from these methods was then be applied as photoanode in DSSC. The physical and optical properties of the synthesized TiO_2 were characterized and they were correlated to the electrical properties of constructed solar cell.

2. Experimental procedure

The ilmenite concentrate used for the synthesis of TiO_2 came from Bangka islands with a size of -200 mesh. The caustic fusion process was conducted by mixing ilmenite with NaOH in a mole ratio of 1:4 at a temperature of 450 °C for 120 minutes. Caustic fusion products were in the form of greenish sodium salts, and then leaching process was carried out with water to remove excess NaOH. The process was continued by acid leaching using 8M H_2SO_4 at 90 °C for 90 minutes. The product produced from acid leaching was $TiOSO_4$ solution, which is a precursor for the synthesized TiO_2. The TiO_2 powder was subsequently synthesized using sol-gel method. $TiOSO_4$ precursor solution was hydrolyzed by adding distilled water with $TiOSO_4$:aquades ratio of 1:6. Next, the pH of the solution was increased by adding a 10% NH_4OH weak base until the pH reached 1.0 at 90 °C for 2 hours until a white precipitate of $TiO(OH)_2$ gel was formed. After settling for 24 hours the washing was conducted to neutralize the pH, then dried and calcined at a temperature of 400 °C for 1 hour to produce crystalline TiO_2 powder. TiO_2 powder was characterized using x-ray diffraction (XRD), scanning electron microscopy-energy dispersive x-ray spectroscopy (SEM-EDS), gas sorption and UV-Vis diffused reflectance spectroscopy (UV-Vis DRS). The TiO_2 synthesis was carried out at the Laboratory of Solid Oxide System, Department of Metallurgical Engineering, ITB.

TiO_2 paste for DSSC photoanode was made by mixing 0.26 grams of TiO_2 powder with 0.1 mL terpineol and triton as much as 0.2 mL as a binder solution, then stirred until thoroughly mixed. Fluorine-doped tin oxide (FTO) substrate was washed with DI water mixed with Tipol solution, deionized (DI) water, and isopropyl alcohol (IPA). DSSC photoanode was coated with a blocking layer using a screen printer, after that it was dried at 120 °C for 10 minutes, and then sintered at 500 °C for 30 minutes. TiO_2 paste was coated on top of the blocking layer using a screen printer and then dried at 120 °C for 10 minutes and sintered at 500 °C for 30 minutes. After the substrate was coated with TiO_2 paste, $TiCl_4$ treatment was then conducted by soaking the substrate in $TiCl_4$ solution at 70 °C for 30 minutes. Once completed, the substrate was rinsed with DI water and sintered at 500 °C for 30 minutes. The next process was immersion of the substrate with Z907 dye solution for 24 hours. After the immersion process was completed, the substrate was rinsed using ethanol. The thickness of the photoanode layer was approximately 22 ± 0.89 μm. DSSC counter electrodes were made by depositing platinum paste (Pt) using a screen printer and then dried at 120 °C for 30 minutes and sintered at 450 °C for 30 minutes. The finished photoanode and counter electrode were then assembled, followed by electrolyte EL-HPE injection. The next step was to seal the holes on the counter electrode using sealants and aluminum foil to prevent electrolyte leakage. DSSC performance was evaluated by measuring the current density-voltage (J-V) characteristics under sun simulator with an intensity of 50 mW/cm^2 equipped with AM 1.5G filter and incident photon-to-current conversion efficiency (IPCE). The fabrication process and DSSC characterizations were carried out at the Research Center for Electronics and Telecommunications, Indonesian Institute of Science.

3. Results and discussion

The chemical compositions of Bangka ilmenite, i.e. a concentrate that was used as the original material for the TiO_2 synthesis, was analyzed using x-ray fluorescence (XRF) characterization and the results are shown in Table 1. The composition of Bangka ilmenite was dominated by TiO_2 (50.80%) and Fe_2O_3 (25.80%) and a small number of impurities such as MgO, Al_2O_3, SiO_2, SO_3, MnO, and others. The XRD diffraction pattern of the synthesized TiO_2 powder in Figure 1 shows that the synthesized TiO_2 powder had an anatase phase with an average crystal size of 9.55 nm. Anatase phase was indeed expected in this study because of more photoactive than rutile [7]. In synthesizing TiO_2 from ilmenite concentrate, impurities especially Fe are highly avoided so that the synthesized product has a high purity of TiO_2. The diffraction pattern of the synthesized TiO_2 powder herein did not visibly show any peaks that correspond to the presence of Fe_2O_3. The peaks of Fe_2O_3 were at 2θ, 24.32°; 33.30°; 35.78°; 41.04°; 49.62°; 54.24°; 62.60°, and 64.16° [8]. This is possibly because the Fe_2O_3 content was too small as shown in Table 2 through energy dispersive x-ray spectroscopy (EDS) testing. Fe impurity would cause a decrease in the value of J_{sc} DSSC. That is because Fe impurities would make agglomerated granules and reduce the crystallinity of TiO_2 powder, whereas J_{sc} was not only influenced by the amount of dye absorbed by TiO_2 powder, but it was also influenced by the grain size and crystallinity of TiO_2 [9].

Table 1. Chemical composition of Bangka Ilmenite

Compound	Amount (%)
TiO_2	50.80
Fe_2O_3	25.80
V_2O_5	4.88
MnO	2.90
Al_2O_3	0.95
SiO_2	0.73
SO_3	0.63
Y_2O_3	0.53
P_2O_5	0.40
Nb_2O_5	0.39
ZrO_2	0.16
MgO	0.08
ZnO	0.06
Cl	0.01

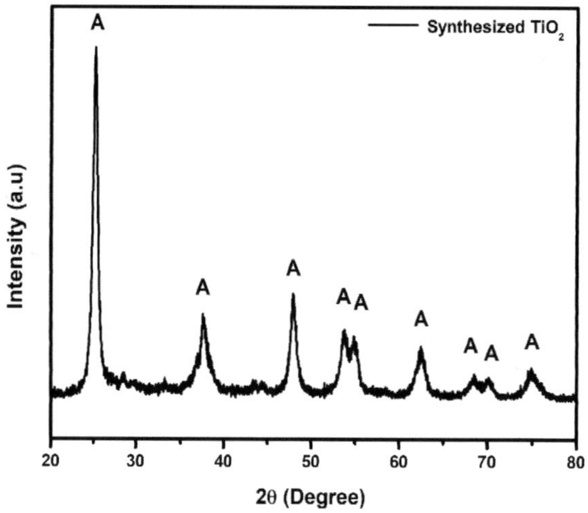

Figure 1. XRD pattern of synthesized TiO_2 powder

Figure 2(a) shows the morphology of the synthesized TiO_2 powder from the ilmenite concentrate. Morphology of synthesized TiO_2 powder was in the form of spherical granules with an average particle size of 0.53 μm. For the implementation in DSSC application, spherical structure is the most widely used because it has a high surface area so that the amount of dye absorbed would increase and allow as much light to enter the surface as possible [10]. The morphology of the synthesized TiO_2 powder shows that there are plenty of pores that would allow the dye to be absorbed well into TiO_2, as well as adsorbing more electrolytes containing redox mediators so it could facilitate the passage of electrons easily [11]. In Figure 2(a), it can also be seen that the particles agglomerate to a larger size due to the presence of Fe deposits in TiO_2 which were evenly distributed and appeared to be homogeneous. The EDS test was conducted to determine the elemental content of the sample shown in Table 2 and Figure 2(b). The EDS result shows that the synthesized TiO_2 powder contained Fe impurities, which was likely to contribute to the yellowish color pigment observed in the synthesized TiO_2 powder.

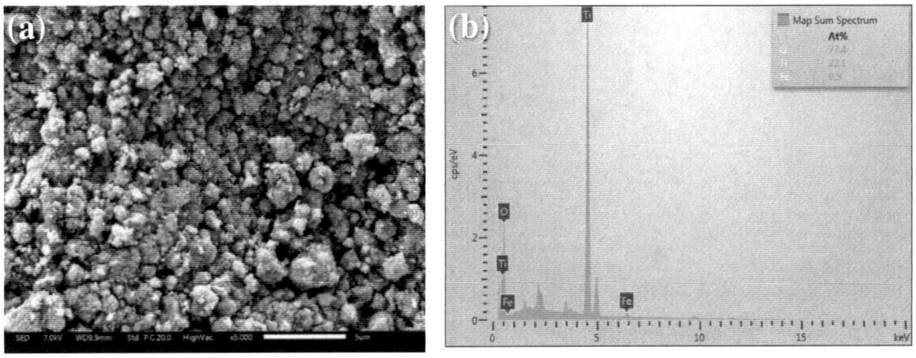

Figure 2. (a) SEM image and (b) EDS spectrum of synthesized TiO_2 powder

Table 2. Composition of synthesized TiO_2 powder

Element	wt %
O	53.33
Ti	45.33
Fe	1.13
Total	100

The value of energy bandgap of the synthesized TiO_2 powder was estimated from UV-Vis DRS characterization data using the Kubelka-Munk method as shown in Figure 3. The synthesized TiO_2 powder value had an anatase phase with an energy bandgap value of 3.0 eV, where it was smaller than most commonly reported energy bandgap value for anatase phase, i.e. 3.2 eV [4,5]. One of the possible reasons is because the synthesized TiO_2 powder did not have nano-sized granules, where the decrease in grain size would cause greater energy band gap value from the material [12]. In addition, the decrease in the band gap value of the synthesized TiO_2 was attributed the presence of Fe impurities. The increase in the amount of Fe in TiO_2 has been reported to decrease the energy bandgap value of TiO_2 [13].

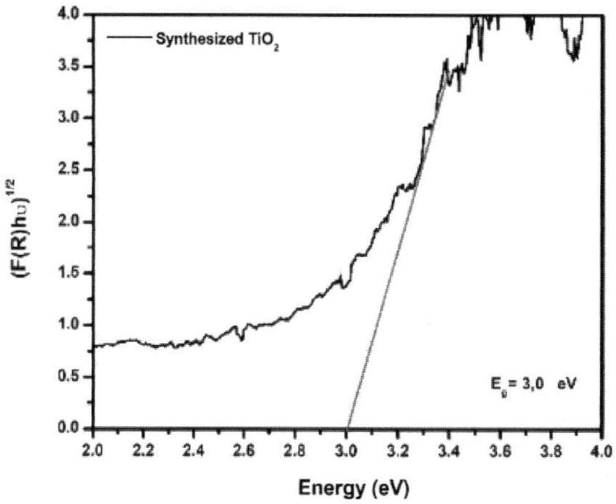

Figure 3. Plot of transformed Kubelka-Munk as a function of energy. The red line was used to estimate the band gap energy of synthesized TiO_2 powder.

Gas sorption analysis was used to determine the Brunauer-Emmett-Teller (BET) surface area of the synthesized TiO_2 powder. The surface area of the synthesized TiO_2 powder was 138.39 m^2/g. Surface area is an important property in DSSC application. The greater the surface area, the amount of dye absorbed would increase so that the DSSC efficiency would increase. Besides, the porosity of the synthesized TiO_2 powder was also calculated from the BET testing data, giving a value of approximately 0.65. Porosity also affects DSSC performance. The smaller the porosity value, the more increase the absorption coefficient of light, so that the current and voltage produced would increase. According to Ni *et al*, good porosity value used for DSSC is around 0.41 because when the porosity was 0.41 the maximum power value generated was greater than the others [14].

The performance of DSSC with the synthesized TiO_2 photoanode was compared to the performance of DSSC with commercial TiO_2 (Merck) photoanode. Figure 4(a) shows a graph of current density-voltage (J-V) of DSSC with a variation of the TiO_2 photoanode. DSSC conversion efficiency depended on the value of short-circuit current density (J_{sc}), open-circuit voltage (V_{oc}), Fill Factor (FF), and the power of incoming light (P_{in}). The efficiency of DSSC and Fill Factor DSSC can be calculated using Equations 1 and 2.

$$FF = \frac{V_{max} \times J_{max}}{V_{oc} \times J_{sc}} \quad (1)$$

$$\eta = \frac{P_{max}}{P_{in}} = \frac{J_{sc} \times V_{oc} \times FF}{P_{in}} \times 100\% \quad (2)$$

Table 3. Electrical parameters of DSSC with various photoanode TiO_2

Photoanode	V_{oc} (V)	J_{sc} (mA/cm^2)	FF	Efficiency (%)
TiO_2 Merck	0.68	36.40	0.40	1.99
TiO_2 Synthesis	0.66	9.60	0.61	0.75

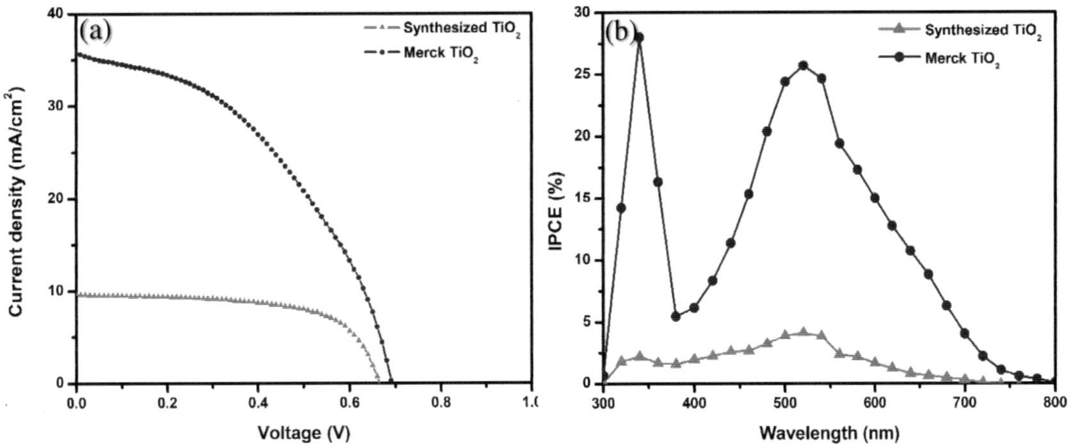

Figure 4. (a) J-V curves of DSSCs and (b) IPCE spectrum of DSSCs

The graph in Figure 4 (a) shows that the performance of DSSC with the synthesized TiO_2 photoanode from Bangka ilmenite was lower than DSSC with Merck TiO_2 photoanode. DSSC with the synthesized TiO_2 photoanode produced conversion efficiency of 0.75%, while DSSC with Merck TiO_2 produced conversion efficiency of 1.99%. The plausible reason for the lower efficiency is because the synthesized TiO_2 powder contained Fe impurities. Although the presence of Fe impurities increased the adsorption of dye molecules, the existence of Fe also caused the TiO_2 granules to agglomerate so that the size became larger [9]. The DSSC parameters with various TiO_2 photoanode are shown in Table 3. The DSSC performance was also tested using incident photon-to-current efficiency (IPCE) to determine the value of the conversion efficiency of photons into electric currents at specific wavelengths. Figure 4 (b) shows the IPCE spectrum of DSSC with variations of the TiO_2 photoanode. It can be seen that most of the visible light conversion into electric by the synthesized TiO_2 occurred at a wavelength range of 380-700 nm. The IPCE spectrum of DSSC with Merck TiO_2 and the synthesized TiO_2 did show any significant differences. The IPCE graph also shows a sharp peak at a wavelength around 340 nm, which was related to the absorption of the energy band gap of TiO_2 [15]. Meanwhile, the second IPCE peak at a wavelength of about 520 nm was related to the maximum absorption of the dye. The fundamental difference between the IPCE spectra of TiO_2 Merck and the synthesized TiO_2 was the large difference in the intensity of the photon-to-current conversion. The higher the value of IPCE, the more number of photons converted to current [16]. IPCE values had increased in anatase DSSC TiO_2 due to high surface area and higher size than other phases of TiO_2, so the dye was well absorbed [17].

4. Conclusions

TiO_2 powder was successfully synthesized from Bangka ilmenite concentrate with a caustic fusion method having an anatase phase, a grain size of 0.53 μm, and a band gap value of 3.0 eV. DSSC with the synthesized TiO_2 photoanode was successfully fabricated with a layer thickness of 22 ± 0.89 μm having J_{sc} 9.60 mA/cm², V_{oc} 0.66 V, FF 0.61, and a conversion efficiency of 0.75%. Although the conversion efficiency of the DSSC with the synthesized TiO_2 photoanode is lower than the DSSC with commercial TiO_2 (Merck), our work demonstrated that ilmenite minerals show great potential to be applied in DSSC. The cost of fabrication of DSSC in Indonesia is expected to be significantly reduced by employing ilmenite-based TiO_2 as photoanode due to its abundant availability.

Acknowledgments

The authors gratefully thank to PT Timah Indonesia for their support in XRF measurements of material Ilmenite and we also would like to thank Materials and Devices for Solar Cells research group at the Research Center for Electronics and Telecommunication, Indonesian Institute of Sciences (P2ET LIPI) for the support in sample preparation and J-V characteristics and IPCE. Loka Penelitian Teknologi Bersih (LPTB) is acknowledged for performing the SEM and XRD analysis.

References

[1] O'Regan B and Grätzel M 1991 *Nature* **353** 737
[2] Hara K and Arakawa H 2003 *Dye-sensitized solar cells, Handbook of Photovoltaic Science and Engineering* (England: John Wiley & Sons)
[3] Ahmad M S, Pandey A K and Abd Rahim N 2017 *Renew. Sustain. Energy Rev.* **77** 89
[4] Diasanayake M A K L, Senadeera G K R, Sarangika H N M, Ekanayake P M P C, Thotawattage C A, Divarathne H K D W M N R and Kumari J M K W 2016 *Mater. Today Proc.* **3** 40
[5] Khataee A and Mansoori G A 2012 *Nanostructured Titanium Dioxide Materials* (Singapore: World Scientific Publishing)
[6] Mulyono J E and Soepriyanto S 2017 *AIP Conf. Proc.* **1805**
[7] Joya M R, Barón-Jaimez J and Barba-Ortega J 2013 *J. Phys. Conf. Ser.* **466**
[8] Reda S M 2010 *Mater. Sci. Semicond. Process.* **14** 417
[9] Liao J Y, He J W, Xu H, Bin Kuang D and Su C Y 2012 *J. Mater. Chem.* **22** 7910
[10] Prima E C, Yuliarto B, Suyatman and Dipojono H K 2016 *Adv. Mater. Res.* **1112** 317
[11] Nowotny J 2012 *Oxide Semiconductors for Solar Energy Conversion* CRC Press.
[12] Soepriyanto S and Aristanti Y 2016 *J. Aust. Ceram. Soc.* **521** 131
[13] Ni M, Leung M K H, Leung D Y C and Sumathy K 2006 *Sol. Energy Mater. Sol. Cells* **90** 1331
[14] Abdullah M H and Rusop M 2013 *Appl. Surf. Sci.* **284** 278
[15] Umale S V, Tambat S N, Sudhakar V, Sontakke S M and Krishnamoorthy K 2017 *Adv. Powder Technol.* **28** 2859
[16] Park N G, Van De Lagemaat J and Frank A J 2000 *J. Phys. Chem. B* **104** 8989

IOP Conf. Series: Earth and Environmental Science 1031 (2022) 012031 doi:10.1088/1755-1315/1031/1/012031

Full-Waveform Inversion in Time Domain by Increased Frequency Iteratively

M J Aparajita[1*] and I S Winardhi[2]

[1] Graduate Program of Geophysical Engineering Study Program, Faculty of Mining and Petroleum Engineering, Bandung Institute of Technology, Bandung, Indonesia
[2]Exploration and Engineering Seismology Research Group, Faculty of Mining and Petroleum Engineering, Bandung Institute of Technology, Bandung, Indonesia

*Corresponding author's email: *madejnanaparama@gmail.com*

Abstract. With seismic imaging techniques using numerical solutions of seismic acoustic wave equations, full-waveform inversion in the time domain can be a tool for estimating high-resolution models from complex geological conditions. However, full-waveform inversion in the time domain has several drawbacks, namely if you determine the initials of the model that is far from the global minimum (the true model) will give a stacked result at the local minimum and to overcome this problem is to use the method of increasing the frequency iteratively. With this method, at least it will keep away the results from a local minimum and close to a global minimum. The result of using the iterative frequency increment method on the full-waveform inversion time domain produces an image that resembles the original synthetic model and produces an updated velocity value that is not too different from the true velocity value of the synthetic model.

1. Introduction

Full-waveform Inversion, or we can say, FWI is the imaging technique of seismic waves that estimates high-resolution models from the subsurface by minimizing the misfit function [1] between modeled and observed data and uses local optimization [2] The FWI method is well to apply in complex geological conditions by simulating the propagation of seismic waves using numerical solutions of seismic wave equations. Minimizing the misfit function itself is optimization algorithm to get a global minimum of the misfit function and the non-linear relationship between seismic data and subsurface model parameters [4].

Research on the FWI method with synthetic data has long been carried out by several universities in the world since 2000. However, this method was first introduced in 1983 by Lailly, then continued by Tarantola in 1984, and in 1986, Mora introduced the method too [5] In 2010, Magrave et al. reviewed developments and proposed the FWI is an iterative cycle that can involve forward modeling, impedance inversion, velocity model updates, and pre-stack migration [3].

2. Data and Methodology

First, we use two synthetic models with different levels of difficulty. Figure 1a shows model 1 with a depth of 474 m and a distance of 1194 m, the model resembles salt diapir. With 4 layer that have velocity in each layer 1000 m/s, 1200 m/s, 1400 m/s and 1500 m/s. In the last model Figure 1b is the Marmousi model with gridpoint intervals dx = 20 m with the lowest velocity of 1500 m/s and the highest velocity of 4700 m/s. For forwarding modeling from Figure 2, we use the acoustic wave equation and use the finite difference time-space domain. In the gradient computation of the objective function from Figure 2, we use the initial velocity as the forward wavefield and use the true velocity as the back-propagation wavefield. In the process of calculating the back-propagation wavefield, there is a forward model that reversed in time.

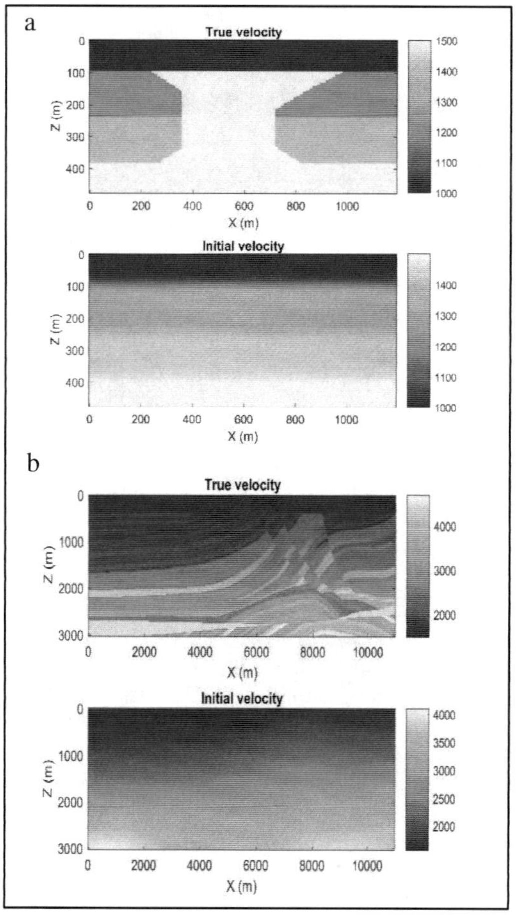

Figure 1. True velocity and initial velocity (a) synthetic model 1 and (b) Marmousi model

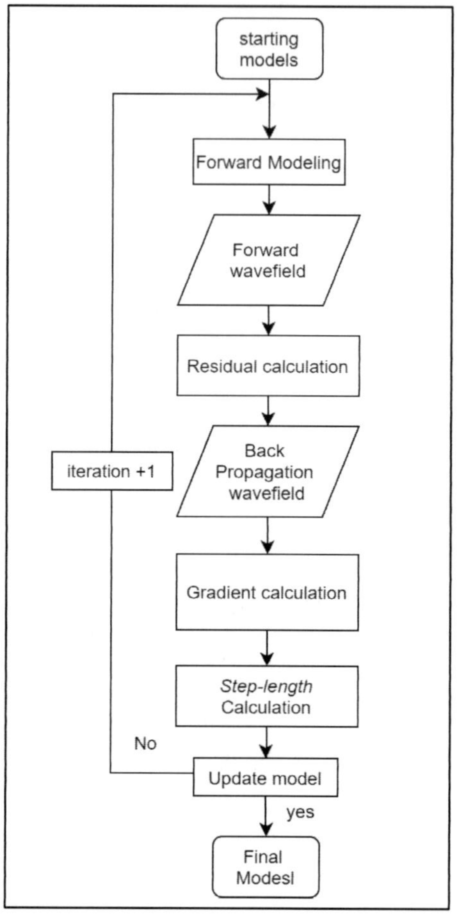

Figure 2. FWI research flow chart

In the calculation of gradient optimization from Figure 2, we use conjugate direction instead of solve the large computation of the Hessian inverse matrix. The step-length calculation uses a back-trace method which aims to eliminate the increased misfit value due to cycle-skipping. The step-length algorithm can be calculated through the following stages.

First, misfit (f_0) calculation,

$$f_0 = s(\boldsymbol{x_0}) - s(\boldsymbol{x_{obs}}) \tag{1}$$

calculate the initial step length (a_0),

$$a_0 = \frac{s(x_0)}{CD} \tag{2}$$

Use back-tracking to get updated step-length calculated,

$$s(\boldsymbol{x_1}) = s(\boldsymbol{x_{obs}}) + a_0 * CD * k \tag{3}$$

Update model,

$$f_1 = s(\boldsymbol{x_1}) - s(\boldsymbol{x}) \, , \tag{4}$$

$f_0 > f_1$, if not, it is repeated by increasing or decreasing the value of k,

$$s(\boldsymbol{x_2}) = s(\boldsymbol{x_1}) + a_0 * CD * k \tag{5}$$

Update model,

$$f_2 = s(\boldsymbol{x_2}) - s(\boldsymbol{x_{obs}}) \tag{6}$$

$f_0 > f_2$, if not, it is repeated by increasing or decreasing the value of k,

$$\gamma = \frac{(f_0 - f_2) + ((f_1 - f_0)}{(2f_0 + 2f_1 - 2f_2)} \tag{7}$$

$$a_n = a_0 \gamma \tag{8}$$

stop if misfit $s(\boldsymbol{x})^{(i)} + \alpha_n G <$ misfit $s(\boldsymbol{x})^{(i)} + \alpha_0 G$. After calculating the step-length (α) and gradient values (CD), the model updated with the following equation:

$$s(\boldsymbol{x})^{(i+1)} = s(\boldsymbol{x})^{(i)} + \alpha CD \tag{9}$$

Stop If the model update matches the true model, otherwise the model update will be repeated until the model update approaches the global minimum or the true model.

3. Result and Discussion

3.1. FWI results for synthetic model 1

There are two FWI processes carried out in the synthetic model 1. First, the synthetic model only uses one frequency (ricker wavelet with frequency 10 Hz) in 1-50 iterations and results can be seen in Figure 3. In 20-50 iterations, there are many multiple effects produced resulting in layer boundaries between layers 2 and 3 that do not match the true model. One of the reasons is the use of model initials are far from global minimums.

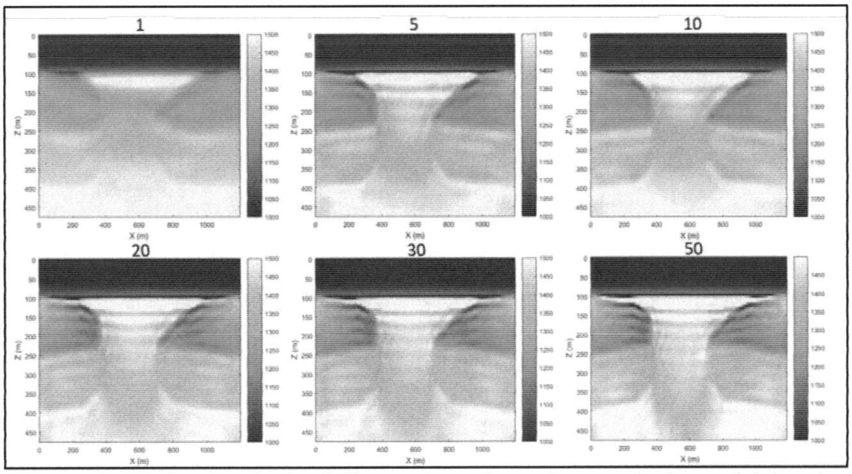

Figure 3. Updated velocity full-waveform inversion synthetic model 1 only uses one frequency (10 hz).

The FWI process carried out in synthetic model 1 is using frequency 5 Hz in iterations 1-10 (frequency changes in every 10 iterations). It's that mean in every 10 iterations used single dominant frequency (on ricker wavelet) and so on until the last iteration is 40 (which uses a frequency of 10 Hz) for the results can be seen in Figure 4. In 1-5 iterations, there is a slower reduction in misfit reduction than the previous process which only uses a frequency of 10 Hz. However, in 10-40 iterations, the resulting model is better than the previous process. With a reduced multiple effect (due to the use of low frequencies) so that the layer boundaries between layers 2 and 3 match the true synthetic model.

For a comparison of the velocity profile generated from the two FWI processes, it can be seen in Figure 5 with the green dotted line as the initial velocity, the black dots are FWI and the blue line is the true model. The effect of wiggle produced in the FWI increased frequency process is iteratively less than the FWI process that only uses one frequency. This is also due to the use of low frequencies in iterations 1-10 which reduce the ringing effect on the shot gather resulting from the forward modeling process. With the minimal wiggle effect on the resulting waves resulting in the level of layer boundary resolution in the FWI increased frequency iteratively better than the FWI process only uses one frequency. So that the resulting misfit value in the FWI increased frequency process is iteratively less than the FWI process only uses one frequency which can be seen in Figure 6.

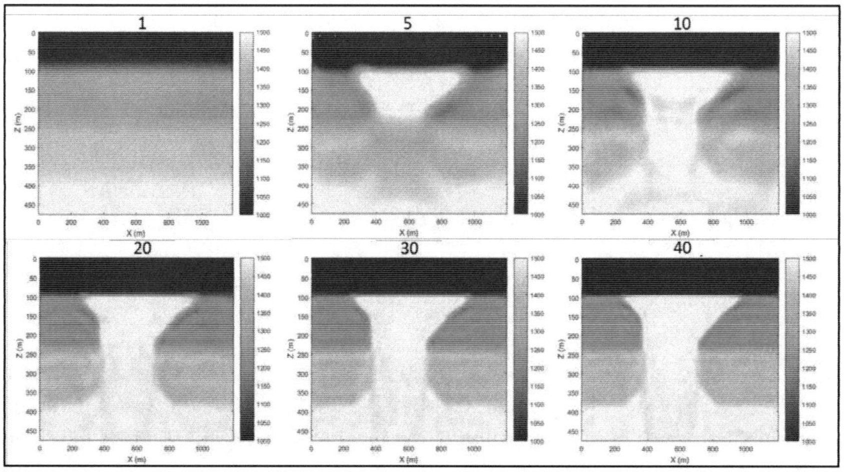

Figure 4. Updated velocity full-waveform inversion synthetic model 1 by increased frequency iteratively (5, 7, 9, 10 Hz)

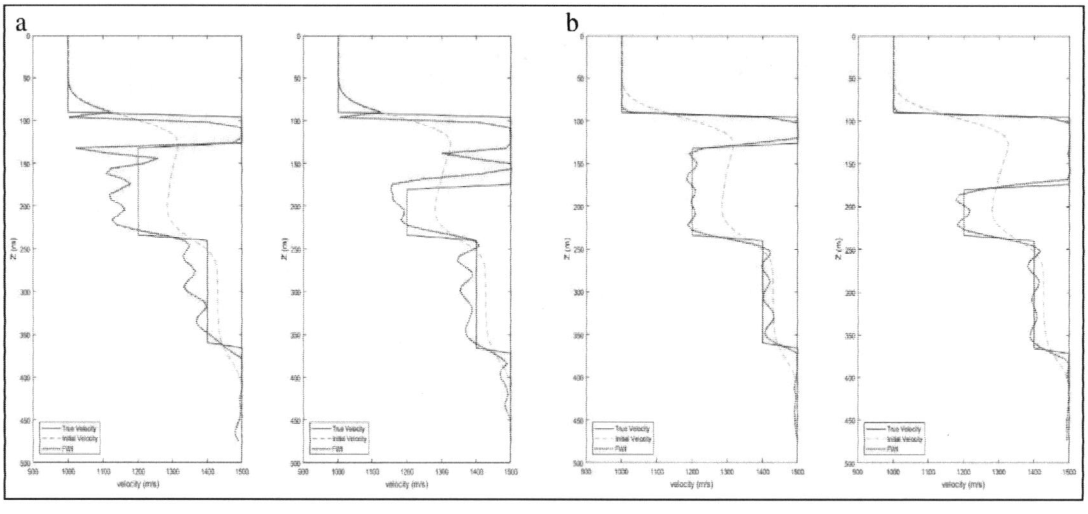

Figure 5. Velocity Profile comparison synthetic model 1 (a) only uses one frequency (10 Hz) at distance 252 m, 900 m and (b) by increased frequency iteratively (5, 7, 9, 10 Hz) at distance 252 m, 900 m.

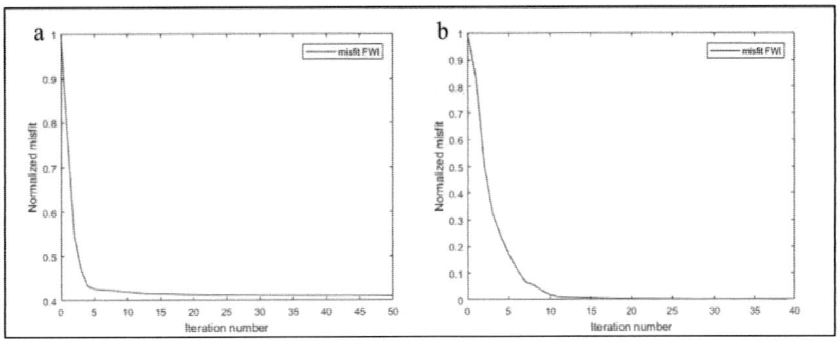

Figure 6. Misfit from (a) synthetic model 1 only uses one frequency (10 Hz) and (b) synthetic model 1 by increased frequency iteratively (5, 7, 9, 10 Hz).

3.2. FWI results for synthetic model Marmousi

After being tested in model 1 and resulting in iteratively better use of the FWI by increased frequency process. And then in testing the Marmousi models. Iterations 1-10 using a frequency of 2 Hz (on ricker wavelet) helps in reducing the resulting local minimum so that in 10 iterations the resulting model has begun to approach the global minimum even though the resulting resolution is still not good.

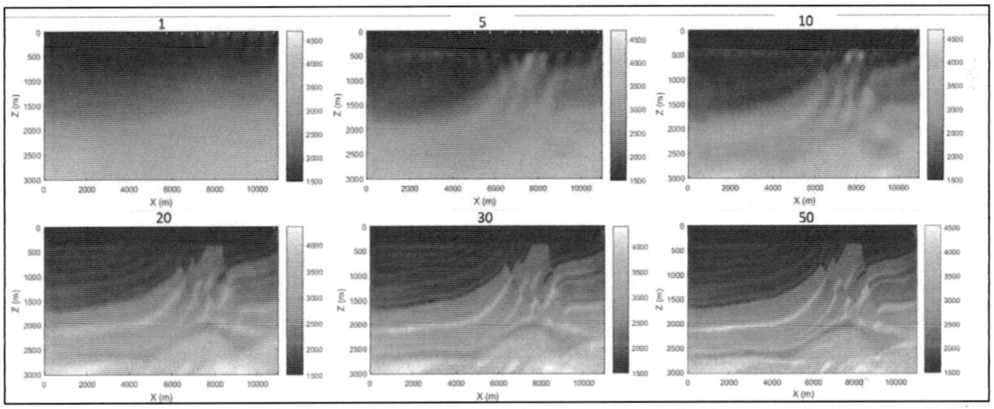

Figure 7. Updated velocity full-waveform inversion synthetic model Marmousi by increased frequency iteratively (2, 4, 6, 8, 10 Hz)

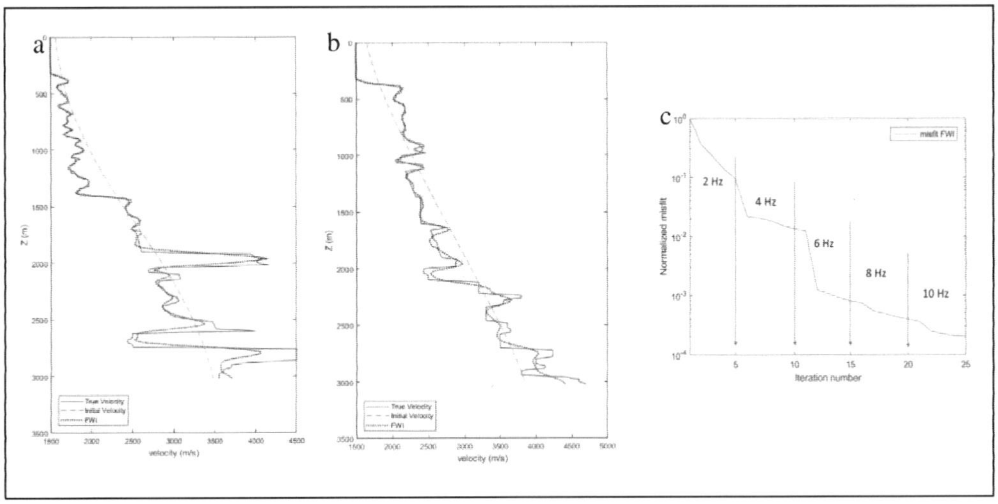

Figure 8. Velocity profile comparison synthetic model Marmousi (a) at distance 4480 m, (b) at distance 8000 m and (c) Misfit from synthetic model Marmousi

IOP Conf. Series: Earth and Environmental Science 1031 (2022) 012031 doi:10.1088/1755-1315/1031/1/012031

To improve the resolution of the result of iteration 10, the frequency is increased in every 10 iterations, so that at iteration 50 the results are very close to the global minimum which can be seen in Figure 7. In Figure 8b (which is the most complex layer in the Marmousi model) using the FWI process by increased frequency iteratively at least gives the velocity profile shape according to the true velocity profile form. The misfit results resulting from this process can be seen in Figure 8c.

4. Conclusion

Our case studies concluded that determining the initial model close to the global minimum on the full-waveform inversion time domain has an important role in simplifying the gradient optimization method to find the global minimum and not stop at the local minimum. Determination of the maximum frequency value on the full-waveform inversion time domain can affect the results of the final interpretation if using too high a frequency on shot gather will get multiple effects which results in the image not being resolved properly. Applying the method of increasing the frequency to the full-waveform inversion of the time domain gives the final velocity model close enough to the true velocity model even though the initial velocity model used is far from the true velocity model, using a frequency of 2 Hz in iterations 1- 10 to reduce the resulting local minimum, then increasing the frequency in every ten iterations will increase the resolution.

References

[1] Galuzzi B G 2017 Modelling and optimization techniques for acoustic Full Waveform Inversion in seismic exploration Universitas Degli Studi Milano PhD Thesis
[2] Jun H 2019 Frequency-domain reflection-based full waveform inversion for short-offset seismic data *Journal of Applied Geophysics* **164** 106–116
[3] Margrave G F, Ferguson R J and Hogan C M 2010 Full waveform inversion with wave equation migration and well control *CREWES Research Report* Vol **22**.
[4] Schuster G T 2017 Seismic Inversion Investigations in Geophysics Series No **20**
[5] Xu S, Wang D, Chen F, Zhang Y and Lambare G 2012 Full Waveform Inversion for Reflected Seismic Data 74th EAGE Conference & Exhibition incorporation SPE EUROPEC 2012 Copenhagen.

IOP Conf. Series: Earth and Environmental Science 1031 (2022) 012032 doi:10.1088/1755-1315/1031/1/012032

Potential REE-Li resources in lateritic granite in Indonesian granitoid formations: A review

Syafrizal[1], P Rasma[2], and AYA Hakim[1]

[1] Earth Resources Exploration Research Group, Bandung Institute of Technology, Bandung, Indonesia

[2] Mining Engineering, Faculty of Mining and Petroleum Engineering, Bandung Institute of Technology

Email: periskarasma@gmail.com

Abstract. The long-term growth trajectory of numerous industries remains dependent on rare-earth elements and lithium (REE-Li) resources, even if China began restricting the supply of REE. As critical raw minerals, REE-Li resources have become a sensitive issue, with the preservation of reserves framed are concerns over protecting the country's sovereignty. REE-Li bearing minerals are surmised to have originated from granite parent rocks. The granite belt in Indonesia is extensive on the Bangka Belitung Islands, West Kalimantan, and West Sulawesi, areas associated with high levels of radioactivity. This study discusses a conceptual analysis on the presence of REE-Li in granitic rock and lateritic granite based on previous studies both in general and in the given area. The region is mineralized with REE-Li bearing minerals in granite, which are absorbed onto clay minerals from weathered granite.

1. Introduction

The Indonesian government is committed to developing high-tech processing infrastructure and has passed legislation to regulate the mining of radioactive energy (Undang-Undang Cipta Kerja No.11, 2020, Article 9), which stipulates that state-owned enterprises may collaborate with private companies to engage in mining associated-radioactive minerals. Rare-earth elements and lithium (REE-Li) are associated-radioactive minerals crucial for developing the high-tech technologies behind Li-batteries, hybrid cars, and magnet alloys – provided these are supported by the availability of raw materials of REE-Li raw materials.

REE-Li deposits can be found in granitoid rock and its weathering products. During the weathering of granitoid, which is mineralized with REE-Li bearing minerals, the REE-Li ions are absorbed onto clay minerals (e.g., kaolinite, smectite, and illite), enriching specific layers within the soil profile [1]. Typically, radioactive elements (e.g., thorium and uranium) are considered impurities because of the accompanying negative issues and impact on mineral exploration and processing. Thus, for REE-Li mining, it is preferable to choose lateritic granite with low radioactive content, which is also easily leached [2]. However, there has been little systematic study of REE-Li in lateritic granite rocks in Indonesia despite the extensive distribution of granitoids across the country. Therefore, this review explores the presences of REE-Li in clays associated with granitoid and speculates on the potential discovery of new deposits in lateritic granite – a direction in which efforts need to be intensified.

2. Characteristics of potential REE-Li in weathered granite

REE-Li is known to be richest during the differentiation process of magma in mantle melts. REE-Li can also be richly concentrated in granitoids, which accumulate as mineral accessories such as titanite, allanite, apatite, monazite, xenotime, zircon, amblygonite, lepidolite, petalite, and zinnwaldite, because these elements tend to remain melted until the final stage of magma differentiation [3 - 4]. However, the main minerals present in granite are mega crystals of quartz, biotite, plagioclase, k-feldspar and hornblende [5].

Ionic radii affect the partitioning of trivalent rare-earth elements (REE^{3+}), which decreases with an increase in atomic number because the ionic radius of the ions becomes larger (Figure 1) and their adsorption is regulated by characteristics such as solution pH, ionic strength, and the clay minerals present [6]. REE deposits via ion-adsorption occur in tropical climates through the following processes: (a) REE are washed off from granites by groundwater, (b) a bold layer of laterite zones with clay-rich content develops above the granites, and (c) the mobilized REE ions are gradually absorbed onto the clays in the lateritic granite zones [7].

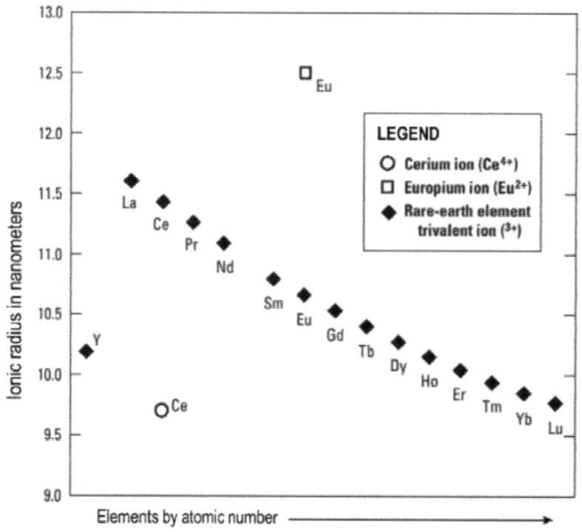

Figure 1. Graph of the ionic radius of rare-earth ions ($^{+3}$), cerium ($^{+4}$), and europium ($^{+2}$) [7]

The REE ion-adsorption deposits from weathered granite comprise the REE accumulation zone in the lower profile, and the REE leached zone is in the upper profile. The REE leached zone exhibits a positive Ce anomaly resulting from the oxidation of Ce^{3+} to Ce^{4+}. Nonetheless, the REE accumulation zone, specifically the thick REE ion-adsorption zones, exhibits a negative Ce anomaly due to the other REE^{3+} ions that are specially adsorbed onto clays. Consequently, the negative Ce anomaly in the ion-adsorption ores has a positive correlation with the ion-exchangeable REE concentration, and this anomaly is a useful indicator in geochemical exploration [2].

Lithium was leached easily during weathering even though sedimentary environments have no sedimentary minerals enriched lithium either than clay minerals. In minerals, univalent lithium can replacement to a degree for divalent magnesium and iron, but its smaller ionic size does no longer allow its substitution to any substantial volume with chemically associated extra-plentiful alkalis, sodium, and potassium. All common cations can dislodge Li+ from base exchange material; consequently, base-exchange reactions tend to deliver lithium into solution [8].

3. REE-Li ion adsorption in weathered granite

In southern China, concentrations of REE ion-adsorption deposits vary roughly from 500 to 2000 parts per million (ppm), with lower concentration in the parent granites or bedrocks [9]. The largest REE deposits are typically found in the B horizon of the lateritic granite zone. In addition to having the highest concentration of REE ion-adsorption deposits, the B horizon also has an abundances of clay minerals. In the upper zone of the lateritic granite profiles, the A horizon is enriched with organic matter and clay minerals, while the C horizon comprises fragments of half-weathered granites and clay minerals.

Lateritic granite has been identified in Jiangxi, China, with REY (REE + Y) grade deposits varying from 143 to 1228 ppm [2]. However, Mamasa and Palu granite enriched with REE ion-adsorption deposits are typically found in the B horizon, with REE concentrations of up to 314 ppm. Studies on this region have reported that the REE contents are dramatically depleted in the A horizon, while the B horizon has deposits approximately 70% to 85% richer than the parent granites. These findings indicate that the REE ion-adsorption enrichment occurred in the B horizon [10]. Although REE ion-adsorption deposits in lateritic granite contain low-grade REE minerals (500 - 2000 ppm of REE_2O_3), in comparison with other deposit types, they are of interest as an exploration target because of their extensive areal distribution, low mining costs, and the low Th and U content of the ores in this zone. [11].

Some muscovite-bearing granites have been identified as lithium-enriched granites, including zones enriched with lithium, tin, fluorine, and tantalum. The peak of a biotite-muscovite granite at the Yichun Mine in Jiangxi, China, transforms into muscovite granite and lepidolite granite, which has been mined for lithium and tantalum [12]. In addition to the weathering process removing lithium from Li-bearing minerals, it incorporates the lithium into clay minerals. Previous studies have reported lithium content in clay minerals with concentrations ranging from 7 ppm to as high as 6000 ppm in hectorites [13]. In humid climates, lithium tends to be concentrated in kaolinite in bauxite, in which lithium concentrations of 340 ppm have been recorded [13]. Nonetheless, Li in lateritic granite is not easily identified. According to Ronov et al. [14], kaolinite from granodiorite contains Li in concentrations of about 45 ppm, which increases slightly to as much as 118 ppm for kaolinite derived from amphibolite. However, studies on kaolinites have compared altered schists and granites, with results showing that there is more lithium in kaolinites from granites than that in kaolinites from schists [13].

4. Areas with potential REE-Li sources in Indonesian granitoids

Based on the literature, granitoid rocks that are widespread in Indonesia, especially on the Bangka Belitung Islands and West Kalimantan are generally associated with high radioactive values, while the West Sulawesi area commonly has high radioactive values from volcanic rock formations (Figure 2). According to Gill [15], granitoid rocks are grouped into diorites, tonalites, granodiorites, granites, and alkaline granites based on mineralogy, with specific tectonic conditions producing a distinct granitoid rock.

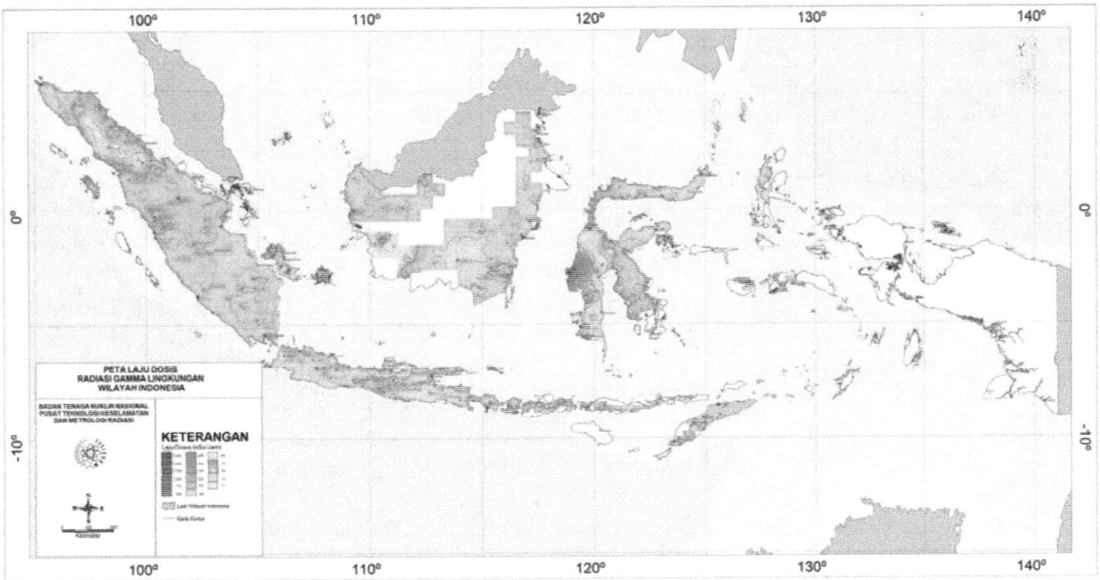

Figure 2. Dose rate map of Indonesian environmental gamma radiation [16]

The Bangka Belitung Islands are composed of main range granitoid and eastern granitoid. Main range granitoid is biotite-monzogranite with an S-type chemical composition associated with Sn, Fe, W, and REE, while eastern granitoid is biotite and biotite-hornblende of monzogranite, which have an I-type chemical composition. West Kalimantan has the Natuna-Sanggau and the Schwaner Mountains complex of granite, granodiorite, diorite, and tonalite. This complex is replete with minerals such as biotite pyroxene, quartz, hornblende, orthoclase, and plagioclase. West Sulawesi has the Mamasa granitic rocks of granite, granodiorite, diorite and quartz monzonite, and has the Palu granitic rocks of monzonite and granodiorite [10]. The main mineral composition includes biotite, hornblende, and quartz, which are grouped into an ilmenite series in the south zone, an ilmenite-magnetite series in the middle zone, and a magnetite series in the north zone. Granitic rocks in West Sulawesi are associated with Cu, Au, Mo, REE, and radioactive elements [17].

Overall, Indonesia has an extensive distribution of granitoids located in a tropical climate; therefore, it has potential REE-Li resources in clay deposits. The weatherable REE-Li bearing minerals are the sources of ion-exchangeable deposits absorbed onto clay minerals.

Acknowledgments
Thanks to reviewers and supervisors for valuable suggestions, and to colleagues who provided crucial information on our research topic.

References

[1] Berger A, Janots E, Gnos E, Frei R, and Bernier F 2014 Rare earth element mineralogy and geochemistry in a laterite profile from Madagascar *Applied Geochemistry* vol **41** p 218–228.
[2] Sanematsu K and Watanabe Y 2016 Characteristics and genesis of ion adsorption-type rare earth element deposits *Economic Geology* vol **18** p 55–79 (Colorado: Soc. of Economic Geologists).
[3] Ng S W P, Whitehouse M J, Roselee M H, Teschner C, Murtadha S, Oliver G J H, and Chang S C 2017 Late Triassic granites from Bangka, Indonesia: A continuation of the main range granite province of the South-East Asian Tin Belt *J. of Asian Earth Sciences* vol **138.**
[4] London D 2018 Ore–forming processes within granitic pegmatites *Ore Geology Reviews* vol **101** p 349–383.
[5] Roselle M H, Ghani A A, Ng S W P, Murtadha S, Oliver G J H, Xiang Q L, and Umor M R 2017 Geochemistry of Bangka Granites, Bangka Island, Sumatera Indonesia *Geo. Soc. of Malaysia* vol **43.**
[6] Coppin F, Berger G, Bauer A, Castet S, and Loubet M 2002 Sorption of lanthanides on smectite and kaolinite *Chem. Geol*ogy vol **182** p 57-68.
[7] Van Gosen B S, Verplanck P L, Seal R R, Long K R, and Gambogi J 2017 Rare-earth elements, chap. O of Critical mineral resources of the United States— Economic and environmental geology and prospects for future supply *U.S. Geological Survey Professional Paper 1802*.
[8] Vine J D 1975 Lithium in sediments and brines-How, why, and where to search: *U.S. Geological Survey Journal of Research* vol **3** p 479-485.
[9] Bao Z and Zhao Z 2008 Geochemistry of mineralization with exchangeable REY in the weathering crusts of granitic rocks in south China *Ore Geology Reviews* vol **33** p 519–535.
[10] Maulana A, Yonezu K, Watanabe W 2014 Geochemistry of rare earth elements in the weathered crusts from the granitic rocks in Sulawesi Island, Indonesia *J. of Earth Sciences* vol **3** p 460-472.
[11] Kanazawa Y and Kamitani M 2006 Rare Earth Minerals and Resources in the World *J. of Alloys and Compounds* vol **408** p 1339-1343.
[12] Schwartz M O 1992 Geochemical criteria for distinguishing magmatic and metasomatic albite enrichment in granitoids—Examples from the Ta-Li granite Yichun (China) and the Sn-W deposit, Tikus (Indonesia) *Mineralium Deposita* vol **27** p 101–108.
[13] Starkey H C 1982 The role of clays in fixing lithium *Geological Survey Bulletin* 1278-F Contributions to geochemical prospecting for minerals p 2-8.

[14] Ronov A B, Migdisov A A, Voskresenskaya N T, and Korzina G A 1970 Geochemistry of lithium in the sedimentary cycle *Geochemistry International* vol **7** p 75-102
[15] Gill R 2010 *Igneous Rock and Processes: a Practical Guide* (West Sussex: Wiley-Blackwell).
[16] Anon 2017 Laporan Kinerja Pusat Teknologi Bahan Galian Nuklir.
[17] Kurniawan A 2014 Geologi batuan granitoid di Indonesia dan distribusinya *Masyarakat Ilmu Bumi Indonesia* vol **1.**

AUTHOR INDEX